This book is a translation of a unique Russian study of fossil plant distributions in the Jurassic and Cretaceous world. The core of the work is the description and assessment of floras of the USSR, China and Japan, currently the largest land area. Information on the floras of this extensive and productive area has hitherto been available only in scattered and sometimes obscure Russian journals. Vakhrameev also summarizes the more familiar Western work and divides the continents into Regions and Provinces illustrating the palaeolatitudinal climatic arrangement of floras. The work deals first with megafossil plants in strictly stratigraphic order, but in many cases also with land plant palynomorphs. The time covered from 200 to 65 million years ago ranges both before and immediately after the main angiosperm radiation from about 130 to 100 million years ago. Vakhrameev's work represents a vast source of new data, which will be of value to any serious student of Mesozoic seed plants.

Jurassic and Cretaceous floras and climates of the Earth

Jurassic and Cretaceous floras and climates of the Earth

V. A. Vakhrameev
Formerly Geological Institute of the Academy of Sciences USSR

TRANSLATED BY
Ju. V. Litvinov

EDITED BY
Norman F. Hughes

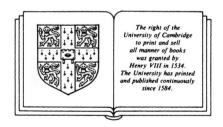

The right of the
University of Cambridge
to print and sell
all manner of books
was granted by
Henry VIII in 1534.
The University has printed
and published continuously
since 1584.

CAMBRIDGE UNIVERSITY PRESS
Cambridge
New York   Port Chester   Melbourne   Sydney

Published by the Press Syndicate of the University of Cambridge
The Pitt Building, Trumpington Street, Cambridge CB2 1RP
40 West 20th Street, New York, NY 10011-4211, USA
10 Stamford Road, Oakleigh, Melbourne 3166, Australia

Originally published in Russian as *Yurskiye i melovyye flory: klimaty Zemli*
by Nauka, Moscow, 1988 and © Nauka, Moscow, 1988

First published in English by Cambridge University Press, 1991
as *Jurassic and Cretaceous floras and climates of the Earth*
English edition © Cambridge University Press 1991

Printed in Great Britain at the University Press, Cambridge

British Library cataloguing in publication data

Vakhrameev, V. A.
Jurassic and Cretaceous floras and climates of the Earth.
1. Climate. Plants. Mesozoic era
I. Title   II. Hughes, N. F. (Norman Francis)
III. [Yurskiye i melovyye flory]
551.76

Library of Congress cataloguing in publication data available

ISBN 0 521 40291 3 hardback

WD

Contents

Preface to Russian edition

This book was completed by Vsevolod Andreyevich Vakhrameev in 1985 and prepared for publication only a few months before his sudden death. He had been working at it for more than ten years and regarded it as the sum total of his efforts. The data interpreted for the first time the Mesozoic history of our planet and the wide range of problems discussed are a total summary of the author's numerous publications on Mesozoic palaeofloristics.

During the last 40 years the scale of Vakhrameev's palaeobotanic and stratigraphic studies steadily increased as the range of his interests expanded and at the same time his scientific weight was also enhanced. Starting with studies of fossil plants, as one of a small number of Moscow students of A. N. Krishtofovich, Vakhrameev soon became a leading specialist in the palaeofloristics and stratigraphy of the continental Mesozoic both in his own country and abroad.

During World War II, Vakhrameev first turned to palaeobotany in an attempt to use floristic data for climatic reconstructions of the conditions of bauxite rise in Western Kazakhstan. After the war he resumed his studies of the stratigraphy of the Cretaceous strata in this region. At first he was guided by A. N. Krishtofovich in his research and description of fossil plants, later he continued on his own. His pioneering discovery of ancient angiosperms in Kazakhstan at that period promoted further study of the problem of their origin. In a paper published in 1947, Vakhrameev suggested that the first angiosperms appeared in subtropical mountain regions. In subsequent publications he adhered to this viewpoint which was later shared by other palaeobotanists.

In 1950 Vakhrameev initiated a detailed phytostratigraphic investigation of the Jurassic and Cretaceous sequences in Siberia. Almost 25 years later he was awarded the Obruchev Prize by the Presidium of the USSR Academy of Sciences for his series of contributions on the Jurassic and Cretaceous floras of Asia, their role in the subdivision and correlation of continental deposits, climatic reconstructions and palaeogeography.

The starting point was Yakutia. Very little had been known about Mesozoic plants in this vast area until the middle of the twentieth century. The only data available were a classification made in the nineteenth century by the Swiss palaeobotanist O. Heer on plants collected in the lower reaches of the river Lena. He dated them as Jurassic. In the pre-war years an extensive Mesozoic uplift was discovered in the mouth of the Vilyui river. However, it was not studied for a long time; even its spatial orientation was erroneously believed to be transverse to the actual one. Vakhrameev and Yu. M. Pushcharovsky explored cross-sections on the right bank of the Lena, north of the mouth of the Aldan; also on the left bank of the Lena in the Ust-Vilyui area (the Sogo-Khai and Oyun-Khai sections), as well as sections in the vicinity of Yakutsk, between the town of Pokrovsky and the Kangalassian Cape. During the same period, Vakhrameev and E. L. Lebedev examined the basin of the middle reaches of the Vilyui and the valley of the Lindya river – a left tributary of the Lena. They succeeded in proving that the Lena–Vilyui sections were made up not only of Jurassic but also of Cretaceous, including Upper Cretaceous, sequences. This abruptly changed the entire conception of the geologic structure of the vast Lena basin. The general picture of the geological map acquired a new aspect. The principally new unified stratigraphic scheme of the Mesozoic in Yakutia and adjacent regions of the southern Siberian Platform now forwarded was accepted by an Inter-Departmental Conference on the Stratigraphy of Siberia. Thus, a stratigraphic basis for the Lena–Vilyui gas-bearing province was created.

In the mid-1950s Vakhrameev concentrated his interests on the eastern areas of the USSR (the Soviet continental Far Eastern area and Sakhalin). A new monograph in this cycle, *Upper Jurassic and Lower Cretaceous Flora of the Bureya Basin and its Stratigraphic Significance* (with M. P. Doludenko as co-author), was devoted to the Mesozoic sections and plants studied in the Bureya and Urgal basins. The monograph substantiated the age of the Upper Jurassic and Lower Cretaceous strata widely represented in the Bureya trough; the correlation of its regional sections with those of the Lena basin was presented by phytostratigraphic methods. The position of the Jurassic/Cretaceous boundary in the continental sequences of Siberia and the Far East was also established. A later publication, *Late Cretaceous Floras of the USSR Pacific Coast: Specific Features of their Composition and Stratigraphic Sites* (1966), clarified the general regularities observed in variations of the make-up of Cretaceous floras in different climatic zones. Papers published in the 1950s and 1960s introduced a considerable number of new taxa, some of which became index ones for identifying and subdividing phytostratigraphic units.

A résumé entitled *Jurassic and Early Cretaceous Floras of Eurasia and Contemporary Palaeofloristic Provinces* (1964) summarized the data published in regional monographs and papers, as well as those obtained from a study of literature sources on foreign countries. It was shown that two major phytogeographic regions existed in Eurasia in Jurassic–Cretaceous times – the Indo-European and Siberian regions – corresponding to the warm-temperate and subtropical climatic belts, respectively. Four provinces were recognized in the first belt (Eurasian, Middle Asian, East European and Indian) and two provinces (Lena and Amur) in the second. Besides maps showing phyto-choria distribution, the work also presented tables and charts showing the distribution of genera and families throughout the Jurassic and Early Cretaceous.

In 1970, Vakhrameev (with co-authors I. A. Dobruskina, Y. D. Zaklinskaya and S. V. Meyen) initiated and prepared for publication a subsequently widely acclaimed monograph *Palaeozoic and Mesozoic Floras of Eurasia and the Phytogeography of that Period*. In addition to the section 'Jurassic and Early Cretaceous floras' Vakhrameev, besides significantly revising and supplementing the text, wrote new chapters: 'Late Cretaceous floras' and 'Botanical–geographical zonation in the past and evolution of the vegetable kingdom'. The latter indicates one of the most important regularities in the evolution of ancient floras, i.e. beginning with the Devonian, each of the three major stages in the evolution of the vegetable kingdom (Palaeophytic, Mesophytic and Cenophytic) falls into two phases traceable throughout Eurasia. In the earlier phase the differentiation of the floras into separate phytochoria is less distinctive. The development of forest vegetation was at its peak. A significant reduction of the arid climate belt facilitated regional stratigraphic correlation. During the later phase, aridization of the climate in the subtropical areas resulted in the extinction of the hydrophilic forms and the spread of new types of plants; this led to an intensification of migration processes and subdivision of the phytochoria. The wide range of problems and extensive factual data presented in this monograph determined its great success both in the USSR and in other countries. In 1978 a revised and supplemented edition was published in German in East Germany (GDR).

The present monograph is a further step in our knowledge of the Mesozoic vegetable kingdom. In the course of previous studies of Eurasian floras, Vakhrameev came to realize the need to acknowledge the existence of continental drift, since otherwise present-day mapping of phytochoria presents serious difficulties, such as the explanation of the unity of the Early Mesozoic Gondwana floras and the similarity in the composition and evolution to the Cretaceous floras on both sides of the South Atlantic. From the fixed continents viewpoint it is also impossible to account for the subtropical type of Mesozoic floras in Greenland, which distinctly differ from coeval floras in the eastern parts of Asia, even from those confined to the present southern latitudes. The author avoided these difficulties by using maps compiled on a mobility basis.

The analysis of the data was also complicated by the different volume of information on the different continents and by the presence of numerous 'blank spots', especially in ancient shield areas. In order to identify the phytochoria and draw their boundaries it was necessary to employ a uniform methodologic and nomenclature approach, aimed primarily at the composition of the dominant taxa ranking from genera to orders. When palaeobotanic data were insufficient, Vakhrameev used litho-logical data in extrapolating boundaries and, especially, in eliciting the degree of environmental aridity or humidity.

Vakhrameev realized that the appearance of one or another phytochoria was determined by floristic complexes and not by a single plant, even if it was found in a large number of localities. He did not share the views of those specialists who derived the names for phytochoria from a single predominant taxon. He found geographic names more acceptable, provided they did not coincide with the names adopted by palaeontologists for biogeographic zoning. He would not apply the largest phyto-chorial unit – kingdom – to his reconstructions of the Mesophytic and Early Ceno-phytic, since he considered that the Jurassic and Cretaceous floras were not as sharply differentiated as the Late Palaeozoic or Cenozoic ones. At the same time, V. A. Vakhrameev remained a firm supporter of the hierarchic system in phytochoria sub-divisions, comprising regions, provinces and smaller units.

The author's consistent examination of the stages in the evolution of Jurassic and Cretaceous floras, performed on a global scale, required additional data on age corre-lation of flora-bearing continental strata in different regions and provinces. For this purpose, Vakhrameev employed a wider than previously used range of biostrati-graphic data on the adjacent marine basins. Here he resorted to such leading groups of Mesozoic fauna as ammonites, inocerams, foraminifera, etc., whose zonal complexes could provide not only a detailed subdivision, but also subglobal tracing of separate zones.

The planetary picture of climatic zonality for each of the Jurassic and Cretaceous epochs and for certain of their smaller units was based primarily on the distribution of typical groups of plants as climatic indicators: warm-temperate, subtropical and tropical. Palaeothermal data were also utilized, lithologic–geochemical criteria and spatial distribution of coal- and bauxite-forming areas were taken into account. On the basis of the combined data Vakhrameev concluded that the current climatic zonality of the Earth was formed only in the second half of the Late Cretaceous. In earlier Mesozoic epochs, the temperate region of the Southern Hemisphere had not yet appeared.

The main virtue of this book is that it presents for the first time a comprehensive picture of the evolution of the Mesozoic vegetable kingdom, but the author also out-lines ways for its further precision and improvement.

M. A. Akhmetiev

Preface to English edition

While working on the concluding part of this monograph concerning phytogeography, global modelling of climate changes and the location of the continents in the Jurassic and Cretaceous periods, Professor V. A. Vakhrameev repeatedly underlined his dream about the translation of his book into English – the language used in most international contacts between scientists. That is why his numerous disciples and colleagues in the Soviet Union met with great enthusiasm the possible publication of his book in Britain by such a well-known scientific publishing house as Cambridge University Press. But for the kind assistance and personal participation of British palaeobotanists Dr N. F. Hughes, Professor W. G. Chaloner, Professor M. C. Boulter and Dr J. Watson, with whom Professor Vakhrameev was in business and friendly relations, this publication would not have been possible.

To make this book more attractive for those who are interested in the Earth's flora history and especially for students – future botanists and palaeontologists, the English version contains some new material not included in the Russian edition. The book is illustrated with palaeontological plates of mostly typical flora of Jurassic and Cretaceous times, and maps, some of which take account of continental drift. The monograph includes a geographic index and a postscript with a small incomplete review of palaeobotanic works which for some reason the author was unable to use while writing his book or which were published after his death.

The book was translated by Ju. V. Litvinov (Vladimir Pedagogical Institute, USSR) with the participation of Dr M. A. Akhmetiev and Dr A. B. Herman (both are from the Palaeofloristic Laboratory, Geological Institute, Academy of Sciences of the USSR). The English text was prepared by Dr N. F. Hughes.

The Postscript was written by Dr M. A. Akhmetiev and E. L. Lebedev. Drs A. B. Herman, M. P. Doludenko and E. L. Lebedev (Geological Institute Academy of Sciences of the USSR) prepared the palaeontological plates of Jurassic and Cretaceous flora and maps. The geographic index was prepared by the widow of the author, Mrs

E. I. Miljutina-Vakhrameeva. Special thanks are due to Mrs K. A. Petchnikova for administrative assistance in preparing the publication.

Cordial thanks must also to to Dr N. F. Hughes (Cambridge University) and Dr Caroline Roberts (Cambridge University Press) for their help in publication of the English version.

English Editor's preface

Publication in English of this work of synthesis by Vakhrameev provides an important opportunity for all palaeobotanists with Mesozoic interests to assess for the first time the relevant floras of the whole world, the work on the extensive Asian constituents having previously been very difficult of access.

The translation and presentation has been done in two stages. The main translation into English was made in Moscow by Ju. V. Litvinov, sadly without the assistance of the author who had by then died. In England we have concentrated on clarification and on consistency of references and tenses, but we have tried to leave the flavour of the book closer to the Russian by not altering such phrases as 'mobilistic maps' into English plate tectonics jargon. We realize that we have not achieved complete consistency of geographic and stratigraphic names because of difficulties through one or more transliterations on slightly different systems, but we have attempted in the time available to remove ambiguities. Russian stratigraphic usage has been left as employed by the author; it will now in any difficult case be possible to follow up into the original detailed references which have all been provided. A set of illustrations of typical Mesozoic plant fossils from the USSR has been incorporated in addition to all the maps and tables of the original Russian edition.

We have received excellent cooperation from Dr M. A. Akhmetiev and his staff in Moscow in attempting to adhere as closely as possible to the known wishes of the author.

Norman F. Hughes
Cambridge

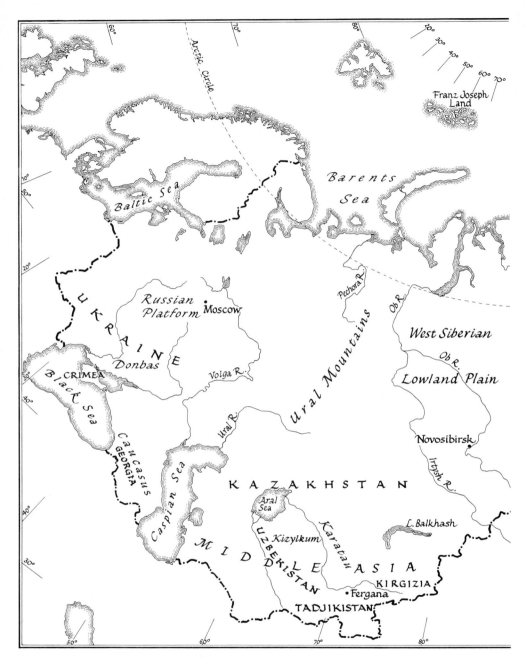

Map of the USSR showing the principal geographic features to which Mesozoic plant fossil localities are referred. Some localities in South America and eastern

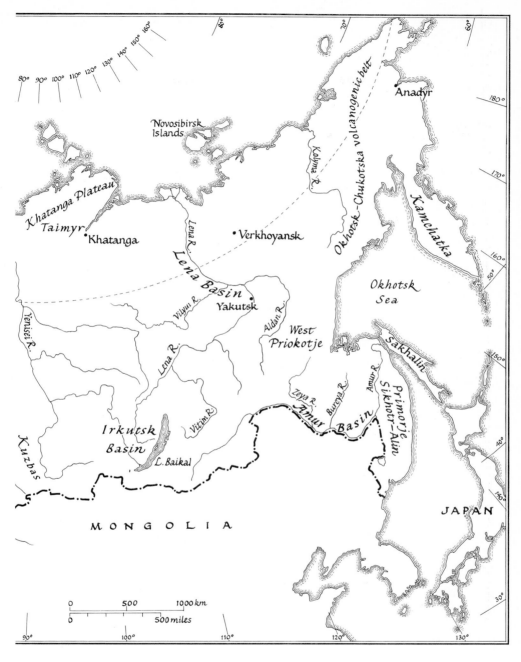

Asia including China are shown on Figs. 2.23 and 3.13 on pp. 85 and 123.

❀ I ❀

Introduction

My earlier book entitled *Jurassic and Early Cretaceous Floras of Eurasia and Contemporary Palaeofloristic Provinces* was published in 1964. Over the past 20 plus years our knowledge of the composition and evolution of the floras of this vast continent has been significantly replenished. What is more important, our previously scarce evidence has been greatly expanded by the studies of the Mesozoic floras of other continents, especially those of the Southern Hemisphere. During this time, not only have new finds of fossil floras been investigated, but also the earlier known floras have been subjected to repeated analysis. Epidermal structure has been examined in many species and genera which play a vital role in ecologic reconstructions. The structure of reproductive organs and of spores and pollen extracted therefrom have also been studied.

Since 1964, research has made considerable strides in the field of palaeoclimatic reconstructions based both on palaeontological (particularly palaeobotanical) and lithologic data. Investigations associated with the measurements of the absolute temperature of marine basins of the Jurassic and Cretaceous periods have also increased, though to a smaller extent. All this has allowed consideration of the phytogeography of the Jurassic and Cretaceous periods as well as the climatic history of this time for the whole surface of the globe. The lithologic data and climatic interpretations based on them were mainly drawn from the works of Strakhov (1960) and Khain, Ronov and Balukhovsky (1975), as well as Ronov and Balukhovsky (1981).

Before embarking on this monograph I had published a number of articles over many years (Vakhrameev, 1975, 1978, 1980, 1981a, b, 1984, 1985; Vakhrameev and Doludenko, 1976) which provide a connecting link between this monograph and the earlier work (Vakhrameev, 1964).

Extending the research scope beyond the boundary of Eurasia, I immediately came across the problem of selection of a geographic basis for compiling maps of phytogeographic zonation. Should this basis be contemporary or should it take into account continental movements? It should be noted that the palaeoclimatic maps in the works mentioned above of N. M. Strakhov, A. N. Balukhovsky, A. B. Ronov and V. E.

Khain were constructed on current evidence and some of these geologists vigorously denied the idea of continental drift.

However, work on the phytogeography of Eurasia already showed the need for recognizing this hypothesis which removed many problems arising from locating phytochoria in the contemporary geographic context. These unsolved issues include the current position of Greenland in the higher latitudes incompatible with the sub-tropic appearance of its Late Triassic and Early Cretaceous floras; similarity of com-position and evolution between the Cretaceous floras of Brazil and Western Africa despite the intermediate existence of the Pre-Albian South Atlantic which was borne out by deep drilling offshore; and identical composition of the Permian and Triassic floras of the southern continents testifying to the existence of Gondwana. To compile maps built on a mobilistic basis we made use of the atlas of Smith and Briden (1977) for the Jurassic and utilized the maps contained in the works by Barron *et al.* (1981).

I deliberately use here the notion of 'continental drift' instead of the more up-to-date 'plate tectonics' or 'tectonics of plates'. The fact is that employing the latter as identical concepts, perhaps implies agreement with the mechanism suggested to explain the drift and the causes behind it. Having no point of view of my own on this issue but recognizing the phenomenon, corroborated by the palaeobotanic evidence, I retain the notion of continental drift. This simply states the fact without accounting for the cause which tectonists and geophysicists are still trying to elucidate.

Here, just as earlier (Vakhrameev, 1964), I apply quite a simple principle of phyto-choria conjugation, i.e. division of regions into provinces. When needed, subregions and subprovinces are identified, with the latter divisible into districts. The largest phytochorial unit available in biogeography, i.e. the kingdom, is avoided in this work since the Jurassic and Cretaceous periods were devoid of distinct and large-scale well-differentiated floras as was the case in the Palaeozoic (Meyen, 1984) or at the present time (Takhtajan, 1978). As a rule, phytochoria are described for each of the epochs of the periods in question. It is only in certain instances when we talk about impover-ished compositions and floras weakly differentiated in time, such as the Jurassic floras of North and South America, that characterization is made age-wise for the entire continent within whole regions rather than for individual provinces.

The position of the regions is established by the climatic belts latitudinally or sub-latitudinally extended. The boundaries between the provinces are determined by the distribution and outlines of marine basins, distances from some regions, and orien-tation and height of mountainous ridges (past and present) separating neighbouring provinces. Impediments to migration due to these reasons are sometimes so significant as eventually to distort the latitudinal or sublatitudinal orientation of inter-regional boundaries; examples are the phytochoria identified from palynologic data for the Late Cretaceous, and also the sector-wise location of the main floras of the Middle Triassic and lower strata of the Late Triassic in Eurasia.

As we see it, such concepts result from inadequate factual material for various climatic belts. It stands to reason that as data are amassed (as was the case for the Late Cretaceous) the sector boundaries will be reduced to those between individual provinces with the top priority being given to the regional differences ascertained through climatic belts having largely latitudinal or sublatitudinal extension.

Differences in the composition of dominant taxa ranking from genus to order serve as criteria for regional identification. Thus, for instance, the Austral and Equatorial regions differ in the Jurassic from the regions of the Northern Hemisphere by the total absence of *Czekanowskia* and poorly developed *Ginkgo* (both regarded as separate orders). It will be noted that in the systematics of Mesozoic plants genera are directly merged into orders in many cases as identification of distinctly delineated families remains a matter for the future.

We distinguish between the following phytochorial regions that existed in the Jurassic and Cretaceous periods (from north to south): Siberian–Canadian (for the Jurassic merely Siberian), Euro-Sinian (including the Cretaceous of the USA), Equatorial and Austral (Notal). These regions correspond to the climatic belts of the moderately warm, subtropical (Northern Hemisphere), and tropical and subtropical (Southern Hemisphere) climates. The last-mentioned, up to the middle of the Late Cretaceous, stretched as far as the Antarctic coast, although what happened in its central portion is obscure. It was only in the second half of the Late Cretaceous epoch that some cooling ensued, marked by the appearance of the pollen *Nothofagidites*. This was accompanied by the concomitant emergence of a belt of moderately warm climate in the Southern Hemisphere within which the Antarctic region came to be situated as a counterpart of the Siberian–Canadian region in the Northern Hemisphere. Initially, in a number of contributions published before the mid-1970s (Vakhrameev, 1964, 1975; Vakhrameev *et al.*, 1970), I singled out the Indo-European region (subsequently split into the Euro-Sinian and Indian subregions). I was compelled to do this due to the major differences observed between the Jurassic floras of Europe, Central Asia and China, on the one hand, and the coeval floras of India. This fact, coupled with the consequent palaeogeographic reconstructions with due regard to continental drift that placed India in the Southern Hemisphere in the Jurassic as well as in the Upper Palaeozoic, made me raise the rank of the Euro-Sinian from subregion to region and consider India as a province of the Austral region.

Apart from the names for the main phytochoria of the Jurassic and Cretaceous which we suggested (the 'Siberian region' was put forward for the Jurassic as far back as 1944 by V. D. Prynada), there are certain names given by other researchers. Thus, Takhtajan (1966) followed by Budantsev (1983) suggested that two regions should be singled out for the Upper Cretaceous of Eurasia; a more northern one, Boreal, similar to our Siberian–Canadian, and a more southerly unit, Ancient Mediterranean, on analogy with our Euro-Sinian. These terms, reflecting adequately the geographic position, would appear to bring out no objections, but the same names were adopted a long time ago by palaeozoogeographers studying the peculiarities of distribution of marine invertebrates. The southern boundary of the Boreal region identified according to these organisms encroaches (in the south) on southern England, France and West Germany, i.e. those regions where subtropical plants typical of the Euro-Sinian region were growing. The Boreal palaeofloristic region boundary passes farther north. Therefore, by accepting these terms we should have to specify what was involved each time, i.e. palaeozoological or phytogeographical region.

As far as configurations are concerned the closest analogues to the regions we singled out are the provinces based on the analysis of palynologic material and suggested by Brenner (1976) for the Middle Cretaceous. He outlined four provinces situated from north to south: North Laurasian, South Laurasian, North Gondwana and South Gondwana. However, to adopt these names for the middle part of the Cretaceous period would hardly be correct since Gondwana as a unit did not exist at this time. Gondwana's disintegrated parts represented the continents well known to us and referred to as South America, Africa, Australia and Antarctica.

A phytogeographic zonation on the strength of palynologic data was recently initiated by Herngreen and Khlonova (1981). Just as did Brenner, they identified large phytochoria directly termed provinces with these names being changed from one age interval to another. For the purpose the authors used both the geographical terms which are quite often abbreviations (ASA for African–South American, or WASA for West African–South American) and the names of predominant plants (province Palmae, province *Nothofagidites*). We think it is also necessary to introduce at least two-stage subdivision of phytochoria (region, province).

It is undesirable to change names, especially those of the largest phytochoria and regions, when going over from one epoch to another or even from period to period. Phytochoria should preserve their names even if the inhabited area somewhat changes its outline and if the composition of taxa characterizing this phytochoria is altered in the process of evolution. The basic ecological requirements of the main floristic elements of a large phytochoria should be unaltered. It is only natural that with time the smaller scale phytochoria (provinces, subprovinces) will disappear or be replaced as a result of major changes of the Earth's relief. All the same, renaming regions should only be necessitated by the most significant restructuring of the floras of the Earth such, for example, as occurred during transition from the Palaeozoic to the Mesozoic. Compliance with this rule will contribute to building a clearer panorama of the Earth's floral evolution in the geological past.

We also believe that names for phytochoria should be geographical rather than based on the terms of predominant plants since the areas of the latter undergo substantial changes with geologic time. Thus, the names for the main phytochoria of South Eurasia suggested by Krassilov (1972 and later), i.e. flora *Phoenicopsis*, flora *Ptilophyllum* and flora Pentoxylales are hardly admissible. The genus *Phoenicopsis* in the Early and Middle Jurassic prevailed not only in Siberia but also occurred in abundance in Central Asia and even in Iran, i.e. within the Euro-Sinian region according to our terminology. *Ptilophyllum* is also common for this region. It was not until the Late Jurassic that the areas of these two genera came to be differentiated strictly enough because at this epoch the distribution of *Phoenicopsis* did not cross the boundary of the Siberian region.

Another rather cumbersome problem is correlation between the floras *Ptilophyllum* and Pentoxylales. The representatives of the latter are known only in the Southern Hemisphere (according to mobilistic concepts India was situated there in the Jurassic) while *Ptilophyllum* was widespread both in the Southern and in the subtropical areas of the Northern Hemisphere. *Ptilophyllum* is quite widely dispersed in India and in the Australian region whence the representatives of Pentoxylales came to be known. Is it right to single out the flora *Ptilophyllum* in such cases even if the area of *Ptilophyllum* encroaches on that of Pentoxylales? Identifying regions and provinces by geographical names we take as *principium divisionis* for this or that floristic complex rather than any single predominant taxon which, as we have shown earlier, may also be encountered in the neighbouring phytochoria.

Consideration of the phytogeographical maps of the geological past indicates that the majority of these maps were based on studies of the macrofossils of some plant or other. It was only later that the wider scale research in palynology allowed incorporation of palynologic data to be relied on in compiling maps. Palynologists started making their maps solely on the basis of dissemination of some taxa of spores and pollen, largely leaving aside the evidence from the investigations of plant macrofossils. This is due to the fact that the correlation of taxa according to macro-remains, on the one hand, and according to spores and pollen, on the other, is made difficult by the lack or extreme fragmentariness of our knowledge of correlational features in various organs of plants. This has led to the emergence of palynophytogeographic maps compiled by Samoilovitch (1977), Zaklinskaya (1977), Srivastava (1978), Herngreen and Khlonova (1981), Batten (1984) and others.

In this work in identifying phytochoria and reconstructing palaeoclimates we tried to employ specific distribution of various taxa of spores and pollen over and above the macrofossils. Special importance in palaeoclimatic reconstructions should be attached to the quantitative content of the pollen of *Classopollis* in palynological spectra. This pollen is produced by Cheirolepidiaceae widespread in the subtropics and tropical areas and particularly well developed in arid climates.

Let us consider a very important issue regarding the role played by phytogeographical schemes in resolving the geological and especially stratigraphic problems. The differentiation of floras on the Earth's surface in past geological periods was far greater than that of the marine fauna which, naturally, makes it more difficult to correlate continental deposits going by the plant remains contained in them. This is particularly true for long distances. The basic unit which preserves more or less uniform floral composition and similar sequence of floristic assemblages in time is the province or, more rarely, the region as we understand it.

The strata occurring throughout the area of the entire province wherein continental deposits of a particular division or of the entire system will largely be developed, include the biostratigraphical units rather than members. These units will be characterized by palaeofloristic assemblages (including palynological ones) which should be named horizons. The latter may consist of a score of members identified according to the lithological composition and replacing each other laterally but containing the same floristic assemblages. Cases in point are the horizons singled out for the Lower Cretaceous of the Lena province by Kirichkova and Samylina (1978) and for the Lower and Middle Jurassic by Genkina (1977). The floristic assemblages or floras characterizing the strata traced within large regions sometimes reaching the proportions of a province were tentatively termed stratofloras by Samylina (1974).

Since the floras of many phytochoria (especially regions) may differ considerably from one another in taxonomic composition it is necessary to find methods permitting the establishment of coeval status of continental formations occurring in two neighbouring phytochoria. To achieve this correlation one should select sequences situated in the boundary regions between two phytochoria. Such sequences are usually composed of continental deposits with plant macrofossils. Taxa concurrently typical for both phytochoria may be encountered in such sequences. For instance, typical representatives of the East Asia province (which is part of the Euro-Sinian region) occur in the sequences of the southern part of the Bureya basin (belonging to the Amur province of the Siberian–Canadian region) together with the forms inherent in the latter. Such zones with transitional compositions of floras or faunas are referred to as ecotones.

The second, though not less important, method for correlating plants by remains in the sequences situated in different phytochoria is one based on elucidating the influence of climatic changes on vegetation cover (the climatostratigraphic method). Fairly major climatic changes are of global nature, i.e. they are manifested simultaneously (though, as a rule, not to the same extent) in different parts of the Earth. Climatic changes involving both cooling and warming as well as moistening and aridization affect most of all the surface plants altering the appearance of vegetation cover over vast areas. Many extinct and living plants can serve as good indicators of climate along with certain types of sedimentary rocks.

The global climatic changes produce the most pronounced effect on the floras spread in the zones of conjugation of the subtropical and moderately warm climatic belts, i.e. in the ecotones. Taking as an example the peculiarities of development of the Late Palaeozoic flora, Meyen (1984) noted an interesting regularity observed for the floras of the Jurassic and Cretaceous. According to him the increase of temperature in a moderately warm belt is accompanied by the invasion of certain subtropical elements. However, no full replacement of the moderately warm subtropical vegetation occurs during this short time. On the contrary, during cooling the vegetation of the subtropical belt retreats in a body being ousted by the moderately warm or moderate one.

Similar phenomena can be observed in the histories of the Jurassic and Cretaceous floras. Thus, during the increase of temperature occurring in the Toarcian in Siberia, the moderately warm flora was fully replaced by the subtropical flora. However, the former received certain subtropical elements (some Bennettitales, thermophilic ferns belonging to the taxa earlier growing to the south; the amount of the pollen *Classopollis* produced by the Cheirolepidiaceae also increased) which made up a unique type of mixed flora mainly preserving its moderately warm appearance. Thus, no complete restructuring of this flora occurred. During the subsequent cooling at the beginning of the Middle Jurassic the southern thermophilic elements quickly disappeared from the floras of the Siberian region.

As we see it, such a process might have taken place only in the wake of an insignificant change of moisture content. If warming was accompanied by major aridization, as was the case in the Northern Hemisphere on the borderline between the Middle and Late Jurassic (Vakhrameev and Doludenko, 1976), the southern peripheral fringe of the moderately warm belt witnessed extinction of the moisture-loving vegetation, giving way to the drought-resistant plants therefore making up part of the subtropical flora. This process is obvious from replacement of the hygrophilic Middle Jurassic flora of the Middle Asia province and the south of ·the Siberian region (Kazakhstan) by the drought-resistant flora of the Late Jurassic. The aridization of climate drastically changed the composition of these floras. It brought about an almost complete disappearance of the Czekanowskiales as well as the majority of the Ginkgoales and *Nilssonia*, dramatically reducing the variety and the number of ferns growing in this territory. The process proved to be irreversible since, during the moistening of climate in the second half of the Early Cretaceous (Aptian–Albian), Czekanowskiales, Ginkgoales and *Nilssonia* failed to return to the south. Quite the reverse, they retreated still further to the north and north-east of Asia. These events can be especially clearly observed from the study of the sequence of South-East Karatau (Southern Kazakhstan)containing the vestiges of both the Middle and Late Jurassic floras.

The list of literature placed at the end of this monograph primarily includes works that appeared during the last 15 to 20 years. The contributions, especially articles, published earlier can be found in my earlier book (1964) as well as from a joint work (Vakhrameev *et al.*, 1970). The latter also comprises a history of the evolution of views on isolation and distribution of phytochoria of the Jurassic and Cretaceous in Eurasia.

In conclusion, I would like to thank A. B. Herman, M. P. Doludenko, E. L. Lebedev, A. I. Kirichkova, I. Z. Kotova, V. S. Lutchnikov and V. A. Samylina who gave me the opportunity of studying some of their materials. I am especially indebted to the senior assistant, K. A. Pechnikova, who retyped the whole text, selected cards for the list of literature and made the bulk of the drawings for this book.

❀ 2 ❀

Jurassic floras

2.1 Early and Middle Jurassic

In this overview I shall consider the Early and Middle Jurassic floras together due to their close inter-relation and virtually the same sets of phytochoria. The majority of localities of the floras of this age are to be found in the territory of Eurasia where they were investigated in greatest detail. In contrast, North America (except for the south of Mexico) illustrates an extreme paucity of floras of this age represented by very rare localities; this does not permit identification of separate phytochoria even as regions. In Africa, South America and Australia one can fairly confidently outline the Equatorial and Austral (Notal) regions. However, it is premature to single out provinces, although there are some grounds to do so.

These continents are therefore described in sections devoted to the composition and development of floras belonging to all three epochs of the Jurassic period since there is insufficient material to characterize the flora of each epoch of the Jurassic. These sections will be placed at the end of the chapter considering the Jurassic floras of the entire planet.

In comparison with the Late Jurassic and Early Cretaceous, the Early and Middle Jurassic epochs featured less differentiation of climate, generally a greater moisture content and a minor thermal gradient, as can be seen from the materials of Eurasia which are the best-studied palaeobotanically. This entailed uniformity of flora whose composition remained more or less the same over vast areas with slow evolution in time, and difficulties in differentiating continental deposits of this age on a palaeo-botanic base. These features were inherited from the climate of the Late Triassic epoch whose flora had developed very gradually.

In the Early and Middle Jurassic of Eurasia rapid coal formation occurred involving both the Siberian and the bulk of the Asiatic part of the Euro-Sinian region. In the Middle Jurassic the Equatorial region was taking shape with floras known from North Africa and South Mexico. The Austral (Notal) region is distinctly traceable in the Southern Hemisphere. According to the data quoted by Ilyina (1985), Victoria Land and the Antarctic exhibited Early Jurassic palynofloras with abundance of *Classopollis*

8

resembling the coeval palynofloras of Australia and South America and suggesting absence of an equivalent of the Siberian region in the Southern Hemisphere.

It is still not clear what was the climatic belt of the northern part of North America, i.e. Canada. It would seem logical to regard it as a continuation of the Siberian region as was the case in the Early Cretaceous for which the Siberian-Canadian region is identified. However, the absence in Canada of localities with macrofossils of Jurassic plants does not permit the drawing of any conclusion.

Ilyina (1985), who studied the composition of the Jurassic palynofloras of Canada from the literature data, was of the opinion that the palynofloras studied (e.g. that of the Hettangian) are closer to those of northern Europe which, as is known, were situated in the northern part of the Euro-Sinian region, while their relationship with the palynofloras of the Jurassic of Siberia is confined to some taxa only. If we consider the mobilistic maps of Smith and Briden (1977), compiled according to the palaeomagnetic data, we shall see that in the Jurassic northern America was shifted with relation to its present position somewhat to the more southern latitudes whereas the north-east of Asia displaced towards the north. It might well be that this is responsible for the relatively warmer climate of Canada as compared with Siberia and the similarities between Canada's palynofloras and those of Europe.

2.2 Siberian region

In the Early and Middle Jurassic this region occupied almost the whole of the Urals except for their southern part, Kazakhstan, all of Siberia, Mongolia, and western, northern and north-eastern China. It also embraced the north-eastern half of the European part of the USSR and, possibly, the northern part of Scandinavia.

Despite the vast territory of this region, it is still not possible to identify in it well-delineated phytochoria of a smaller rank, i.e. provinces, although there are certain differences between the compositions of floras of various regions. The list of works describing the Early and Middle Jurassic floras of the Siberian region can be found in the monographs of Vakhrameev (1964) and Vakhrameev et al. (1970). Out of the later and fuller publications of note are the contributions by Vlassov and Markovich (1979a, b) citing the most complete lists of the Jurassic plants for the South Yakutsk basin, the monograph by Ilyina (1985) devoted to the depiction of the Jurassic spores and pollen of Siberia and to the development of palynofloras during the Jurassic, and the work of Kirichkova (1985) reviewing the Jurassic flora of the Lena basin and Teslenko (1970) describing the Jurassic floras of western and southern Siberia and Tuva.

From the point of view of systematic variety the flora of the Early and Middle Jurassic of the Siberian region appears far inferior to the coeval flora of the Euro-Sinian region. The Siberian region flora numbers about 60 genera and correspondingly about 120 species while in the Euro-Sinian region the number of genera runs into the 200s whereas that of species reaches 500 over this period of time. Thus the flora of the Euro-

Sinian region is more than twice superior to that of the Siberian region as far as the number of genera and species is concerned. It is, then, only natural that there is virtually no possibility of identifying the exact number of taxa for some phytochoria due to different quantities of this or that taxon adopted by different palaeobotanists. However, despite more than 20 years that have elapsed since the publication of my earlier work this correlation has not changed.

The principal core of the Siberian flora of the Early and Middle Jurassic is constituted by Ginkgoales, Czekanowskiales and conifers represented by the woody forms as well as the herbaceous, relatively low-growing ferns. Among Ginkgoales the representatives of the genera *Ginkgo*, *Baiera* and *Sphenobaiera* predominate. The leaves of Ginkgoales are usually dissected into lobes, the most widespread being *Ginkgo* ex gr. *sibirica*. *Ginkgodium* finds are far rarer although the independence of this genus is doubted by some palaeobotanists.

Czekanowskiales are represented primarily by the genera *Phoenicopsis* and *Czekanowskia*. The co-occurring female fructifications *Leptostrobus* belong to the latter (Krassilov, 1972a; Harris, Millington and Miller, 1974). Czekanowskiales possess narrow, as well as linear (*Czekanowskia*) or ribbon-shaped (*Phoenicopsis*) leaves drawn into brachyblasts which are seasonally dropped by these plants. The shoots of *Phoenicopsis* and *Czekanowskia* are often found strewing the bedding planes within coal-bearing deposits corroborating the seasonal nature of the leaf-fall. Ginkgoales, such as, for example, *Sphenobaiera*, also dropped their shoots seasonally.

The monotony of outward morphology of both Ginkgoales and Czekanowskiales without investigating the epidermal structure of their leaves inhibits the development of the systematics of these groups of plants. In recent years research has been undertaken into epidermal structure in the USSR (Doludenko and Rasskazova, 1972; Krassilov, 1972a; Samylina and Kirichkova, 1973; Kirichkova and Samylina, 1979, 1983). This is especially true of the Czekanowskiales, the greater part of whose area of occurrence is situated in the territory of the USSR. Recently a new genus *Leptotoma* Kirichkova and Samylina was isolated within this order. The epidermal characterization of both the well-known and the later established species according to the features of the epidermal form of the new species allows updating of their diagnosis. This is of major importance for developing the stratigraphy of the carbonaceous Early Cretaceous and Jurassic deposits of Siberia in the first place, because the species identified with due regard to the epidermal structure have a narrowed stratigraphic distribution.

The pollen of Czekanowskiales remains so far unknown. Probably, this pollen, just as that of Ginkgoales, cycads and Bennettitales, belongs to the monocolpate type and possesses a smooth exine. This plain structure fails to permit the discernment of the pollen of this group of gymnosperms under a light microscope.

Ancient Pinaceae and *Podozamites* were predominant among the conifers. At this time the contemporary genera of the family of conifers that evolved perhaps from the Late Cretaceous, did not exist. The Jurassic forerunners of current Pinaceae possessed a combination of features observed now in the prevalent genera of this family. The Jurassic and Early Cretaceous Pinaceae were represented by the shoots (*Pityocladus*), isolated needles (*Pityophyllum*), cones (*Pityostrobus*), cone scales (*Pityolepis*) and winged seeds (*Pityospermum, Schizolepis*). Bisaccate pollen appeared widespread, belonging mainly to the Pinaceae.

Along with Pinaceae, *Podozamites* are quite abundant being represented by four to five species. These are to be found both in the shape of single leaves and, often, as foliated shoots. *Podozamites* seem to have been branch-dropping plants. The shoots of conifers with the scaly (*Brachyphyllum*) or short awl-shaped leaves were encountered only in two habitats in the south of the Siberian region (Eastern Kazakhstan and northern China).

Remains of ferns are mainly represented by the sterile fragments of complicated pinnate leaves belonging to the form-genus *Cladophlebis* whose species number in the Jurassic reaches 40. The species of this genus demand revision with a view to reducing them since minor differences between close species may often depend on what part of the double or triple-pinnate fronds some fragmentary trace or other belongs to. Towards the end of the Early Jurassic the fern of the genus *Raphaelia* came into existence (usually represented by *Raphaelia diamensis*). Both sterile and rarer fertile pinnae are found (the Vilyui depression, Bureya basin). The study of their sporangia by Vassilevskaya and Pavlov (1967) and Krassilov (1978) showed their association with *Osmunda*; Krassilov suggested that the species of this genus should be directly referred to the genus *Osmunda*.

In the Early Jurassic the fertile fragments of the fern genus *Coniopteris* appear. The number of their species increases towards the end of the Liassic achieving maximum variety as early as the Middle Jurassic. The sterile leaves of similar structure are usually attributed to the form-genus *Sphenopteris*. *Hausmannia* is to be mentioned as a representative of more rarely found genera. Rare findings of the ferns *Clathropteris* and *Dictyophyllum* as well as bi-pinnate *Hausmannia* are to be noted in company with some Matoniaceae (*Phlebopteris*) and Marattiaceae (*Marattiopsis*). As Teslenko (1970) and Ilyina (1985) found out, these originate mainly from the deposits of Kuzbass and the Irkutsk basin comparable with the marine Toarcian of northern Siberia and their emergence is due to the warming that took place at this time. The representatives of these genera are common elements of the Euro-Sinian region. In south-eastern Kazakhstan (the Kenderlyk trough located to the south of the town of Zaissan), Turutanova-Ketova (1962) described a new genus, *Kenderlykia*, based on the sterile leaves bearing a close resemblance to the leaves of bi-pinnate ferns.

The Cycadales and Bennettitales are far from numerous in the Early Jurassic floras of the Siberian region and the findings of their fossils are rare. They are represented by two species of *Nilssonia*, one species of *Anomozamites*, and several species of *Pterophyllum* (4) and *Taeniopteris* (3). *Ptilophyllum* increases and then rapidly vanishes in the Toarcian, being an important element in the Jurassic flora of the Euro-Sinian region. Equisetales grew in humid areas being represented mainly by *Equisetites* and

Neocalamites. The latter scarcely continues into the Middle Jurassic and its presence in the Jurassic deposits may be taken as a convincing proof of its association with the lower part of this system. *Equisetites beanii* appears as the largest and yet the rarest occurrence among the horse-tails.

Judging by the palynologic data (Ilyina, 1985), the carbonaceous sequences contain numerous spores of sphagnum mosses (*Stereisporites*). According to palynologic research the club-moss-like plants were also quite widespread. Lycopodiaceae are represented by the spores *Lycopodiumsporites*, and Selaginellaceae by *Uvaesporites* and *Perotriletes*. The rarity of the fragmentary findings of the foliated stalks of Lycopodiaceae is probably attributable to the fact that the creeping stalks of these plants do not lend themselves to transportation but rather decay on the site of growth. *Lycopodites tenerrimus* from the lower reaches of the Jurassic of the Irkutsk basin appears to be a case in point (Prynada, 1962). *Phyllotheca sibirica* was also encountered there just as in the Karaganda basin.

The Toarcian age occupies a special place in the evolution of the Early Jurassic floras of the Siberian region. It is a long time since remains of thermophilic plants (*Thaumatopteris schenkii* and *Ptilophyllum* sp. as well as *Dicroidium* sp. (?) uncommon for Siberia) were discovered (Samylina and Yefimova, 1968) from the marine deposits of the Toarcian characterized by ammonites outcropping in the middle stream of the Kolyma river. But the fullest evidence tracing the Toarcian thermophilic plants in Siberia was obtained by Ilyina (1985). She has identified the palynoflora which includes the spores *Matonisporites*, *Marattisporites*, *Klukisporites*, *Dictyophyllum* and *Contignisporites* as well as the enhanced content of the pollen *Classopollis* (Fig. 2.1). The remaining predominant part of plant remains is represented by the species typical for the Siberian region. The Toarcian marine deposits of the middle reaches of the Vilyui river are especially rich in southern elements. This assemblage features a combination of forms disseminated in the Siberian region along with the representatives of the Euro-Sinian region. Apart from the Vilyui basin, Ilyina detected this palynoflora in the eastern portion of the Yenissei–Hattang trough and in the east of Taimyr. The maximum of migrants from the Euro-Sinian region occurs in the Lower Toarcian (the zone of *Harpoceras facifer* and the lower half of the zone of *Dactylioceras athleticum*). The find by Kirichkova (1985) of *Ptilophyllum sibiricum* in the basin of the Vilyui river should be dated at the same level. This genus is characteristic of the subtropical and tropical areas of the Jurassic and Early Cretaceous.

A palynoflora rather similar to the above in composition containing the spores of thermophilic ferns mentioned above was discovered in the uninterrupted series of the Jurassic coal-bearing basins of the south of Siberia (the Ossinovsk member of Kuzbass, the Idansk member in the lower portion of the Periayansk suite as well as the Ust-Baleisk member in the upper strata of the Cheremkhovsk member of the Irkutsk basin, the upper strata of the Pereyaslav member of the Kansk basin, etc.). Over and above the spore–pollen floras, these sequences exhibited the macrofossils of thermophilic ferns: *Clathropteris elegans*, *C. obovata*, *Phlebopteris polypodioides*.

In earlier publications (1964, 1970) I referred the continental analogues of the Toarcian of south Siberia to the lower part of the Middle Jurassic relying on the presence in the floras of these deposits of several species of *Coniopteris* as well as on the occurrence of the enumerated bi-pinnate and Matoniaceae ferns in the Euro-Sinian region both in the Lower and Middle Jurassic (the Caucasus, Middle Asia). However, the detailed palynologic investigations by Ilyina (1985) who compared the palyno-floras from northern Siberia extracted from the faunistically characterized deposits with the palynofloras from the continental formations of southern Siberia compelled me to change my original point of view and agree with V. I. Ilyina and Y. V. Teslenko.

Reviewing again the sequences of the Jurassic of Kazakhstan I find that the Sorkusk suite developed in the Karaganda basin may be deemed to be analogous to the Toarcian stage in this region. It is composed mainly of sandstones with intercalations of conglomerates and aleurolites coloured red-brownish and greenish-grey. Deprived of coal-bearing beds the Sokursk suite contains the subadjacent carbonaceous

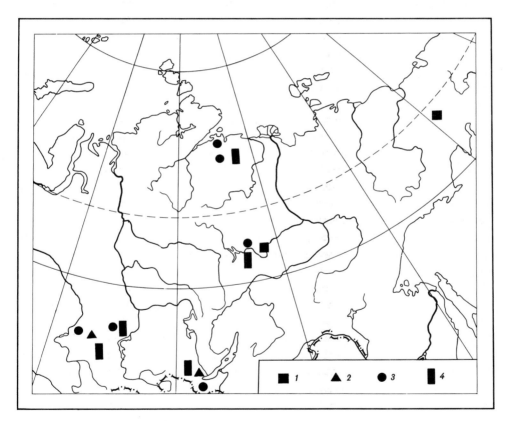

Fig. 2.1 Localities of thermophilic plant remains in Jurassic Toarcian deposits. Megafossils: 1, *Ptilophyllum*; 2, *Clathropteris*. Spores: 3, *Marattisporites scabratus* and *Klukisporites variegatus*; 4, Dipteridaceae (according to Ilyina, 1985).

deposits of the Dubovsk suite dated as Early Jurassic and is overlain by the carbon-aceous Mikhailovsk suite comprising a variety of *Coniopteris* permitting dating as Middle Jurassic.

Possibly, the Uzunbulak (banded) suite singled out in the sequences of the basin of the Alakul lake (south-east of Kazakhstan) and made up of the thin interlayers of aleurolites, argillites and sandstones is also of Toarcian age. This suite has an under-lying carbonaceous Alakul suite with numerous *Neocalamites* indicative of Liassic age. The Uzunbulak suite is overlain by the Katusk suite which is probably of Middle Jurassic age. The reason for dating these carbon-deprived suites as Jurassic is the fact that coal formation ceases in the Toarcian in south Siberia (the Ilansk suite in the Kansk–Achinsk basin, the Idansk member of the Peri-Sayan suite of the Irkutsk basin, the variegated member of the Tersuk suite in the Kuznetsk basin). The palynofloras of all these (Ilyina, 1985) correlate with the Toarcian palynofloras of northern Siberia. Unfortunately, the spores and pollen of the aforesaid suites of Kazakhstan were studied long ago and cursorily, which only enables us to date them as Lower Jurassic without further refinement. The position in the sequence between two carbonaceous suites and the lithologic appearance provide convincing reason for comparing these coal-devoid members with the Toarcian deposits of Siberia. The remains of plants discovered from these, including *Neocalamites*, suggest their classification with the Liassic flora of the Siberian region.

The climate of the Siberian region in the Toarcian may be characterized as moder-ately subtropical, i.e. intermediate between the genuine subtropical climate of the Euro-Sinian region and the warm-temperate climate of the Siberian region prevail-ing in the Early and Middle Liassic. It is noteworthy that the Toarcian warming brought about a dramatic reduction of coal formation mentioned earlier. The diminishing quantity of spores of sphagnum mosses is indicative of this, too. Perhaps this phenomenon is associated with the drying of swamps which indicates somewhat, though not greatly, enhanced dryness of climate. The same is corroborated by the emergence of variegated rocks in some member sections. Yet the nature of floras which, in spite of the cropping up of thermophilic species, were dominated by the Siberian elements (Ginkgoaceae, Czekanowskiaceae and ancient Pinaceae) does not allow identification of floras of the Euro-Sinian and Siberian regions and suggests the formation of a single region. Rather we are entitled to speak of the penetration to the north of certain subtropical elements owing to a short-term warming. During the transition from the Early to the Middle Jurassic, the boundaries of the Siberian region remained virtually unchanged but the composition of the flora has somewhat altered. According to V. I. Ilyina, thermophilic elements enumerated above disappeared at this point of time or, perhaps, even earlier, between the Early and Middle Toarcian, due to the cooling of climate.

The cycadophytes of the Middle Jurassic yielded the following: *Anomozamites lindleyanus, Ctenis* sp., *Nilssonia* cf. *acuminata, N. tenuissima, N. villosa, Pterophyllum* sp. *Heilungia* (*H. mongolensis*) was found in the Vilyui depression and in Mongolia but apart from it no other cycadophytes were detected in these districts in the Middle Jurassic. The principal, although scarce, finds of cycadophytes are confined to the basins situated in the south of Siberia. In the Middle Jurassic, *Neocalamites*, *Phyllotheca* and *Annulariopsis* virtually disappear.

The thin-stalked *Equisetites lateralis* (= *E. ferganensis*) emerges among the horse-tails. The amount and variety of *Coniopteris* markedly increases, reaching its maximum in the Bajocian. The remains of *Raphaelia*, represented in the main by *R. diamensis*, are more and more frequently encountered, as are the representatives of the genera *Klukia*, *Osmundites*, *Gleichenites* and *Todites* as well as bi-pinnate forms (save for *Hausmannia*). Matoniaceae and Marattiaceae are absent. *Sagenopteris* appears in the Siberian region for the first time. Ginkgoaceae, Czekanowskiaceae and ancient Pinaceae continue to prevail. The difference between the Early Jurassic and the Middle Jurassic species of these groups can only be established after study of the epidermis. Examination of the outward morphology alone usually does not permit delineation of the change of species during the transition from the Early to the Middle Jurassic.

2.3 Euro-Sinian region

This vast region, extending from western and southern Europe through the southern half of the European part of the USSR, Central Asia and southern China up to the coast of the Pacific Ocean, is divided into three provinces: European, Middle Asiatic and East Asiatic. The climate of this region was subtropical. The northernmost floral habitats of the Euro-Sinian region are to be found in Scotland, southern Sweden, Donbas, east of Smolensk within the Volga river bend near Kuibyshev, southern Urals, Issykkul, and in the province of Szechwan in China (the basin of the Yangtse river).

We shall dwell on some general features characteristic of the floras of the Early and then the Middle Jurassic of the entire region. For the Lower Jurassic numerous ferns of the families Dipteridaceae, Matoniaceae and Marattiaceae are typical. The first of these families is usually represented by the species *Clathropteris elegans*, *C. meniscoides*, *C. obovata*, *Dictyophyllum acutilobum*, *D. nathorstii*, *D. nilssonii*, *Thaumatopteris schenkii* and *T. brauniana*. The Matoniaceae feature *Phlebopteris braunii* and *P. polypodioides*, while Marattiaceae exhibit *Marattiopsis hoerensis* and *M. muensteri*. During the transition to the middle Jurassic some of these species vanish but the following persist: *Clathropteris obovata*, *Phlebopteris polypodioides*, *Marattiopsis muensteri* and *M. hoerensis*.

The family Osmundaceae is represented by *Osmundopsis plectrophora* and *Todites princeps* that replaced *Raphaelia* whose distribution was restricted to the Siberian region. *Stachypteris* emerges as a representative of the family Schizeaceae. Its first appearance was pinpointed for the Liassic in south China. *Klukia* is another genus of the same family appearing from the Middle Jurassic. Dicksoniaceae comprise the genera *Coniopteris* and *Eboracia*. The former is represented in the Early Jurassic by sporadic species, attaining its peak development in the Middle Jurassic when the number of species of this genus does not exceed three. In the Siberian region *Eboracia* is represented by the endemic species *E. kataviensis* known only in the Bureya basin situated in the zone of ecotone between the Siberian and the Euro-Sinian regions. The Equisetales include *Neocalamites* and a diversity of horse-tails whose largest forms are *Equisetites beanii*. The dissemination of the genus *Sagenopteris* in the Early Jurassic epoch appears to have been confined to the region under consideration only. This region yielded such finds as microsporophylls (*Caytonanthus*) as well as cupules with ovules (*Caytonia, Gristhorpia*) belonging to the order Caytoniales just as the leaves described under the generic name of *Sagenopteris* (Fig. 2.2).

The floras of the Euro-Sinian region of both the Early and Middle Jurassic epochs are characterized by the abundance of Bennettitales, cycads and cycadophytes of unestablished systematics (*Taeniopteris, Cycadites*). A number of Bennettitales were employed by us as indicators of subtropical and tropical climates outlining the extent of the northern boundary of the Euro-Sinian region. They include such genera as *Zamites, Sphenozamites, Otozamites. Dictyozamites, Pseudocycas* and *Ptilophyllum*. The rest of the Bennettitales, such as *Anomozamites, Pterophyllum* and *Nilssoniopteris*, as well as the form-genera like *Taeniopteris*, are encountered in the Siberian region, too, but they are far more numerous and diverse in the Euro-Sinian region.

Nilssonia are at a peak among the cycads. Micro- and megastrobili of these plants are described under the generic names of (respectively) *Androstrobus* and *Beania*. Dorato-phyllum, *Ctenis, Pseudoctenis* and *Paracycas* are encountered more seldom. Except for *Paracycas*, the representatives of these genera are known in the Siberian region as well.

Pachypteris, Cycadopteris, Ctenopteris (= *Ctenozamites*) and *Stenopteris*, lacking in the Siberian region, are spread among pteridosperms in the Euro-Sinian region. It has been established that the largest of these genera, *Pachypteris*, represented by several species, grew in the coastal zone of marine basins because its vestiges are always con-nected with the shallow deposits of such basins. The evidence available shows that remains of other pteridosperms are also associated with such deposits. It is not by chance that *Pachypteris* appears to be most widespread in the Late Jurassic which is the epoch of a vast transgression with a coastline of complicated configuration and a number of islands (particularly in Europe).

Ginkgoales (*Ginkgo, Baiera* and, rarer, *Sphenobaiera*) are widely represented, with Czekanowskiales being less extensive. The distribution of these groups within the Euro-Sinian region is far from uniform. The Middle Asia province is especially rich in them. In the Middle Jurassic the number of Ginkgoales and especially that of Czekanowskiales within the European province is drastically decreased.

The conifers of the Euro-Sinian region appear to be more diverse in comparison with those of the Siberian region. Apart from the ancient Pinaceae and *Podozamites*, particularly well developed in the latter region, wide areas became inhabited by the conifers whose shoots are covered by scaly or awl-shaped needles, probably belonging to the families Cheirolepidiaceae and Araucariaceae (*Cheirolepis, Brachyphyllum, Pagiophyllum*), and to *Elatocladus* as well as the cones *Elatides*.[1] The family Podozamitaceae is said to be linked with the seed scales *Cycadocarpidium* encountered in the Liassic but vanishing in the Middle Jurassic although the leaves *Podozamites* occur in the upper strata of the sequence. The leaves of the genus *Ferganiella* spread in the adjacent area of the Euro-Sinian and Siberian regions (Middle Asia, Southern Siberia, Far East) are structurally close to Podozamitaceae. The seed scales *Swedenborgia* are also characteristic of the Early Jurassic.

2.3.1 European province

This province embraces western and southern Europe as well as Greenland, southern Sweden and the greater half of the European part of the USSR including the Smolensk and Moscow regions; I refer the Caucasus here, too. References to numerous works devoted to the description of the floras of this and other provinces of the Euro-Sinian region published before 1970 can be found in the monograph by Vakhrameev *et al.* (1970). References to contributions that appeared in print later are obtainable from the list of literature in this monograph.

Let us deal with the definition of the boundary between the Triassic and the Jurassic which is established most distinctly in the sequences of Greenland and southern Sweden. Harris (1931, 1932a, b, 1935, 1937) was the first to establish the change of floras in eastern Greenland which, as was shown later from the research in the southern part of West Germany, took place between the Rhaetian and the Lower Liassic. The boundary between these is justified by the change of marine fauna (Gothan, 1914, 1935; Weber, 1968). *Lepidopteris ottonis*, the key form of the upper strata of the Triassic (Rhaetic), disappears at this point of time followed by *Camptopteris spiralis, Dictyophyllum exile, Ptilozamites nilssonii, Pterophyllum schenkii, P. ptillum, Cycadocarpidium erdmannii*, etc. The supplanting assemblage features *Thaumatopteris schenkii, Marattiopsis hoerensis, M. muensteri, Todites princeps, Dictyophyllum dunkeri, D. muensteri, D. nilssonii, Phlebopteris angustiloba, Pterophyllum subaequale, Clathropteris meniscoides, Swedenborgia cryptomeroides*, etc.

The Early Liassic witnesses the emergence of a characteristic megaspore of the lycopsid *Lycopodites scottii*. Layers with *Lepidopteris ottonis* get separated into the zone

---

[1] Cones of the genus *Elatides*, established by Heer (1876) are assigned to the family Pinaceae with which we should, perhaps, agree. It has been proved that certain female cones, investigated in detail seated on leaved shoots described from the Jurassic and Cretaceous under the name of *Elatides*, actually refer to the family Taxodiaceae (*E. bommeri* Harris, *E. asiatica* (Yokohama) Krassilov, *E. williamsonii* (Lindl. and Hutt.) Nathorst, *E. harrissii* Zhou, etc.). Morphologically the cones of these plants are quite different from those of *Elatides* from the Irkutsk basin and, therefore, they should probably not be described under this generic name. On the genus *Elatides* see the article by M. P. Doludenko and Y. U. Kostina (1987) – Ed. (M.A.A.).

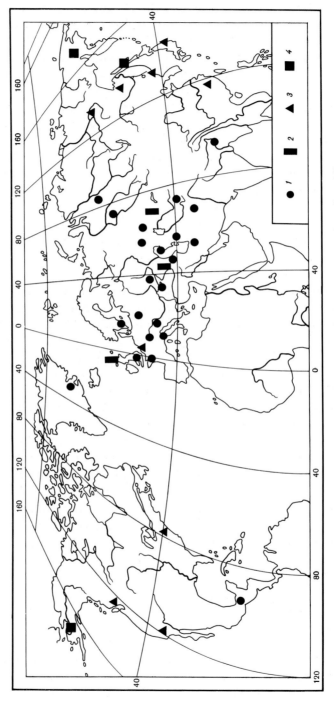

Fig. 2.2 Localities with *Sagenopteris* remains (Caytoniales). 1, Early and Middle Jurassic; 2, Late Jurassic; 3, Early Cretaceous; 4, Late Cretaceous. Female reproductive organs together with leaf compressions were found in the localities of Yorkshire (England), Scoresby Sound (Greenland), Poland, and Southern Far East (USSR).

of *Lepidopteris* while those with *Thaumatopteris brauniana* are in the zone of *Thaumatopteris*. Their analogues, as we shall see later, are traceable throughout the Euro-Sinian region and were recently established in China.

The age of the layers with *Thaumatopteris* in West Germany and Sweden is defined as Early Hettangian according to presence of the overlying strata containing *Schlotheimia angulata* (the upper zone of the Hettangian). In Greenland and in Romania these layers are located under the Pliensbachian whereas in Hungary they are under the Sinemurian. Judging by the composition of the plant remains found in eastern Greenland, southern Sweden, the southern part of West Germany, Hungary and Poland, the vegetation of these areas in the Early and, in places, in the Middle Liassic was dominated by Marattiaceae, Matoniaceae and especially Dipteridaceae ferns with large ramified fronds, as well as by certain Bennettitales. Ginkgoales (*Ginkgo, Baiera, Sphenobaiera*) occupied a very small place, still less Czekanowskiales. Both these latter groups had a woody appearance. Remains of conifers are rare and generally belong to Cheirolepidiaceae and Podozamitaceae. Going by the rarity of ancient Pinaceae remains, the latter played a subordinate role.

Close inter-relation of the continental deposits occasionally containing intercalations and (in Hungary, Romania and Poland) strata with marine formations lodged, as a rule, in the roof of the continental beds suggests availability of this vegetation association in the littoral lowlands including deltas. These lowlands used to be quickly inundated by sea during the ongoing European transgression that set in as early as the end of the Hettangian and reached its peak in the Toarcian. Over and above this assemblage there was a second one, particularly well manifested in France growing on the fringes of the Massif Central. Usually the remains of the plants of this association are to be found in the coastal deposits of the marine basin surrounding this ancient massif and containing Hettangian invertebrates.

Unlike the first association the second one comprises far more pteridosperms (*Ctenopteris, Cycadopteris*) as well as leaf remains classified as *Thinnfeldia* but, possibly, pertaining to *Pachypteris* (Doludenko, 1969). As distinct from the first association, Bennettitales include a multitude of *Otozamites*, although *Cycadites, Nilssoniopteris* and *Pterophyllum* appear to be rare occurrences. Cycads are represented, just as in the first association, by a few *Nilssonia* and *Pseudoctenis*. Ginkgoales are encountered more seldom (*Ginkgo, Baiera*), while Czekanowskiales have not been recorded at all. There are plenty of conifer remains which are mainly represented by shoots with scaly or awl-shaped needles the majority of which seem to belong to Cheirolepidiaceae and some, possibly, to Araucariaceae. Remnants of probable Cheirolepidiaceae were detected, too, represented by female cones of Cheirolepidiaceae (*Cheirolepis*). Remains of ferns, Ginkgoales and, particularly, Czekanowskiales are encountered here much more rarely than in the first vegetation association. Judging from the composition of ferns and Bennettitales occurring in both the assemblages they may be considered coeval.

A bennettito-coniferous association was probably growing on drained and relatively dry slopes gently descending to the sea. Boggy lowlands in these areas were apparently rather restricted, hence the paucity of ferns and lack of coal intercalations (Vakhrameev *et al.*, 1970). Habitats with a merged type of plant remains occur occasionally. I would like to dwell in more detail on habitats with plant vestiges of younger Late Pliensbachian (Domerian) age situated in the Venetian Alps. Rocks containing vegetation remains are represented by thin interspaced limestones and clays. Thin lenses of coal situated on the same stratigraphic level are observed therein. Above and below this level marine fauna remains can be detected in limestones. Imprints are usually perfectly preserved which is due to the fine grain structure of limestones resembling the Upper Jurassic limestones of Solnhofen (Bavaria) and the dolomitized limestones of the fresh-water lake Karatau in the south of Kazakhstan belonging to the upper part of the Upper Jurassic (Kimmeridgian?) and containing the remains of various animals and insects alongside plant fossils. Deposition of rocks with plant remains in the Venetian Alps appears to have proceeded in the calm shallow waters of the lagoon encircled by vegetation. The latter was dominated by conifers with rigid shoots covered with scale-like and awl-shaped leaves belonging to *Brachyphyllum*, *Dactylethrophyllum*, *Pagiophyllum* and *Elatocladus*. The conifers encompassed numerous but, probably, shorter Bennettitales along which *Otozamites* (6 species) and *Sphenozamites* (2 species) prevailed. *Ptilophyllum*, absent in the Hettangian and Sinemurian floras, was emerging. The ferns (*Hymenophyllites*, *Phlebopteris*) and horsetails (*Equisetites*, *Phyllotheca*) represented by rare fragments seem to have inhabited moist areas. Given the prevalence of Bennettitales with narrow leaves and scale-leaved conifers which probably belonged to the Cheirolepidiaceae, the climate was subtropic and dry. It is probable that the vegetation association of the Venetian Alps stood very close to the Hettangian floras of France mentioned above as far as growing conditions and systematic compositions were concerned.

The differences between floral compositions in northern Italy and those of southwest Germany can hardly be accounted for by the disparity in age between these floras. These differences were certainly brought about by the diverse ecological conditions of growth. The floras of the Hettangian in West Germany and southern Sweden, Romania, Hungary and Poland were associated with humid habitats (as indicated by the abundance of ferns and the occurrence of coals) whereas the majority of floras of France and the flora of northern Italy were growing under semi-arid conditions on drained coastal surfaces. It is interesting to note that the localities of the remains of *Nilssonia* as well as of Ginkgoales and rare Czekanowskiales tend largely towards humid habitats; the same is applicable to Podozamitaceae.

The Early Jurassic floras of the Caucasus, which we classify as associated with the European province, are represented only by rare habitats located in the northern Caucasus in the basin of the Baksan and Kuban rivers as well as in the vicinity of the Lok pass (Svanidze, 1971). They are related to the coal-bearing Hettangian and Pliensbachian formations containing preferentially the remains of horse-tails (*Neocalamites*), ferns (*Marattiopsis hoerensis*, *Dictyophyllum nilssonii*, *Phlebopteris polypodioides*), Ginkgoales, Czekanowskiales and rare cydadophytes (*Anomozamites*,

Nilssonia). Conifers are dominated by Podozamitaceae and ancient Pinaceae (*Pityo-phyllum, Schizolepis*). In the tufagenic deposits of the Domerian (the rivers Eshkaton, Tarakul-Tube, Chechek-Tokahassu) the composition of the flora changed dramatically as compared with the Pliensbachian. Here researchers failed to discover *Thaumatopteris schenkii, Hausmannia rara* or *Phlebopteris caucasica*. Czekanowskiales (*Czekanowskia rigida* and *Phoenicopsis* ex gr. *angustifolia*) as well as *Podozamites lanceolatus* were detected.

Despite its relative scarcity in taxa the Early Liassic flora of the northern Caucasus exhibits a certain difference from those of western Europe. This difference consists in extreme paucity and sometimes absolute lack of Bennettitales and in frequent occurrence of Ginkgoales and Czekanowskiales which approximates it with the flora of the Middle Asia province virtually adjoining the northern Caucasus. The composition of ferns coincides with that of the European flora but is far more impoverished, although this might have been caused by inadequate sample collections. In describing the floras of the Siberian region we saw that the Toarcian age was marked by a warming phase expressed in the penetration of the thermophilic elements to the north.

In western and southern Europe the Toarcian age witnessed a transgression; therefore the continental deposits of this time are virtually non–existent. *Neocalamites* sp., *Otozamites reglei, O. bucklandi* and *Ginkgo* ex gr. *digitata* have been found in the marine shallow deposits of France and Hungary while *Coniopteris* ex gr. *hymenophylloides*, *Ptilophyllum cutchense* and *P. acutifolium* were present in the Caucasus (river Knukh, basin of the river Kuban). The only relatively rich habitat of the Toarcian flora associated with the bituminous marly limestones is situated in the highland Budoshi in Chernogoria (Yugoslavia). There the following were identified (Pantič, 1981): *Equisetites columnare, Coniopteris* sp., *Caytonia* sp., *Pachypteris* sp., *Zamites* sp., *Otozamites beanii, O. gramineus, O. tennantus, Ptilophyllum pectinoides, P.* cf. *pecten, Brachyphyllum crucis, Pagiophyllum kurri, Elatides williamsonii, Podozamites lanceolatus*. This flora illustrates an almost complete absence of ferns and an abundance of thermophilic Bennettitales (*Zamites, Otozamites, Ptilophyllum*) as well as the shoots of conifers with scaly or awl-shaped leaves. The presence of *Pachypteris* that grew on sea coasts is quite striking. All this is indicative of the subtropical semi-arid climate of the Toarcian age in the Balkan peninsula which is consistent with the Toarcian warming phase so distinctly manifested in the Siberian region.

A more distinct conclusion on the Toarcian warming can be inferred from the review of palynologic data. These data show that in the Lower and Middle Liassic of the south of the European part of the USSR only individual grains of the pollen *Classopollis* (produced by Cheirolepidiaceae) are encountered. However, the average content of the pollen *Classopollis* rises up to 25–40%. In some samples, such as the Toarcian in the northern Caucasus (Yaroshenko, 1965), in the north-western part of Donbas (Semyonova, 1966), and in the Dniepr–Donetsk hollow (Shramkova, 1963) it is as high or higher than 50%. The earlier review of the locality in Chernogoria

provides no evidence on percentage content of the pollen *Classopollis* but points out
its partial presence. As was shown by my earlier work (Vakhrameev, 1970a, 1980) the
enhanced contents of the pollen *Classopollis* is indicative of climatic warming whereas
the proportions of 60–90% points to its aridity. However, it is also necessary to take
into account the lithologic composition of the deposits whence the samples were
collected. The samples from carbonaceous deposits that formed even under the conditions of subtropical or tropical but humid climate usually contain a significantly
smaller amount of the pollen than the coeval sequences originating in semi-arid and
arid climates. We shall revert to this issue later when considering Late Jurassic time.

Thus, the Toarcian optimum is easily traceable almost throughout Eurasia including India which was at the time situated in the Southern Hemisphere.

Taking up the review of the Middle Jurassic floras we shall directly address the flora
of Yorkshire (north-eastern England) minutely scrutinized by T. M. Harris who published the results of his investigations in five monographs (Harris, 1961, 1964, 1969,
1979; Harris *et al.*, 1974). This flora features all major groups of Jurassic plants represented by approximately 150 species.

The delta series connected with vegetation remains is underlain by Upper Aelenian
rocks (the zone of *Ludwigia murchisonae*) and is overlain by the Lower Callovian. The
upper portion of the series is separated from the middle by layers of marine origin with
Teloceras bladgeni which is a key form of the upper zone of the Middle Bajocian.
Consequently, the deposits of the delta series may be referred to the Bajocian or
Bathonian.

Considering the flora of Yorkshire, Harris came to the conclusion that no significant alterations occurred in its composition throughout the vertical profile. Therefore, this flora may be regarded as a single entity. It comprises mosses, rare lycopsids
and horse-tails suggestive of a humid sea climate. The horse-tails show the relics of
Annulariopsis and *Neocalamites* as well as *Schizoneura*.

Ferns are represented by *Todites, Osmundopsis, Klukia, Stachypteris, Dicksonia,
Coniopteris, Eboracia, Kylikipteris, Matonidium, Phlebopteris, Selenocarpus, Clathropteris,
Dictyophyllum* and *Hausmannia*, as well as by a number of species of *Cladophlebis*. These
feature the key form of the Jurassic – *Dictyophyllum rugosum*, the appearance of the
genus *Klukia* and the abundance of the species of *Coniopteris* (6 species). Pteridopserms
are represented by *Pachypteris, Stenopteria* and *Ctenopteris*.

Pachypteris papillosa (Thomas and Bose) Harris, whose anatomy has recently been
studied by Harris (1983) is looked upon as a large-sized bush forming mangrove
thickets along river mouths inundated by tides. Caytoniales are represented by leaves
(*Sagenopteris* microsporophylls (*Caytonanthus*), megasporophylls (*Caytonia*)).

Cycads and Bennettitales are abundant. The former are represented by diverse
species of *Ctenis, Pseudocycas, Paracycas,* various *Nilssonia* and the *Nilssonia*-associated
reproductive male organs of *Androstrobus* and female organs of *Beania*. Findings of
Bennettitales include *Anomozamites, Nilssoniopteris, Zamites, Otozamites* (the latter
being represented by 14 species), *Dictyozamites, Pterophyllum, Cycadolepis, Ptilophyllum,*

Williamsonia, Weltrichia and *Bucklandia*. A variety of Ginkgoales such as *Ginkgo, Baiera, Sphenobaiera, Eretmophyllum* and *Pseudotorellia*, are also to be found.

Czekanowskiales display several species of *Czekanowskia, Leptostrobus* and, possibly, the microstrobili of *Ixostrobus* belonging to this order. Conifers are numerous with *Brachyphyllum, Pagiophyllum, Geinitzia, Elatides, Cyparissidium, Pityanthus, Pityocladus, Schizolepis, Elatocladus, Torreya* and *Podozamites* as well as the representatives of such rare genera as *Marskea* Florin and *Poteridion* Harris.

The deltaic nature of host deposits and the presence of three members of marine origin indicates that the vegetation of Yorkshire occupied a littoral band. Its composition points to a humid subtropical climate. The humidity is corroborated by abundant ferns and horse-tails and the presence of mosses and lycopsids. The subtropical climate is borne out by the diversity of *Otozamites* and the presence of other Bennettitales as well as by a large number of the shoots of *Brachyphyllum* and *Podozamites* and rarity of the ancient Pinaceae remains.

The Yorkshire flora has been most extensively studied and is known to be rich in the species of the Middle Jurassic flora of the European province and, probably, of the entire Euro-Sinian region, and may be held up as a standard for the flora that grew under the conditions of humid and subtropical climate. It is this type of flora that was prevalent in the Middle Jurassic in this region.

A relatively rich Aalenian flora which was, however, less carefully studied than the Yorkshire one is known from Dagestan (Vakhrameev and Vassina, 1959; Vassina and Doludenko, 1968). Almost all genera detected in it are known from the Bajocian–Bathonian floras of Yorkshire but the specific composition turned out to be naturally different. The specific feature of the Dagestan flora is the abundance of *Otozamites* and presence of *Phoenicopsis* which brings it closer to the coeval floras of the Middle Asia province. But it is already in the Bathonian (Figs. 2.3, 2.4, 2.5) of Georgia (Twarkchelli, Tkibulli) that *Otozamites* crops up just as in Central Asia. *Pachypteris* is familiar in both the floras which is in full conformity with their habitat on littoral lowlands where coal-formation was a primary feature and was at intervals interrupted by transgressions leaving marine sediments.

The Middle Jurassic flora of Sweden, France, Sardinia, Poland, western Ukraine, the Volga area, Donbas and the southern Urals, briefly described in the work of Vakhrameev (1964), as well as new finds in Romania (Dragastan and Barbulescu, 1977–78), comprise the same genera as the flora of Yorkshire. Certain discrepancies may be ascertained only in comparing the specific composition which is due to the presence of local species.

The characteristic features of all these floras (including the northern Caucasus and the Transcaucasian (Figs. 2.6, 2.7, 2.8, 2.9, 2.10) amount to the wide distribution of *Equisetites beanii, Klukia exilis, Dictyophyllum rugosum* and various *Coniopteris*, constant presence of *Ptilophyllum* and, in the majority of floras, abundance of *Otozamites*. Availability of *Pachypteris* is also a specific feature, as is the extreme rarity of the finds of *Phoenicopsis* against a relatively higher frequency of *Czekanowskia*.

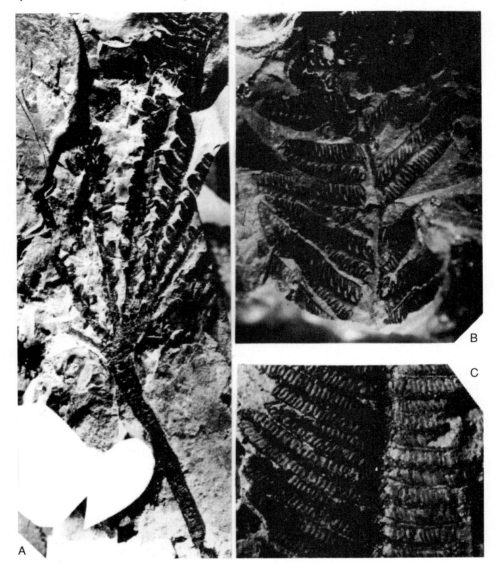

Fig. 2.3 Plant fossils of Bathonian age from Georgia. A–C, *Matonidium goeppertii* (Ettingshausen) Schenk; A, × 2; B, × 5; C, × 5.

Fig. 2.4 Plant fossils of Bathonian age from Georgia; natural size. A, *Paracycas harrisii* Doludenko; B, C, *Ptilophyllum okribense* Doludenko and Svanidze.

Fig. 2.5 Plant fossil of Bathonian age from Georgia; natural size; *Pachypteris speciosa* (Ettingshausen) Andrae.

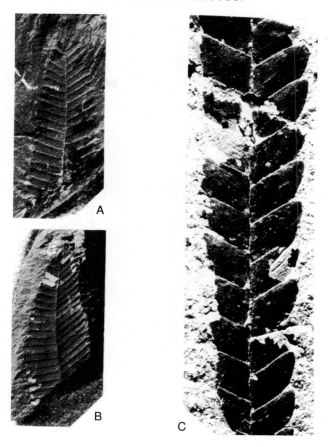

Fig. 2.6 Plant fossils of Callovian age from Georgia; natural size unless otherwise stated. A–C', *Ptilophyllum caucasicum* Doludenko and Svanidze; C, × 3.

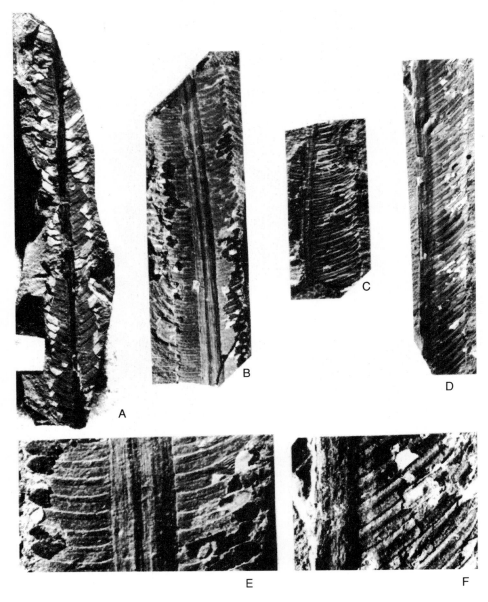

Fig. 2.7 Plant fossils of Callovian age from Georgia; natural size unless other-wise stated. A–F, *Paracycas brevipinnata* Delle. C, D, × 3.

A

B

Fig. 2.8 Plant fossils of Callovian age from Georgia; natural size unless otherwise stated. A–B, *Paracycas raripinnata* Doludenko. B, × 3.

Fig. 2.9　Plant fossil of Callovian age from Georgia; natural size; *Pachypteris lanceolata* Brongniart.

Fig. 2.10 Plant fossils of Callovian age; natural size unless otherwise stated; *Pachypteris lanceolata* Brongniart. A, from Georgia; B–C, from north Caucasus; C, × 3.

The list of Middle Jurassic floras includes, among others, the flora of Poland whose habitat is located west of Krakow on the northern slope of the Swentokshisk mountains earlier dated by Raciborsky as Liassic. However, the presence in it of the typical *Klukia exilis* confirmed by the research of Harris (1977) refers it to the Middle Jurassic. Up to now this species was encountered in a number of mutually very remote habitats (England to the Caucasus) only in the Middle Jurassic. Recently *Cycadopteris* sp. was discovered in a borehole core in the deposits of the Bathonian–Lower Callovian west of Smolensk (near the town of Safonovo). Representatives of the thermophilic genus do not spread north of the boundary of the Euro-Sinian region which allows delineation of this boundary within the limits of the Russian Platform. The second reference point is Samarskaya Luka whence V. D. Prynada (1937) identified a small flora including *Phlebopteris* sp. and *Elatides curvifolia*.

2.3.2 Middle Asia province

The Middle Asia province is situated in the south-western projection of the Eurasian continent. Its northern boundary passes south of Karaganda, Baikonur and Turgay. The southern boundary passes roughly along the northern coast of the Jurassic sea washing the Eurasian continent from the south. The habitats in northern Iran and Afghanistan are the most southerly of all known localities of the Middle Asia province.

The border with the East Asia province is rather general and is thought to pass east of Tsaidama. The western boundary extends through the Caspian Sea. Of course, all these boundaries are quite general since there are gradual transitions between the floras of the Middle Asia province and its surrounding phytochoria (except for the southern boundary passing along the sea basin).

The longest border of the Middle Asia province is in the north separating it from the Siberian region. Sea failed to penetrate here throughout the Jurassic period. The migration of temperate elements of the Siberian region southwards proceeded across this largely general border. Such elements included Ginkgoales, Czekanowskiales and ancient Pinaceae which, as we shall see later, are especially numerous in the floras of the northern part of the Middle Asia province.

A characteristic feature of the floras of the Middle Asia province in Liassic time is the presence of a number of species of *Cladophlebis* unknown in the European province but gaining access to the Siberian region. The Middle Jurassic is characterized by a number of *Coniopteris* species either keeping within or only slightly crossing its boundary. Many of the species possess finely dissected tiny pinnules. These species include principally *Coniopteris angustiloba* (= *C. simplex* (Lindl. and Hutt.), *C. furssenkoi*, *C. nerifolia*, etc.). Another feature of the Middle Jurassic floras is an uncommon diversity of *Nilssonia* (Fig. 2.11). In certain localities the number of species of this genus runs to 8 to 10 or even 15. Possibly, this number may be somewhat exaggerated due to the extremely fragmentary conceptualization on the part of some palaeobotanists. Finds of *Otozamites* are numerous. *Sagenopteris* is found in almost every locality both in the Early and Middle Jurassic. *Ferganiella* appears to be a typical genus widespread both in the Siberian and Middle Asian regions.

The composition of floras within the province under consideration changes markedly first and foremost from south to north, as observed by Sykstel (1954). This differentiation compels us to single out and separately describe three subprovinces. The Transcaspian subprovince comprises, among other things, the southern floras of Iran; Hissar embracing Darwaz, the southern slope of the Hissar ridge and the Kugitang ridge. The habitat of the southern flora of Afghanistan should be referred to the same subprovince although rather generally. The third subprovince – Fergana – encompasses Fergana and its current surrounding highlands as well as the basins of the lakes Issykkul and Somkul.

The Transcaspian and Hissar subprovinces were inhabited by vegetation assemblages covering lowland surfaces adjacent to marine basins. In the Bajocian and particularly in the Bathonian these were now and then flooded by the transgressing sea which interrupted accumulation of carbonaceous series. The Fergana subprovince accommodated intermontane depressions, the sea being kept away from this area throughout the Triassic and Jurassic periods. Coal accumulation was interspersed only by the uplifts of the surrounding highlands which brought about coarser terrigenous formations devoid of coals in the sequences. The same processes occurred in the other two subprovinces (Vakhrameev, 1969).

2.3.2.1 TRANSCASPIAN SUBPROVINCE

Only the youngest (possibly, the Toarcian) floras of the Early Jurassic are known within the borders of this subprovince. In Mangyshlak they are associated with the variegated Kokalinsk member while in Tuarkyr also with the multicoloured deposits of the uppermost strata of the Lower Jurassic. Ferns are rather few within this flora (Burakova, 1963; Kirichkova and Kalugin, 1973). They are represented by rare *Coniopteris* appearing for the first time in the section, as well as by individual species of *Hausmannia*, *Dictyophyllum*, *Clathropteris*, *Phlebopteris*, *Marattiopsis* and more diverse *Cladophlebis*. The latter are noted for such species as *C. suluktensis* and *C. aktaschensis*, which are typical representatives of the Middle Asia province.

The flora of this subprovince is rich in Bennettitales and cycads. The former are represented by several species, i.e. *Anomozamites*, *Nilssoniopteris*, *Ptilophyllum* and *Pterophyllum*. Finds of *Williamsonia* and *Williamsoniella* are numerous. No representatives of the genus *Otozamites* were found and this was probably not by chance. *Nilssonia* are diverse among the cycads (Fig. 2.11). Ginkgoales (*Ginkgo* and *Sphenobaiera*) and Czekanowskiales (*Czekanowskia* and *Phoenicopsis*) are quite common. Ancient Pinaceae (*Pityophyllum*) coexist with *Elatocladus* (= *Taxocladus*) and *Podozamites*.

In the Mangyshlak–Kokalinsk member remains of *Pachypteris lanceolata* were found (Kalugin and Kirichkova, 1968). The Tuarkyr member was established to contain

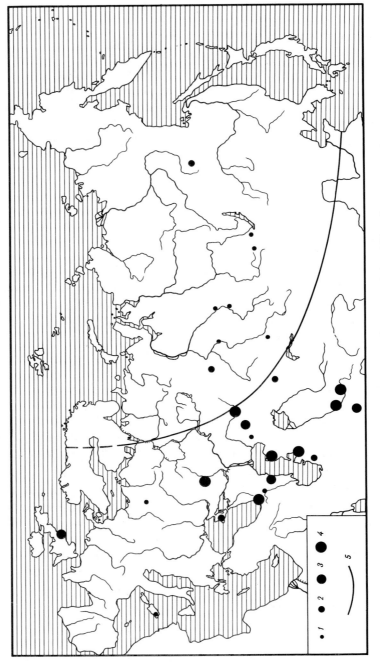

Fig. 2.11 Localities of Middle Jurassic floras containing *Nilssonia*. 1, one species; 2, two to three species; 3, four to five species; 4, six or more species; 5, boundary between the Siberian and Euro–Sinian regions.

none. No identifiable remains of plants were detected in the basin of the river Emba in the Lower Jurassic.

Palynologic data show that the content of the pollen *Classopollis* within the Transcaspian subprovince appreciably increased in the Toarcian (in Mangyshlak it fluctuates from 1.5 to 50%, in Tuarkyr from 0.8 to 8.5%, in Usturt from 18 to 86%). During transition to the Aalenian the content of this pollen was drastically reduced. High proportions of the *Classopollis* pollen coupled with the richness and variety of Bennettitales as well as the variegated colouration of rocks is strongly suggestive of a semi-arid climate for the Toarcian in the Transcaspian area indicating that here, just as in the Caucasus and the Siberian region, the Toarcian was marked by climatic warming.

The remains of the Middle Jurassic plants of the Transcaspian subprovince are familiar from all three stages of this division which enables us to give quite an adequate characterization of the flora of this epoch. The most ancient Middle Jurassic flora dated as the Aalenian is linked with the Tonashinsk suite of Mangyshlak, the lower half of the coal-bearing suite of Tuarkyr and the lower part of the sand-to-clay formation of the basin of the river Emba. It is characterized by the presence of horse-tails including the large-sized *Equisetites beanii* as well as by the significant increase in the specific composition of the fern *Coniopteris* including the species with small finely dissected pinnules such as *Coniopteris angustiloba*, *C. furssenkoi*, *C. neriifolia*, *C. simplex* and *C. vialovae*. It was only in the Aalenian that *Coniopteris margaretae* was encountered. Apart from *Coniopteris* the ferns were found to include rare *Hausmannia* and *Marattiopsis* (relics of the Early Jurassic flora), *Phlebopteris*, *Clathropteris* and *Dictyophyllum* as well as several species of *Cladophlebis*.

Bennettitales are represented by *Anomozamites*, *Nilssoniopteris*, *Pterophyllum* and *Ptilophyllum*. The latter comprises a new species, i.e. *Ptilophyllum ketovae* Kirichkova (Kirichkova, 1976b), possessing narrow laminae and very short segments diverging from the rachis at a right angle. The typical imprints found both in Mangyshlak and in Tuarkyr include narrow features with tiny segments whose edges are sort of hemmed. Originally these imprints were described as *Tyrmia cingularis*. Later, after revealing the invalidity of the genus *Tyrmia*, this species was classified by A. I. Kirichkova as belonging to the genus *Otozamites*. Mention of this species here is due to its being the key form of the Aalenian in the Transcaspian subprovince.

Nilssonia are not numerous. The generic composition of Ginkgoales, Czekanowskiales and conifers does not change appreciably when compared with the Toarcian. Along with the ancient Pinaceae (*Pityospermum*, *Pityophyllum*), *Brachyphyllum* and *Pagiophyllum* are encountered. The Bajocian in the Emba river is considered to incorporate the upper part of the sand–clay formation, in Mangyshlak – the Karadiirminsk suite, in Tuarkyr – the upper portion of the coal-bearing formation.

Just as in the other subprovinces of the Middle Asia province, the Bajocian here was the time of the greatest variety and abundance of *Coniopteris* (up to 6 species) and *Nilssonia* (up to 9 to 10 species). The latter, more often than not, is represented by

Nilssonia vittaeformis. Klukia exilis emerges being a characteristic form of the Middle Jurassic promulgated from Middle Asia to England and occurring *par excellence* in the Bajocian. Another intrinsic form of the Middle Jurassic is *Dictyophyllum rugosum*, the last representative of this genus which was nearly extinct. Isolated finds of *Marattiopsis hoerensis* may be regarded as relics.

The Bajocian of the Transcaspian subprovince is believed to include an extremely small number of Bennettitales. Thus, in the Emba river a single species of *Pityophyllum* was found (Baranova and Kirichkova, 1972; Baranova, Kirichkova and Zauer, 1975), in Mangyshlak two species of *Anomozamites* were encountered (Kalugin and Kirichkova, 1968), while in Tuarkyr only *Pterophyllum tietzei* was detected (Burakova, 1963). The composition of Ginkgoales, Czekanowskiales and conifers remained roughly the same as the Aalenian. We should note the absence of *Czekanowskia* in the floras of the river Emba and in Mangyshlak, whereas *Phoenicopsis* persisted everywhere. The rare occurrence of Bennettitales and the abundance of *Coniopteris* and *Nilssonia* suggest a slight climatic cooling as well as significant humidization in the Bajocian. The latter is corroborated by the wide dissemination of coals.

The Bathonian age is characterized by the onset of a transgression that reached its maximum development as early as the first half of the Late Jurassic. Therefore, the continental deposits of the Transcaspian subprovince witness the appearance of members of coastal and marine formations. The composition of the plants whose remains were discovered in the Bathonian becomes tangibly impoverished. Reduction is obvious as far as the specific variety of *Coniopteris, Cladophlebis* and *Nilssonia* is concerned. Uncommon Bennettitales are represented by the rare finds of *Ptilophyllum, Pseudocycas* and *Williamsoniella. Ferganiella* was prominent in the Emba river. In the Tuarkyr, *Pachypteris lanceolata* makes its appearance, testifying to the proximity of sea coast.

An abrupt decrease (Emba) and, in places, complete disappearance of Ginkgoales and Czekanowskiales coupled with increasing numbers of the shoots of *Brachyphyllum* and *Pagiophyllum* as well as the greater content of the pollen *Classopollis* in the spore–pollen spectra indicate the onset of climatic drying. The Middle Jurassic flora of the southern Urals (south of the town of Orsk), exhibiting an uncommon variety of *Nilssonia* and an abundance of Ginkgoales and Czekanowskiales, is close in composition to the flora of the Transcaspian subprovince. *Leptostrobus* is encountered amid Czekanowskiales. This locality is situated on the northern peripheral edge of the Transcaspian subprovince.

In the south the floras of northern (Elburz) and south-eastern Iran may be referred to the Transcaspian subprovince. The former flora was described by Sadovnikov (1977), Barnard (1965, 1967) and Kilpper (1964). The flora of south-eastern Iran (Kerman) was determined by T. A. Sykstel (Polyansky, Safronov and Sykstel, 1975). The lists of both the Liassic and Middle Jurassic flora cited in the above-mentioned works are not different in principle from those for Mangyshlak and Tuarkyr which were quoted above.

The Liassic is characterized by the absence of *Thaumatopteris schenkii* as well as by a variety of the species of *Dictyophyllum, Phlebopteris, Marattiopsis* and *Todites*, some of which ascend to the middle Jurassic. The lower part of the Liassic reveals several species of *Pterophyllum. Otozamites* is encountered very rarely and there is a single find of *Dictyozamites*. Isolated species of *Ginkgo* and *Sphenobaiera* are found. *Scoresbya dentata* has been detected.

Czekanowskia ex gr. *rigida* and *Phoenicopsis* ex gr. *angustifolia* were found in the Elburz and Kermadesk districts, both in the Lower and Middle Jurassic rocks. These appear to be the southernmost finds of Czekanowskiales. Conifers are made up of *Podozamites, Cycadocarpidium* (only in the lower part of the Liassic), *Elatides* and *Brachyphyllum*. An almost complete absence of ancient Pinaceae (*Pityospermum, Schizolepis*, etc.) is an outstanding feature.

The transition to the Middle Jurassic witnesses the emergence of *Klukia exilis* and *Eboracia lobifolia*, a diversity of *Coniopteris* (including the ones with the finely dissected pinnules), two species of *Pityophyllum* and diverse *Nilssonia*. Sadovnikov (1977) pointed to the presence of *Pachypteris* in Elburz. The peculiarities of the Transcaspian subprovince viewed as a whole are thought to comprise the occurrence of *Pachypteris* in the Liassic and Bathonian of the various areas signalling the proximity of a littoral line; as well as the presence of a number of species in common with the European province and especially the Caucasus. These are *Coniopteris margaretae, Dictyophyllum rugosum, Marattiopsis hoerensis, M. muensteri, Phlebopteris polypodioides, P. muensteri*, etc. *Otozamites cingulatus* seems to be an original species so far not identified outside the subprovince. Another feature is the lack of diversity and scarcity of Ginkgoales and an almost complete disappearance (from the Aalenian or the Bajocian) of *Czekanowskia*. It is noted only in Iran in the Middle Jurassic. *Phoenicopsis* is present on a large scale both in the Liassic and the Middle Jurassic.

2.3.2.2 HISSAR PROVINCE

Its boundaries outlined above reveal the most ancient floras of the Liassic (Luchnikov, 1967, 1987), known from Darwaz (the basin of the river Obiniou) and the southern slope of the Hissar ridge (Guliob). From the Jurassic section of Darwaz (the first complex) the representatives of the following genera were collected: *Todites, Clathropteris, Anthrophyopsis, Anomozamites, Zamites, Sphenozamites, Otozamites, Pterophyllum, Pseudoctenis, Nilssonia Taeniopteris* (many) and *Macrotaeniopteris*. Conifers are represented by the shoots of *Brachyphyllum, Pagiophyllum* and *Elatocladus*. A striking feature is the abundance of Bennettitales, including *Otozamites*, as well as of the rarely encountered *Sphenozamites*. There are many scale-leaved conifers (*Brachyphyllum*) or conifers with awl-shaped or short bent leaves (*Pagiophyllum, Elatocladus*). Ferns are exceptional. The lower part of the Liassic or even the uppermost of the Triassic is suggested by a find of *Anthrophyopsis*. Ferns (*Todites, Phlebopteris, Hausmannia, Cladoplebis*) are far more numerous in the lower part of the Jurassic sequence in the upper reaches of the river Guliob. V. S. Luchnikov noted the presence of *Coniopteris*, which normally is not found until the second half of the Liassic. Bennettitales are dominated by various species of *Pterophyllum* (6 species). Isolated Ginkgoales were found (*Ginkgo*, *Sphenobaiera*). *Pityophyllum* is present in the midst of conifers along with *Brachyphyllum* and *Elatides*. Conifers are tantamount to 25 to 30% of the plant vestiges collected from both Darwaz and Guliob in the lower portions of the Jurassic section.

We appear to be dealing with two oryctocoenoses reflecting the composition of two vegetation associations. The first of these (Darwaz) inhabited the area under the conditions of a dry climate indicated by the rarity of ferns, absence of Ginkgoales and abundance of *Brachyphyllum* and *Pagiophyllum*. The second one grew under more humid conditions (frequent ferns and absent Ginkgoales). The humid conditions in which the vegatation of Guliob was growing are also indicated by finds of plant remains in a thin lens of carbonaceous clays deposited in the upper part of the bauxite-bearing member of the Tashkutansk suite.

Complexes more impoverished but richer in composition were detected in a number of other localities on the southern slope of the Hissar ridge (Sayed, Dibod, Oksu). Some of these revealed Czekanowskiales. The remains of Bennettitales are everywhere predominant over cycads represented by *Nilssonia* and, more rarely, by *Ctenis* and *Pseudoctenis*. *Neocalamites* were found, too.

The second and younger complex of the flora is represented in the section of the coal deposits of Tashkutan and Shargun (the southern slope of the Hissar ridge) as well as in the deposit of Fan-Yagnob. In Tashkutan one can observe the bedding of host deposits stratigraphically above the rock member containing the plant remains of the first complex. It is dominated by various ferns including Dipteridaceae (*Clathropteris obovata*, *Thaumatopteris elongata* and *Hausmannia leeiana*), Matoniaceae and Marattiaceae. Ferns referred to the form-genus *Cladophlebis* are widely disseminated. At the same time the number of cycads is reduced. This particularly applied to Bennettitales; *Otozamites*, *Zamites*, *Sphenozamites* and *Anthrophyopsis* vanish altogether. Rare *Coniopteris* were found. Ginkgoales and Czekanowskiales are encountered quite often.

The third complex was established by V. S. Luchnikov in the coal-bearing deposits of Shargun, Mianadu, Sayed and Tashkutan. A diversity of *Cladophlebis* whose number of species reach eight are prevalent among the ferns. The *Cladophlebis* include certain species typical for the Middle Asia province (*Cladophlebis aktashensis*, *C. magnifica*, *C. suluktensis*). The number and the specific variety of Dipteridaceae and Matoniaceae are sharply diminished. Out of these occasionally one encounters *Phlebopteris*, *Clathropteris* and *Hausmannia*. Up to two or three species of *Coniopteris* are sporadically encountered. The cycadophytes are represented only by isolated species of *Anomozamites*, *Pterophyllum*, *Ctenis* and *Taeniopteris*. Ginkgoales and Czekanowskiales are not numerous; they were not detected at all in the deposit of Mianadu.

Conifers are represented in abundance. Over and beyond the prevalent *Brachyphyllum* and *Pagiophyllum*, *Elatocladus* and *Pityophyllum* appear to be numerous while *Podozamites* are rare. In certain habitats, e.g. in Fan-Yagnob, various *Neocalamites* occurred along with some species of *Ginkgo, Sphenobaiera, Czekanowskia* and *Ferganiella* with the number of species of *Podozamites* increasing.

The fourth complex confined to the terrigenous formation deprived of coals is well represented in Darwaz (the river Gring), on the southern slope of Hissar (Luchob and Hodjamaston) and in the Kugitangtau ridge (the section of Kampyrtube). *Neocalamites* are virtually absent in it, yet *Equisetites* are present including *E. lateralis* (= *E. ferganensis*). Among the ferns most prevalent are the large-pinnule species of the genus *Cladophlebis* whose number of species in certain localities attains 10 to 12. Two to three species of *Coniopteris* are constantly present including sometimes *Eboracia lobifolia*. *Hausmannia* and *Marattiopsis* frequently appear. The increasing number of species of *Nilssonia* (up to three to five) and appearance of *Ptilophyllum* is another striking feature. Isolated *Anomozamites, Pterophyllum* and *Taeniopteris* are preserved. Always present are certain species of Ginkgoales and a few Czekanowskiales. The number of conifers is drastically diminished with *Brachyphyllum* and *Pagiophyllum* vanishing altogether in their midst.

Here we ought to dwell on the relationship among the four complexes singled out by V. S. Luchnikov for Darwaz and the Hissar ridge, on the one hand, and the floras characterizing the phytostratigraphic horizons established somewhat earlier by Genkina (1979) for the Jurassic of the east of Middle Asia on the other. The first and the second complexes identified by Luchnikov (1987), on the whole correspond to the flora of the Tashkutin horizon. The third complex is referred by Luchnikov to the Pliensbachian, while Genkina dates it as the Pliensbachian–Toarcian (the Shargun horizon). The fourth complex is considered by Luchnikov to belong to the Toarcian and one cannot but agree with him on this score.

As we have already seen when reviewing the Jurassic floras of the Siberian region, a phase of climatic warming has been established for the first half of the Toarcian (Ilyina, 1985). This time is marked by the enhanced proportions of the content of the pollen *Classopollis* produced by Cheirolepidiaceae whose shoots are described under the form-generic names of *Brachyphyllum* and *Pagiophyllum*. Besides, this epoch in Siberia witnesses the emergence of thermophilic ferns (Dipteridaceae, Marattiaceae and Matoniaceae). If we proceed from climatostratigraphic principles then the Early Toarcian should correspond to the third complex of Luchnikov while the Late Toarcian to the Aalenian corresponds to the fourth complex.

The third complex, as was indicated above, is rich in the shoots of *Brachyphyllum* and *Pagiophyllum* whose presence is suggestive of a warming phase, whereas in the fourth complex, which supplants the third one, these conifers disappear. This change occurs in Siberia on the boundary of the Early and Late Toarcian. At the same point of time the Middle Asia province sees thriving *Coniopteris* and *Nilssonia* earlier represented by one to two species while Bennettitales become more infrequent. Therefore, the fourth complex is most probably of the Late Toarcian to Aalenian age and can be compared with the complex of the Vandob horizon (Genkina, 1979). It is likely that subsequently researchers will manage to separate the Aalenian deposits proper from the Late Toarcian ones going by changes in floristic composition. So far no criteria for such differentiation have been ascertained.

The flora of the Bajocian is composed of plant remains gathered in the first place in the coal-bearing strata on the eastern slope of Kugitangtau laid immediately over the marine deposits of the Upper Bajocian (the zone *Parkinsonia parkinsoni*). Age analogues of the Bajocian (the Sherdjan horizon of R. Z. Genkina) are known in the northern part of Surkhantau and the southern slope of the Hissar ridge, as well as in Darwaz (the upper part of the Gring suite).

A specific feature of this flora is the maximum development of the fern *Coniopteris* whose species number up to 15. Among these many forms with small, finely dissected pinnules are encountered, i.e. *Coniopteris furssenkoi, C. neriifolia, C. zindanensis, C. vialovii, C. pulcherrima,* etc. *Klukia exilis* and *Klukia westii* appear. *Cladophlebis* are numerous although researchers have so far failed to identify their species typical for the Bajocian. One to two species of *Marattiopsis, Clathropteris* (usually *C. obovata*) and *Phlebopteris* are retained. *Dictyophyllum rugosum*, characteristic of the Middle Jurassic of Europe and the Transcaspian subprovince, is not found. Following a prolonged lapse (the Pliensbachian–Aalenian) a few *Otozamites* appeared again. It was typical to observe two to three species of *Ptilophyllum* and certain species of *Anomozamites* and *Pterophyllum*.

Just as for *Coniopteris*, the Bajocian age appears to be a thriving period for *Nilssonia* whose number of species runs to 20. Such a diversity was never encountered in any other phytochoria of the Jurassic period. It is probable that this figure is somewhat exaggerated owing to the major morphologic variability of the leaf lamella of *Nilssonia*, but even so, the frequency of occurrence of *Nilssonia* in the Bajocian and their variety are very significant.

The fullest sections of the Bathonian continental deposits containing plant remains are situated in Darwaz and the basin of the river Fan-Yagnoba (Luchnikov, 1973, 1982), as well as in the Yakkabag mountains (Gomolitsky, 1968). R. Z. Genkina identified them as the Shelkan horizon due to the fact that it is impossible for this as well as for other continental strata to be shown to correspond completely to the stage whose stratotype is identified for marine formations. In the east of Tadjikistan, in Darwaz, the Bathonian is represented by the variegated deposits, possibly of deltaic origin devoid of coals. In the western areas the variegated coloration vanishes. In the sequence of Fan-Yagnob the Bathonian is composed of coal-bearing deposits passing up in the section into shallow-water marine formations (the Kukhimalek suite) classified as upper Bathonian (Luchnikov, 1982). In the Yakkabag mountains and south of them in the sequences of Surkhantau and Kugitangtau, the carbonaceous deposits placed in the marine upper Bajocian are overlain by the marine strata of the Bathonian. The Bathonian ushers in a new stage of development of the Jurassic floras, paving the way for a transition to the original floras of the Late Jurassic.

By and large the Bathonian flora experiences a certain suppression attributable to the onset of the aridization of climate. The specific diversity of *Coniopteris* and

Nilssonia as well as of *Cladophlebis* is reduced. These, coupled with Cheirolepidiaceae (*Brachyphyllum*, *Pagiophyllum*), form the background for the Bathonian flora. *Ptilophyllum* is constantly encountered while *Otozamites* (particularly *O. graphicus*) are found more frequently in the Bajocian. Ginkgoales (*Ginkgo*, *Sphenobaiera*) are not numerous. Czekanowskiales virtually disappear. Abundance of the shoots of *Brachyphyllum* and *Pagiophyllum* should be emphasized, whereas the remains of ancient Pinaceae (*Pityophyllum*) are rare. *Pachypteris lanceolata* growing along the coasts of sea basins was detected in Darwaz and the south-western slopes of the Hissar ridge. It makes its appearance within the Hissar subprovince only in the Bathonian; this is probably due to the transgression of this time advancing from the south-west.

The Hissar subprovince occupying the south-western part of Middle Asia and Afghanistan exhibits a richness in Bennettitales including *Otozamites* (mainly in the Liassic and Bathonian) and *Ptilophyllum* (largely in the Middle Jurassic). In comparison with the Transcaspian subprovince the number of species that are also common in Europe is on the decrease. At times Cheirolepidiaceae (*Brachyphyllum*, *Pagiophyllum*) which are especially common in the Liassic and Bathonian become quite widespread. Ginkgoales and Czekanowskiales, just as ancient Pinaceae, were not numerous.

2.3.2.3 FERGANA SUBPROVINCE

The Jurassic deposits of continental origin are exceptionally developed within the Fergana subprovince's boundaries because even the powerful Jurassic transgression failed to make inroads here. The flora associated with the most ancient deposits of the Jurassic is well represented in the basin of the lake Issykkul. The composition of the flora indicates its Early Liassic (the Dgil suite) and Middle Liassic (the Aksai suite) age, with the uppermost parts of the sequence possibly belonging to the Toarcian.

According to the evidence obtained by Genkina (1966) the Early Liassic saw the development of a variety of *Neocalamites* and *Equisetites* including *N. issykkulensis*, as well as the endemic genus *Dzergalanella* resembling *Schizoneura*. The ferns are represented by the following genera, *Marattiopsis*, *Todites*, *Hausmannia*, *Clathropteris*, and by various *Thaumatopteris* and *Cladophlebis*. An interesting phenomenon is the emergence of *Raphaelia* which is a representative of the Siberian flora co-occurring with *Todites* disseminated in the Euro-Sinian region. The Bennettitales exhibit only two species of *Anomozamites* and the cycads only one species of *Nilssonia*. On the other hand, *Ginkgo*, *Sphenobaiera* and *Podozamites* are very diverse. *Czekanowskia* and *Phoenicopsis* are numerous. Two species of *Cycadocarpidium* were detected.

The classification of this complex as belonging to the Liassic rather than to the Rhaetic is based on the presence of *Todites princeps*, the form familiar from the zone containing *Thaumatopteris* and absent in the Rhaetic (the zone of *Lepidopteris*). This classification is also supported by the available variety of *Cladophlebis*, and Ginkgoales and Czekanowskiales scarcely represented in the Rhaetic sedimentation (the Koktuisk suite) of Issykkul. The younger complex referred to the Middle Liassic–

Lower Toarcian reveals an increased diversity of Bennettitales and cycads represented by *Anomozamites* (2 species), *Pterophyllum* (2), *Nilssonia* (1) and *Pseudoctenis* (1). The possibility of classifying the upper part of the sequence as Lower Toarcian is suggested by the disappearance of coals in the second half of the Aksai suite and the reduction of the number of *Neocalamites* and horse-tail remnants, as well as by the emergence of the shoots of *Pagiophyllum*. In the analogues of the upper parts of the Aksai suite (the rivers Ichki-Djerges, Djergalan, Tyup) the pollen *Classopollis* crops up whereas the host rocks acquire a variegated coloration. These peculiarities are likely to have been brought about by the warming and certain drying of the climate and temporally correspond to the phase of warming established for Siberia by Ilyina (1985).

The richest Middle Jurassic flora of the Fergana subprovince originates from the Fergana ridge. Prior to the detailed research by Genkina (1977, 1979), the bulk of this flora whose remains came out of the Tuyyk and Chaartash suites was regarded as Early Jurassic in age. This error did not allow the drawing of a well-judged correlation between the Jurassic deposits of the Fergana ridge and other sequences of Middle Asia.

R. Z. Genkina believes that the Tuyyk suite approximately conforms to the Aalenian, the Chaartash suite to the Bajocian, and the Zindan suite (which had always been referred to the Late Jurassic) to the Bathonian. All three suites, i.e. the entire Middle Jurassic, are characterized by the prevalence of the ferns *Coniopteris* and *Cladophlebis* although their specific variety is on the ebb downward in the sequence. The majority of the species are represented by the forms described from the Jurassic of Middle Asia usually confined in their distribution to this territory. Fourteen species of *Cladophlebis*, including *C. aktashensis*, *C. bidentata*, *C. nifica*, *C. sulkata*, *C. suluktensis*, *C. stenolopha*, *C. undulata* and *C. zauronica*, are quoted for the Aalenian (the Tuyyk suite). Also present are five species of *Coniopteris*, including *C. angustiloba*, *C. kumbelensis* and *C. spectabilis*, also established on the basis of the material from Fergana. Other species comprise *Todites princeps*, *Osmundopsis plectrophora* and *Eboracia lobifolia*, as well as *Phlebopteris muensteris* and *Clathropteris obovata*. The Equisetales species show a probable relict, i.e *Lobatannularia* and the frequently met *Neocalamites* and *Equisetites* sp.

Bennettitales are represented by isolated species of *Anomozamites* and *Pterophyllum*. *Nilssonia* are numerous (8 species) as well as *Ctenis* (3 species). A more common occurrence is *Taeniopteris*. Various *Ginkgo*, *Baiera*, *Sphenobaiera* and *Podozamites*, as well as *Ferganiella*, are frequently encountered. Czekanowskiales are, as usual, represented by *Phoenicopsis*, *Czekanowskia* and *Leptostrobus*, and the ancient Pinaceae by *Pityophyllum*, *Pityocladus*, *Schizolepis*; *Brachyphyllum* appears rarely.

In the Bajocian flora (the Chaartash suite) the diversity of *Coniopteris* rises (9 species), while that of *Cladophlebis* is reduced (8). *Klukia exilis* makes an appearance; this is the form encountered primarily in the Bajocian. Other newcomers are *Ptilophyllum acutifolium* and a rare fossil for the Fergana subprovince – *Otozamites* (1 species). In other respects the changes in the composition of plant remains from the Chaartash suite as against the Tuyyk one are minor and insignificant.

The Bathonian flora (the Zindan suite) follows the pattern of other localities of the Middle Asia region, undergoing impoverishment in terms of specific composition rather than of genera. Of the major groups, Czekanowskiales totally disappear. The shoots of *Elatocladus* and *Brachyphyllum* are encountered in almost all the sequences of the Zindan suite. The vanishing of Czekanowskiales and the extensive occurrence of *Elatocladus* and *Brachyphyllum* are likely to be due to climatic drying which is a characteristic feature of the Bathonian age. The drying is especially well pronounced in the sequence at Shuraba (south Fergana) wherein the Bathonian deposits are deprived of coals followed by impoverished flora and emerging red beds.

A typical feature of the Fergana subprovince incorporating the habitats not only from the eastern but also from northern and southern Fergana is paucity of Bennettitales both in the number of taxa and in their occurrence. The variety of Bennettitales decreases north of the Hissar subprovince down to the basin of Issykkul. Ferns, *Cladophlebis* and *Coniopteris*, become quite diverse with many of the species being local. It is noteworthy that *Pachypteris* is totally lacking because the Fergana subprovince was not inundated by the sea.

The influence of the Siberian region is quite obvious from frequent occurrence of Czekanowskiales and Ginkgoales up to the Bathonian. The great extent of the land borders between the Siberian region and the Fergana subprovince contributed to exchanges between individual taxa rendering this boundary vague and setting up a transition zone (ecotone).

2.3.3 East Asia province

This province occupies central and south China and Japan. The zone of ecotone between the Siberian and Euro-Sinian regions embraces the southern Far East Sea region wherein a Liassic flora was recently established. The Early Jurassic floras of China highlight a richness both in plant taxa found and in the number of habitats. For a long time researchers failed to separate the Liassic continental deposits from the uppermost parts of the Triassic (Rhaetic–Noric) although individual findings (*Lepidopteris* sp.) of this age had been known long ago just as the localities of the typical Liassic flora.

It is only recently that Zhou (1983) singled out in the sequences of the province Khunan two adjacent horizons with plant remains of the Rhaetic and Liassic age in the midst of continental beds. The lower of these, correlating with the zone of *Lepidopteris ottonis* from western Europe and Greenland, contains a less rich flora which, however, exhibits such characteristic forms as *Lepidopteris, Anthrophyopsis, Ptilozamites, Drepanozamites, Pterophyllum ptilum* and *Cycadocarpidium erdtmanii*.

In the upper half of this continental formation a rich flora was collected, represented by 72 species belonging to the genera *Equisetites, Neocalamites, Marattiopsis, Dictyophyllum, Clathropteris, Stachypteris, Gleichenites, Coniopteris, Todites, Ctenozamites,*

Sagenopteris, Scoresbya, Ctenis, Nilssonia, Otozamites (9 species), *Ginkgoites, Sphenobaiera, Pseudotorellia, Ixostrobus, Ferganiella, Brachyphyllum*, etc. This list is devoid of the key form for the Lower Liassic, i.e. *Thaumatopteris schenkii*, but the stratigraphic position of this horizon situated above the one containing *Lepidopteris* coupled with the presence of such genera as *Stachypteris, Gleichenites, Todites* and even *Coniopteris* enables it to be referred to the Liassic with confidence.

The most important species helping to determine the age are those co-occurring with the species from the zone of *Thaumatopteris* in western Europe and Greenland, viz. *Equisetites scanicus, Gleichenites nitida, Ctenis stewartensis* and *Scoresbya dentata*. It is very likely that this continental formation may correspond not only to the Lower but also to the whole of the Liassic. Unfortunately, the Chinese palaeobotanist Zhou (1983) does not give the distribution of the species described throughout the sequence. The availability of younger deposits, i.e. Middle and, possibly, Upper Liassic, is displayed by the presence of *Coniopteris* and even *Stachypteris*; the latter usually appears from the Middle Jurassic onwards. More to the south, in the provinces of Khunan and Guandan, strata of marine origin emerge containing ammonites *Schlotheimia* and *Arnioceras* suggesting a Hettangian–Sinemurian age. Probably these beds partially replace the continental deposits of the Liassic.

A typical Liassic flora is also known from the lower half of the Syantsi suite developed in the eastern peripheral area of the Szechwan depression (Khubei province). Here the variegated formation of the Upper Triassic containing *Avicula arcuata* and other bivalves includes terrigenous sediments of the Syantsi suite having a continental origin. Hence, Sze (1949) described plant remains typical for the Liassic. We shall indicate a few, characteristic of this age: *Neocalamites carrerei, Clathropteris meniscoides, Phlebopteris* cf. *polypodioides, Dictyophyllum* cf. *nathorstii* and *D.* cf. *nilssonii*. These are followed in diversity by Bennettitales represented by *Anomozamites, Nilssoniopteris*, several species of *Otozamites, Pterophyllum* and *Ptilophyllum*, as well as *Sphenozamites* sp. and *Zamites sinensis.* There are many cycads including *Nilssonia*. Ginkgoales (*Ginkgo* and *Baiera*) are not numerous.

An impression of *Phoenicopsis* aff. *speciosa* was encountered. Conifers include several species of *Elatocladus* and *Pityophyllum* as well as *Podozamites lanceolatus* and *Swedenborgia cryptomeroides*. The latter is characteristic of the Lower Liassic of Greenland and Sweden. Another species of *Swedenborgia* (*S. dentata*) found in the western part of the Szechwan depression and in Khunan province is known not only in the Liassic of Greenland and Sweden but also in Fergana (Kok-Yangak).

The presence of many common forms throughout Greenland, western Europe, Central Asia and south China tells the story of a large-scale migration of a number of plants along the broad belt located north of the Tethys Ocean. The genus *Ferganiella*, absent from western Europe, is a link between the Middle Asia and East Asia provinces.

Northwards and back the migration was restricted by the belt of temperate climate whose southern border passed somewhere in the vicinity of Beijing. Quite a few forms crossed this boundary, as was shown above. For instance *Phoenicopsis* aff. *speciosa* was detected in Szechwan having probably migrated from the north. However, for the majority of Bennettitales, such as *Zamites*, *Otozamites* and *Sphenozamites*, this boundary, judging from the known habitats of the flora, proved to be an insurmountable barrier. This also applies to the majority of Dipteridaceae (except for *Hausmannia*), Matoniaceae and Marattiaceae ferns which reveals their thermophilic nature.

The ferns belonging to the genera *Coniopteris*, let alone the form-genus *Cladophlebis*, Bennettitales (*Pterophyllum*) and cycads (*Nilssonia*), as well as Ginkgoales (*Ginkgo*, *Baiera*) and Podozamitaceae, spread to the north and south of this province.

Turning to Japan, whose Early Jurassic floras were quite recently investigated by Kimura and Tsujii (1980a, 1981, 1982, 1983, 1984), we see that it has a largely subtropical appearance and should fall within the East Asia province. The flora of Japan presents *Neocalamites*, several local species of *Equisetites*, *Marattia* (= *Marattiopsis*) *asiatica*, five species of *Todites*, *Osmundopsis nipponica*, *Dictyophyllum kotakiense*, *Thaumatopteris elongata*, *Hausmannia* sp., *Clathropteris meniscoides* and several species of *Cladophlebis*.

Bennettitales are represented by *Otozamites fujimotoi*, *O. neiridaniensis*, *Pterophyllum* ex gr. *propinquum* and several species of *Ptilophyllum* including two local species. The cycads feature *Ctenis kaneharai* and *Pseudoctenis nipponica*, as well as several species of *Nilssonia* (see Fig. 2.11). In the midst of cycadophytes whose systematics were not then established the described species comprise *Cycadites* cf. *saladinii*, *Ctenozamites sanranii* and a number of *Taeniopteris* including *T. gracilis* and *T.* cf. *jourdyi*.

Ginkgoales are represented by *Ginkgo* ex gr. *sibirica* and three species of *Sphenobaiera*. The representatives of Siberian flora are also present, i.e. *Phoenicopsis* ex gr. *angustifolia* and *Czekanowskia* ex gr. *rigida*. Conifers comprise *Elatocladus* spp., *Stogaardia spectabilis* and the imprints of the leaves of *Podozamites* defined up to the generic status. Many new species so far unknown outside Japan are quite conspicuous. This suggests Japan's relative isolation from the continent of Eurasia although, according to geologic data, Japanese islands did exist in the Jurassic and Japan was directly adjoining Eurasia. This specific peculiarity allows identification of Japan, just as China, as belonging to different subprovinces of the Euro-Sinian region. Another striking character is the admixture of the elements from the Siberian region (*Phoenicopsis angustifolia* and *Czekanowskia rigida*) which is explained by the proximity of the southern boundary of this region.

The plant remains discovered and described (Krassilov and Shorokhov, 1973) from the Earlier Liassic of the southern Far East Sea area should be referred to the typical ecotone flora. Here, in the basin of the Petrovka river discharging into the Ussuriysk bay from the east as well as in the tributaries of the river Sibichuan, plant remains were detected in the wash-out of the lower Triassic and overlain by marine shallow-water sediments with *Harpax laevigatus* and *Uptonia* sp. of the Middle Liassic. Their composition points to the Early Liassic age which allowed V. A. Krassilov to compare them with the zone of *Thaumatopteris schenkii* in Greenland and western Europe. The typical forms of the Euro-Sinian region, such as *Neocalamites hoerensis*, *Marattiopsis hoerensis*, *Clathropteris elegans* and *Cycadocarpidium swabii*, go side by side with Czekanowskiales (*Czekanowskia rigida* and *Phoenicopsis angustifolia*) inherent in the Siberian region. Apart from the forms enumerated, representatives of the following genera are encountered: *Sagenopteris*, *Ctenis*, *Nilssonia*, *Pterophyllum*, *Taeniopteris*, *Ginkgo*, *Sphenobaiera*, *Podozamites* and *Elatocladus*. It should be pointed out that not a single thermophilic species of the Bennettitales group was found in the southern Far East Sea area although *Otozamites* can be met in Japan.

We have ascertained that the generic composition of the Liassic flora of the East Asia province is very close to that of the European and Middle Asia provinces. Exceptions to this rule include several rarely occurring genera from south China unknown outside. These are *Vittifolium*, resembling *Pseudotorellia*, and *Sinophyllum*, belonging to cycadophytes. If the specific compositions of all three provinces are compared they can be shown to be quite different from each other in the presence of species occurring solely within the boundaries of each of them. We have also mentioned the great specific discrepancy between the Liassic floras of Japan and China. The specific variety of *Otozamites* should be singled out as a characteristic feature of the Liassic flora of China which is far less rich than that of the other two provinces.

South China, rich in the localities of the Liassic floras of subtropical appearance, has so far failed to exhibit remains of Middle Jurassic plants. Localities of the floras of this age emerge somewhat to the north. They outcrop west of the isthmus of the Shandun peninsula, west of Beijing (the Beijing Sishan) and in the loop of the river Khuan-Khe. The Middle Jurassic flora is represented here by several species of the ferns *Coniopteris* and *Cladophlebis*, isolated *Clathropteris* and *Raphaelia diamensis*, as well as by various Ginkgoales, Czekanowskiales and ancient Pinaceae. Certain species of *Elatocladus* and *Podozamites* are encountered, too. The remains of cycads and Bennettitales belong to the genera *Anomozamites*, *Pterophyllum*, *Nilssonia* and *Ctenis*, i.e. the genera encountered both in the Siberian and Euro-Sinian regions (see Fig. 2.11). A find of *Cycadites* appears to be an exception because this is generally known only in the Euro-Sinian region (Chen Fen *et al.*, 1980). Thus, China has until now failed to furnish remains of the subtropical flora of the Middle Jurassic age typical for the Euro-Sinian region. If we compare the maps of the localities of the Early and Middle Jurassic floras it will be obvious that the known units from this part of China have also been referred to the Siberian region according to the abundance of Ginkgoales and Czekanowskiales and absence of certain Bennettitales (*Zamites*, *Otozamites*, *Ptilophyllum*) and a number of Matoniaceae and Marattiaceae ferns. It will be remembered that only south China can record the typical Liassic floras of subtropical appearance containing the above-mentioned groups of plants fully compatible with the floras of western parts of the subtropical belt that extended throughout western Europe, the south of the European part of the USSR, the Caucasus and Middle Asia farther embracing China. The Sinemurian floras at present known from China (the basin of the river Khuan-Khe) may be somewhat remotely regarded as a transition from the floras of the Siberian region to those of the Euro-Sinian one, i.e. the floras of an ecotone, but subtropical elements identified in them so far are very few.

2.4 Late Jurassic

The onset of Late Jurassic time is marked by significant climatic events expressed in the overall warming of climate in the Northern Hemisphere resulting, in particular, in the boundary between the Siberian and the Euro-Sinian regions being advanced northwards in the east of Eurasia by at least 15 to 20° latitudinally from the river Khuan-Khe in northern China up to northern Mongolia.

The general warming encompassed marine basins too. Thus, Arkell (1956, p. 615) writes: 'Throughout the Middle Oxfordian the main coral belt covered the southern portion of Central Europe and the coral reefs in Yorkshire were situated 20° north of the northernmost of the current coral reefs in the Gulf of Suez, the Bermudas and Japan'. Vast transgressions that occurred repeatedly in Europe, Western Siberia and Middle Asia were conducive to the warming. At the same time dramatic aridization of climate took place within a broad belt crossing south England, France, Switzerland, western Ukraine, the Caucasus, Middle Asia, Mongolia and western China. The whole broad belt is noted for sedimentation of evaporites in disjunct basins.

Tadjikistan and the adjacent areas of Turkmenia and Uzbekistan as well as, in places, the northern Caucasus show gypsum associated with the deposits of rock and some-times potash salts. Evaporites become extremely disseminated in the upper Jurassic in the west and south of the USA (New Mexico, Arizona, Colorado, Wyoming, Montana, South Dakota). The uplift at the end of the Jurassic led to the formation of a vast internal lake whose multi-coloured sediments merge into the Morrison suite containing but few plant remains. Habitats of the late Jurassic floras are not encoun-tered in the rest of the United States save for rare instances.

The change in the composition of the continental deposits in Asia at the point of time described is quite noteworthy. During the Early and Middle Jurassic numerous basins (both internal and open into marine realms) housed deposition of grey-coloured, primarily coal-bearing sediments. At the beginning of Late Jurassic time this situation changed. A number of littoral depressions sunk and were invaded by the sea (west Georgia, Mangyshlak, Tuarkyr, Tadjik depression, West Siberian lowland), an event which is signalled by the replacement of the coal-bearing with marine deposits. The internal depressions of Middle Asia (Fergana), Kazakhstan, Siberia (Chulym-Yenisei, Kuznetsk, Kansk, Irkutsk basins) and China (except for its north-eastern part) accommodated sediments of variegated and red colours supplanting accumulation of coal-bearing formations. Owing to this the flora of these aridization-stricken areas drastically changed its composition as will be shown below.

In the Southern Hemisphere the bulk of splitting Gondwana remained land con-stituting in the main an area of displacement. Thus, on the vast territories of the Afro-

South American continent, continental deposits of this age are almost unknown except for northern Africa. The sea occupied the Andean geosyncline and the coastal areas of eastern and northern Africa and Australia.

2.5 Late Jurassic of the Siberian region

In the Late Jurassic this region embraced the extreme north of Scandinavia, the Pechora basin, the northern part of Western Siberia, Eastern Siberia, and the north-east of the USSR. As already mentioned, its southern boundary shifted as against that of Middle Jurassic time by 15 to 20° due to the warming. Owing to the ensuing aridization the flora of the subtropical belt was subjected to dramatic alterations which will be dealt with below when reviewing the Euro-Sinian region. The flora of the Siberian region did not change significantly. Just as in the Middle Jurassic, the dominating plant groups of this region were the Czekanowskiales, Ginkgoales and conifers; the latter were dominated by ancient Pinaceae. The climate remained moderately warm and humid, as indicated by an abundance of ferns and widely distributed coal sedimentation. The seasonal nature of the climate is revealed from the well-expressed annual rings.

The description of the Siberian region will be started with its eastern portion occupied by the Lena province within which relatively numerous habitats of the Late Jurassic floras are known. More northern areas (north of Siberia and the Pechora basin) are characterized almost exclusively by palynofloras.

2.5.1 Lena province

The Upper Jurassic primarily coal-bearing deposits are most widespread in the Lena basin where they are identified as constituting the Chechum horizon. In the basin of the river Sitoga, in the right tributary of the lower reaches of the Aldan river and on the left bank of the Lena river upstream of Zhigansk, the coal-bearing formations of the upper third of the Chechum horizon are replaced by marine terrigenous sequences defined as the Sitoga and, in the north, the Chonok suites. These suites contain species of *Buchia*, determining their age as the second half of the Late Jurassic (Vakhrameev, 1958; Kirichkova, 1976a, 1979). The lower half of the Chechum horizon, the Djaskoy member, rests in the areas where the upper strata are represented by marine deposits. The main collections of plant remains from the Djasko coal-bearing suite were effected near Djaskoy promontory on the left bank of the Lena river downstream of Zhigansk. The remnants of plants from the upper half of the Chechum horizon, represented here completely by carbonaceous formations, were gathered in the river Markh, the left tributary of the Vilyui river. But, although these collections originate from various parts of the sequence of the Chechum horizon, their composition turned out to be very similar.

Horse-tails are represented by *Equisetites tschetachumensis*, and ferns by *Coniopteris* ex gr. *hymenophylloides*, *Osmundopsis acutipinnula*, *Gleichenia* (*Gleichenites*) *jacutica*, *Hausmannia* sp., *Cladophlebis aldanensis*, *C. serrulata*, *C. denticulata*, *Raphaelia diamensis* and *R. kirinii*. A variety of Ginkgoales were detected, too (*Ginkgo* spp., *Baiera* spp., *Sphenobaiera* spp.), coupled with Czekanowskiales (*Czekanowskia* spp., *Leptostrobus* sp., *Phoenicopsis* sp.) and conifers (*Podozamites* spp., *Schizolepis* spp., *Pityophyllum* spp. and *Pityospermum* spp., as well as *Taxocladus tschetshcumensis* and *Coniferites marchaensis*).

2.5.1.1 SOUTH YAKUTSK BASIN

The second vast area making part of the Lena province is the South Yakutsk basin which was described earlier in detail when considering the Middle Jurassic flora. The Upper Jurassic here comprises (Vlassov and Markovich, 1979a, b) the Kabaktin, Kabatkin–Berkatite and Nerungrin phytocomplexes confined to the Kabaktin, Berkatite and Nerungrin suites.

The species occurring throughout or on a number of levels of the sequence of the Upper Jurassic both in the Aldan–Chulman and in the Tokinsk areas (in the latter they are less diverse) include *Equisetites tenuis*, *E. tschetschumensis*, *Gleichenia jacutica*, *Coniopteris burejensis*, *C.* ex gr. *tyrmica*, *Lobifolia lobifolia*, *Hausmannia* sp., *Cladophlebis* ex gr. *haiburnensis*, *C. serrulata*, *C. williamsonii*, *C. argutulla*. *C. orientalis*, *Raphaelia diamensis*, *R. stricta*, *Taeniopteris* sp., *Nilssonia* spp. and *Yakutiella amurensis*. Ginkgoales are rather varied (*Ginkgo* ex gr. *sibirica*, *Sphenobaiera* ex gr. *czekanowskiana*), as are Czekanowskiales (*Phoenicopsis* ex gr. *angustifolia*, *Czekanowskia* ex gr. *rigida*, *Leptostrobus* sp.) and conifers (*Pityophyllum* sp., *Pityocladus* sp., *Schizolepis* sp., *Elatocladus* spp., *Pagiophyllum* sp., *Brachyphyllum* sp., *Podozamites* spp.).

The Upper Jurassic flora of the Tokinsk area is somewhat poorer than the flora of the Aldan–Chulman area but is comparable with the latter. Here the same floristic complexes are identified, i.e. Kabaktin and Kabaktin–Berkatite.

Recently a rich complex was collected along the rivers Khudurkan, Undytkan and Algoma. This complex belongs to the lower part of the Tokinsk suite comparable with the Kabaktin–Berkatite complex from the Aldan–Chulman area stemming from the upper strata of the Kabaktin suite and the entire Berkatite suite.

The rocks of this stratigraphic level in certain localities contained large numbers of *Raphaelia diamensis*, *Cladophlebis serrulata*, *Equisetites tschetschumensis*, etc., i.e. the forms characteristic of the Upper Jurassic. Upper strata of the Tokinsk suite deprived of definable plant remains appear to be an analogue of the Nerungrin suite in the Tokinsk basin.

The characteristic species for the entire Lena province in the Late Jurassic incorporate *Equisetites tschetschumensis*, *Cladophlebis aldanensis*, *C. serrulata*, *C. orientalis*, *Raphaelia diamensis* and *R. stricta*, as well as the rarely encountered *Coniferites*

marchaensis. This species was found both in the Lena basin (in the rivers Aldan and
Markh) and in the Upper Jurassic of the Amur province, in the basins of the rivers Zeya
and Bureya. All the forms enumerated above vanish at the boundary of the Jurassic and
Cretaceous.

The emergence of a number of thermophilic conifers missing in the Middle
Jurassic attracts attention in the Upper Jurassic of southern Yakutia, particularly the
Aldan–Chulman area. These are various *Elatides*, *Elatocladus* and especially *Brachyphyllum* and *Sciadopitys*. Their appearance along with a somewhat greater diversity of
ferns and cycadophytes seems to testify to a certain warming of climate as compared
the Middle Jurassic. Possibly the more southern position of the South Yakutsk basin
as against the Lena basin gave rise to a more tangible change in floristic composition
on the borderline between the Middle and Late Jurassic since the warming here
proved to be more significant.

At the same time the Lena basin was doubtless isolated from the South Yakutsk basin
in the Late Cretaceous epoch which is borne out by the analysis of the dissemination
of cycads and Bennettitales; probably, this was determined by temperature conditions.
No representatives of Bennettitales have as yet been detected in the Upper Jurassic of
the Lena basin (Kirichkova, 1985). Of cycads the Lena basin provides only the genus
Heilungia. In the Early Cretaceous of the Lena basin both the cycads and Bennettitales
were represented quite broadly for this latitude.

The relative paucity of the Late Jurassic flora in the Lena basin appears to run
counter to the warming pervading the Northern Hemisphere in this epoch and may
be accounted for by the fact that in the Middle Jurassic the bulk of the Lena basin was
occupied by the sea gradually retreating to the north during the Middle and Late
Jurassic epochs. The settlement of plants on the dried surfaces proceeded from the
fringes of the basin which had not been flooded by the sea and whose vegetation was
very poor. The latter phenomenon was generally characteristic of the vegetation of
the Middle Jurassic of Siberia. From the south the Lena basin was encircled by the
Stanovy ridge and in the west by the Patomsk highland, interfering with the penetration of thermophilic elements from the Amur province.

It was only in the Early Cretaceous that this hindrance was overcome; this is
corroborated by the appearance of a number of southern elements, viz. *Onychiopsis
psilotoides*, diverse *Pterophyllum*, *Ctenis*, *Pseudoctenis*, *Neozamites* and *Nilssonia*. Simultaneously, under the conditions of a climate which was warmer as against that of the
Middle Jurassic, local thermophilic species underwent evolution. A number of new
species of *Heilungia* emerged, its relative *Aldania* came into existence and *Ctenis*,
Nilssoniopteris, etc. were differentiated.

2.5.1.2 NORTH–EAST USSR

Of special interest are the localities of the Late Jurassic flora situated in the north-east
of the USSR where the correlation between members is well established. These

members contain plant remains with marine deposits containing a fauna. Thus, in the river Pezhenka, the left tributary of the Big Anuy, a member of continental deposits resting on the marine formations of the Volgian stage with *Buchia* species supplied *Cladophlebis aldanensis*, *Raphaelia diamensis*, *Ctenis anyuensis*, *C.* aff. *borealis*, *Heilungia* cf. *amurensis*, *Nilssonia* sp. and *Phoenicopsis* sp. (Kirichkova and Samylina, 1984).

On the left bank of the Kolyma river in the Lyglykhtap depression the deposits of the Malinovsk suite, laid down in the marine beds with *Buchia* species of the middle part of the Volgian stage, displayed *Cladophlebis aldanensis*, *Ctenis* sp., *Nilssonia* sp. and *Heilungia* cf. *amurensis*. These were found by Paraketsov (1982). Finally, the right bank of the river Omolon contains a number of rocks of continental origin within marine Volga deposits with *Cladophlebis aldanensis*, *C. nebbensis*, *Sphenobaiera* sp., *Phoenicopsis angustifolia*, *Pityophyllum longifolium* and *Podozamites* ex gr. *lanceolatus* (Vakhrameev, 1964). The relationship between the marine and continental deposits confirms that *Cladophlebis aldanensis* is confined in its distribution to the Upper Jurassic and that *Raphaelia diamensis* does not continue into the Early Cretaceous.'

The Late Jurassic floras of the north-east of the USSR are interesting in that they represent a vegetation association consisting of plants which migrated to the coastal land after the Early Volga Sea had retreated. As we have seen above, the composition of plants collected from various localities is quite uniform. It amounts to the Siberian flora-specific Czekanowskiales among which *Phoenicopsis* are prevalent, *Podozamites* and ancient Pinaceae making the tree stands, accommodating the habitats of ferns and certain cycads (*Nilssonia. Heilungia, Ctenis*).

2.5.1.3 WEST OF BAIKAL

The Siberian and Euro-Sinian regions are here quite easily distinguished. Some provinces may be delineated only according to the palynologic data. The only area furnishing the habitats with macro-remains of the Late Jurassic flora is the basin of the river Severnaya Sosva flowing down from the eastern slope of the Sub-Polar Urals. In this area at the base of the Jurassic strata deposited in the weathering outcrop of the Palaeozoic is situated a carbonaceous formation whose thickness is about 200 m divided into three suites (upwards): Yanymanyinsk, Tolyinsk and Otoryinsk. Later the Tolyinsk and Otoryinsk suites were merged into one referred to as the Tolyinsk suite. Some researchers date the latter as the Bathonian–Lower Callovian while others classify it as Lower Oxfordian. The roof contains the marine sediments of Upper Oxfordian age.

Genkina (1960) identified the following from the Tolyinsk suite, primarily from its upper and coal-bearing part: *Equisetites* sp., *Coniopteris hymenophylloides*, *Cladophlebis* cf. *denticulata*, *Nilssonia* ex gr. *orientalis*, *N.* cf. *polymorpha*, *N. vittaeformis*, *Taeniopteris vittata*, *Ginkgo* sp., *Sphenobaiera* cf. *longifolia*, *Phoenicopsis angustifolia*, *Czekanowskia rigida*, *Pityophyllum nordenskioldii* and *Podozamites angustifolius*.

As can be seen from the list this is a typical flora of the Siberian region with all its

components. Of all cycads only *Nilssonia* are present. Firmly established Bennettitales
were not found since the form-genus *Taeniopteris* covers the species with similar
morphology pertaining to both the cycads (*Doratophyllum*) and Bennettitales
(*Nilssoniopteris*). According to the composition the flora, that from the Severnaya
Sosva is very close to the Middle Jurassic floras of the Siberian region. The warming
that occurred on the boundary of the Jurassic and Cretaceous did not affect its composition, probably due to the position of the territory where it was growing, i.e. the
high latitudes slightly to the south of the present polar circle.

2.5.1.4 THE WEST SIBERIAN LOWLAND

The West Siberian lowland is located between the eastern slope of the Urals and the
Siberian Platform. Within its boundaries we know of virtually no habitats with plant
macrofossils. On the greater part of the West Siberian lowland the Upper Jurassic
deposits which are mainly of marine origin never come out on the surface being
known only from boreholes. The composition of plants may be inferred only from the
palynologic evidence and analysis because cores extracted from these deposits contain
virtually no definable remains of plants. As was revealed by the investigations of
palynologists (Rovnina, 1972; Ilyina, 1985), the composition of spores and pollen during the transition from one area to another is rather changeable.

In order to draw a boundary between the Siberian and the Euro-Sinian regions,
both in Western Siberia and in the European part of the USSR, use was made
(Vakhrameev, 1970, 1980) of the pattern of distribution of *Classopollis* pollen whose
quantity rises in the rocks of the Jurassic and Early Cretaceous when moving from the
moderately warm climate to subtropical areas (Fig. 2.12). *Classopollis* pollen is
especially abundant in the deposits formed under the conditions of an arid climate. As
far as the ecological requirements are concerned, ferns and ancient Pinaceae were
some sort of a contrast to the Cheirolepidiaceae producing the *Classpollis* pollen. The
former produced a bisaccate pollen.

In drawing the boundary between the two regions we rely on the percentage content of the *Classopollis* pollen in the spore–pollen complexes. Some palynologists
(Herngreen and Khlonova, 1981) draw the boundary on the basis of an almost complete disappearance of the pollen *Classopollis* which to the north is found only as
isolated grains. However, in this case we fail to reconcile this boundary with the
border drawn in the eastern part of Eurasia on the strength of the emergence of such
Bennettitales as *Otozamites*, *Ptilophyllum*, *Zamites* and similar genera. Earlier it was
mentioned that Bennettitales cannot be differentiated from cycads and Ginkgoales
distinctly according to the pollen evidence and still less can they be divided into
separate genera. Therefore, the boundary between the regions of Western Siberia is
established along a line south of which the content of *Classopollis* pollen surpasses 10
to 15% in the overwhelming majority of samples. This boundary roughly extends
along the latitudinal section of the river Ob.

2.5.1.5 THE KANSK–ACHINSK BASIN

This basin testifies that the southern part of Western Siberia belongs to the Euro-Sinian region. In this basin the Upper Jurassic deposits are represented by the Tyazhinsk suite which overlies the carbonaceous Itatsk suite of the Middle Jurassic. The Tyazhinsk suite is made up of variegated terrigenous rocks in which researchers managed to find *Coniopteris gracillima* and *Pizyospermum gracile*. The spore–pollen complex contains much *Classopollis* pollen (Borgolepov, 1961), up to 80% in some samples; the bisaccate pollen has a subordinate significance. The woody *Xenoxylon latiporosum* is encountered. According to the palynologic data the following areas (palynologic provinces) may be identified within the northern part of Western Siberia and the Sub-Polar Urals entering the Siberian phytochoria. On the eastern slope of the Sub-Polar Urals (the basin of the river Severnaya Sosva) the basis of the palyno-floras of the Upper Jurassic is made up of the spores of *Cyathidites* whose content decreases upwards in the sequence from 40–20 down to 10–20% (Ilyina, 1985). *Gleicheniidites* (up to 15–20% in the Volga stage) and *Dicksonia jatrica* are constantly present. Pinaceae dominate among the pollen being accompanied by *Classopollis* (in

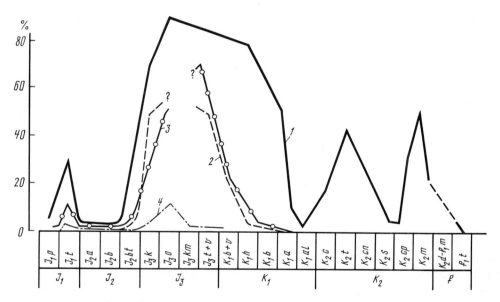

Fig. 2.12 *Classopollis* pollen content curves for Jurassic and Cretaceous deposits. 1, southern regions of the USSR, i.e. Moldavia, Caucasus, Crimea, southern Kazakhstan, Middle Asia (the right part of the curve embracing the Upper Cretaceous was constructed according to the results of the analysis of the samples collected from Middle Asia); 2, central part of the Russian Platform, the Moscow region, the Vyatka-Kama trough; 3, Western and Middle Siberia (Vilyui trough); 4, northern fringe of the Asiatic part of the USSR (Ust–Yenisei and Hattang depressions).

the lowest strata of the Upper Jurassic 5% and in the uppermost, 10–20%). The content of the pollen *Sciadopityspollenites* rises upwards in the sequence of the Upper Jurassic from 10 to 15%. Rovnina (1972) considered this area as the Urals province of the Siberian region.

The Oxfordian–Lower Volgian deposits of the north-west of the West Siberian lowland, i.e. the basin of the lower reaches of the river Ob, contain an abundance of the pollen *Sciadopityspollenites* (more than 50%), and presents the spores of *Gleicheniidites*, *Osmundacites* and *Cyathidites* as well as the pollen *Podocarpidites*, Pinaceae and *Classopollis*. L. V. Rovnina defined this area as the Beryozov–Shaimsk province.

2.5.1.6 UST-YENISEI AREA

The spores of *Cyathidites* and *Osmundacites* and the pollen of Pinaceae are prevalent in the more northerly Ust-Yenisei area. The spores of *Gleicheniidites* prove rare. In the Oxfordian the average content of the pollen of *Classopollis* amounts to 10%, although in rare samples it increases up to 30%. It is common knowledge that the Oxfordian was the time of the greatest climatic warming of the Late Jurassic epoch and this might be the cause of such enhanced proportion of the *Classopollis* pollen in some samples collected in the far north. In the Kimmeridgian the *Classopollis* pollen content is considerably reduced which is concurrent, however, with the increasing amount of the pollen of ancient Pinaceae prevailing in the Volgian stage. In drawing the boundary between the Siberian and Euro-Sinian regions in Western Siberia, which was mentioned above, we are going by the average content of the pollen *Classopollis* taken throughout the sequence of the Upper Jurassic rather than by samples with maximum contents.

2.5.1.7 SUMMARY

Another pattern seems to be taking shape which is indicated by V. I. Iliyna. The pollen of *Sciadopityspollenites* occurring in large amounts in the basin of the lower reaches of the river Ob, is seldom encountered when moving eastwards. The quantity of spores of *Gleicheniidites* and of the pollen *Classopollis* diminish in the same direction. The latter is found in the Upper Jurassic in Eastern Siberia and in the north-east USSR only in the form of isolated grains. Looking at the map (Smith and Briden, 1977) which takes into account the movement of continents it is obvious that Chukotka in the Late Jurassic epoch was located immediately south of the North Pole, i.e. 20 to 25° south of the Pechora basin. This serves as a good explanation of the impoverishment of the spores and pollen of thermophilic plants when moving from the south to the north along the northern fringe of current Asia. The same is also observed for the Early Cretaceous epoch.

The north-westernmost locality of the Late Jurassic floras of Eurasia is to be found in the island Andoe which is part of the Westeralen archipelago in the north of Norway. A detailed history of the investigation of this flora, rather impoverished in its systematic composition, which was carried out by a number of palaeobotanists, is described by Doludenko (1984). The Middle and Upper Jurassic terrigenous littoral deposits rest here on the Precambrian granites.

The plant macrofossils have been collected in the sequences of the Callovian covered by the Oxfordian with *Amoeboceras macrophylla* and other fauna as well as in the Kimmeridgian deposited up in the sequence. This collection comprises undefinable fragments of plants such as *Ginkgo huttonii*, *Ginkgo* sp., *Pseudotorellia heeri*, '*Phoenicopsis*' sp., *Nilssoniopteris* sp., *Sciadopitys macrophylla*, *S. lagerheimii*, *Sciadopityoides nathorstii*, *S. persulcata* sp. and *Pinus* spp. There is a striking abundance of the remains of *Sciadopitys* in such a poor flora. Besides these some layers contain many fragments of *Ginkgo* sp. and *Pseudotorellia heeri* (Manum, 1987).

Judging by the prevalence of conifers featuring enhanced strength and better withstanding of long transportation the host rocks were of marine origin. This is confirmed by the availability of marine fauna in certain members of this formation. This predetermined an almost complete absence of ferns possessing an easily destructible leaf blade. According to the composition and location of this habitat on the map it probably belongs to the ecotone between the Siberian and Euro-Sinian regions.

2.5.2 Amur province

The southernmost part of the Siberian region is the Amur province, taking up the basin of the river Amur and the Transbaikal area. The Late Jurassic flora of the Bureya river basin has been well studied (Vakhrameev and Doludenko, 1961; Vakhrameev and Lebedev, 1967; Krassilov, 1972a, 1973a, 1978). The coal-bearing formation of this basin starts from the Talynzhansk suite overlain by the Urgal suite. Later the Urgal suite was divided according to the lithologic evidence into two (upwards): the Dublikansk and Soloniysk suites (Davydova and Goldshtein, 1949). Initially the flora was gathered from the Soloniysk suite. Its comparison with the Lena basin has revealed that it belongs to the lowest strata of the Lower Cretaceous. Somewhat later remains of plants from the Dublikansk suite were collected which permitted reference to the uppermost strata of the Upper Jurassic (Vakhrameev and Lebedev, 1967).

Originally the Talynzhansk suite was believed to belong to the Upper Jurassic (the Volgian stage?) since it had the underlying Chaganyisk suite of marine origin containing no fauna. It was only stratigraphically below that, already in the Elginsk suite, that bivalves classified as Oxfordian were detected, whereas at the base of this suite Bathonian ammonites were found. However, later in the upper strata of the Chaganyisk suite (certain researchers thought it to be the base of the Talynzhansk one) Bathonian–Callovian *Arctocephalites* were discovered (Markov, Trofimuk and Shcherbakov, 1970). It remains obscure how the Bathonian–Callovian ammonites occur twice in the sequence (the lower strata of the Elginsk suite and the roof of the Chaganyisk suite), being separated by a thick formation of marine deposits with a monoclinal dip.

Let us consider together the floras of the Talynzhansk and Dublikansk suites, primarily because they are close to each other in floristic compositions and due to the fact that there is disagreement over classification of some outcrops with habitats of plant remains.

The Late Jurassic flora of the Bureya river (Figs. 2.13, 2.14) is the richest among the floras of the Amur basin. Liverworts, mosses and lycopods were examined within this flora (Krassilov, 1973a); the composition of mosses is diverse. The species failing to continue into the Lower Cretaceous (Soloniysk suite) exhibit the following: *Coniopteris sewardii, C. tyrmica, Raphaelia diamensis, R. stricta, Cladophlebis aldanensis, C. laxipinnata, C. orientalis, Sphenopteris samylinae, Eboracia kataevensis*. Some characteristic species passing into the Soloniysk suite are present, too, viz. *Haussmania leeiana,*

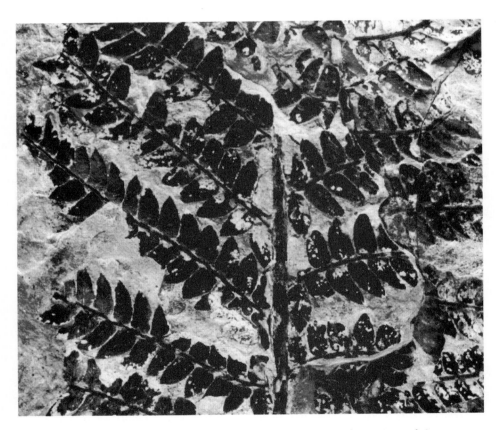

Fig. 2.13 Late Jurassic plant fossil from Bureya river, Amur basin; natural size; *Raphaelia diamensis* Seward.

Coniopteris bureyensis, C. saportana, etc. Bennettitales and cycads are various; the former include *Anomozamites angulatus, Pterophyllum sensinovianum, P. pterophylloides, P. regidium* (earlier this species was singled out by V. D. Prynada as an independent genus *Bureja*), *Pseudocycas polynovii* and *Cycadolepis sixtelae*.

The cycads include very characteristic species such as *Ctenis burejensis, C. angustissima, Heilungia amurensis* (the species spread in the overlying Soloniysk suite, too), *Nilssonia mediana* and *N. schmidtii*. Ginkgoales and Czekanowskiales are represented in abundance (the former include *Ginkgo, Baiera, Sphenobaiera, Eretmophyllum, Pseudotorellia, Umaltolepis* and *Carpolithes*; the latter include *Czekanowskia, Leptostrobus, Phoenicopsis, Staphidiophora* and *Ixostrobus*). *Czekanowskia* and *Pseudotorellia* do not pass over to the Soloniysk suite in the Bureya basin (the Lower Cretaceous). The monograph of Krassilov (1972a) is devoted to the description of these two groups. In particular the monograph cites evidence on the association of *Ixostrobus* with

A   B

Fig. 2.14 Late Jurassic plant fossils from Bureya river, Amur basin; natural size. A, *Sphenopteris samylinae* Vakhrameev; B, *Coniopteris burejensis* (Zalesky) Seward.

Czekanowskia. Conifers are represented by *Podozamites*, *Pityocladus*, *Pityophyllum*, *Elatocladus*, *Sorosaccus*, *Schizolepis*, *Stenorachis* and *Coniferites marchaensis*.

It should be noted that Krassilov studied in detail certain ferns. On the basis of the examination of fertile pinnules occurring together with sterile pinnae defined as *Raphaelia diamensis* he thought it possible to transfer *Raphaelia* to the genus *Osmunda* retaining its former specific name. Prior to this Vassilevskaya and Pavlov (1967) had pointed out that *Raphaelia* belonged to the Osmundaceae but they had not gone as far as Krassilov in identifying the new genus *Osmundiella* established according to the fertile pinnules of *Raphaelia diamensis*. We should no doubt support the view that *Raphaelia* belongs to *Osmunda* but to classify all the finds of this widespread fern directly as belonging to the contemporary genus *Osmunda* is, in our opinion, premature. The isolated fragments of the sterile pinnules found, classified, for that matter, as belonging to several species of *Raphaelia* due to changing morphology, so far does not permit reference of them to the current genus *Osmunda*. To do this it is necessary to conduct a comparative investigation of fertile leaves from different mutually remote areas of Siberia.

The study of fertile pinnae on the basis of the material from the Bureya basin also enabled Krassilov to ascertain a systematic classification of several species of *Coniopteris* within the generic rank. Thus, *Coniopteris burejensis* and *C. arctica* proved to belong to the genus *Dicksonia*, and *Sphenopteris tyrmensis* to the genus *Cyathea*.

The second area enclosing a number of habitats of the Late Jurassic flora within the Amur province is the basin of the river Zeya (Figs. 2.15, 2.16, 2.17, 2.18) situated west of the basin of the river Bureya. The Ayaksk, Depsk and the lower half of the Molachansk suites are dated as the Upper Jurassic (downwards); the Depsk suite is best characterized. Lebedev (1965) managed to determine 51 species from the upper half of this suite. The systematic composition of plant remains collected in both the suites is very similar to the composition of the Late Jurassic floras of the Bureya basin. This should not come as a surprise since in both instances the plant remains are confined to carbonaceous deposits that formed under similar conditions.

The Ayaksk suite is characterized by *Raphaelia diamensis*, *Hausmannia bilobata*, *Cladophlebis laxipinnata*, *C. orientalis*, *Coniopteris depensis* and *Nilssonia schmidtii*. In the Depsk suite these are supplemented by *Raphaelia stricta*, *Coniopteris sewardii*, *C. burejensis*, *Cladophlebis saportana* and *C. serrulata*. Various Ginkgoales are widely disseminated in both the suites, particularly the representatives of *Ginkgo*, with *Pseudotorellia* being present. There are many Czekanowskiales (*Czekanowskia*, *Leptostrobus*, *Phoenicopsis*). Conifers comprise *Podozamites* and *Pityophyllum*; *Coniferites marchaensis* was also encountered.

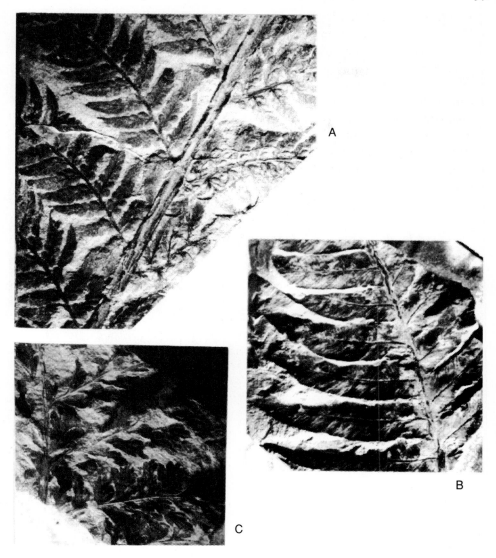

Fig. 2.15 Late Jurassic plant fossils from Zeya river, Amur basin; natural size.
A, *Raphaelia diamensis* Seward; B, *Cladophlebis aldanensis* Vakhrameev;
C, *Coniopteris vsevolodii* É. Lebedev.

Fig. 2.16   Late Jurassic plant fossil from Zeya river, Amur basin; natural size;
Cladophlebis haiburnensis (L. and H.) Brongniart.

A

B

Fig. 2.17 Late Jurassic plant fossils from Zeya river, Amur basin. A, *Nilssonia schmidtii* (Heer) Seward, natural size; B, *Heilungia amurensis* (Novopokrovsky) Prynada, × 0.5.

Fig. 2.18 Late Jurassic plant fossils from Zeya river, Amur basin; natural size.
A, *Phoenicopsis speciosa* Heer; B, *Ginkgo obrutschewii* Seward.

Cycads grow to be more diverse in the Depsk suite being represented by *Heilungia amurensis*, *H. zejensis*, *H. baganoensis* and *Butefeja burejensis* as well as various *Nilssonia*. However, Bennettitales, in particular *Ctenis* and *Pterophyllum*, widely distributed in the Late Jurassic of the Bureya basin and in the basin of the river Zeya, were not encountered.

The Late Jurassic flora is also known in the upper reaches of the River Aldan where it was studied by Dobruskina (1961, 1965a, b). The composition of the flora collected within the Tolbuzin area located on the left bank of the Amur river is very near that from the Depsk suite of the river Zeya. The discoveries here were *Coniopteris saportana*, *C. burejensis*, *Cladophlebis aldanensis*, *Raphaelia* sp., *Nilssonia schmidtii*, *Heilungia amurensis*, *Bufetia burejensis* and *Pseudotorellia ensiformis*, as well as numerous *Phoenicopsis*, *Czekanowskia* and *Podozamites* and rarer Ginkgoales (*Ginkgo*, *Sphenobaiera*). Conifers include *Podozamites* and *Pityophyllum*.

In the Transbaikal area the Upper Jurassic is believed to embrace the Shandoronsk suite now looked upon as a series. It is composed of effusive rocks, various tufogenic formations and, to a lesser extent, sedimentary terrigenous rocks of continental origin. The plant remains detected in the lower and upper parts of the sequence of this series were evaluated at various times by Prynada (1962), Vakhrameev (1964), Teslenko (1968, 1975), Bugdyeva (1983) and others. The comparison of the lists of species from the lower and upper parts of the series (the middle portion is mainly composed of effusive rock containing undefinable plant remains) indicates a great similarity between them. This enables us first to give a general list and then point out the forms encountered only in one of the suites. Different palaeobotanists determined the following from the Shandoronsk suite: *Equisetum* sp., *Raphaelia diamensis*, *Cladophlebis williamsonii*, *C. toungusorum*, *Coniopteris* ex gr. *hymenophylloides*, *C.* ex gr. *burejensis*, *Heilungia iszetujensis*, *Butefia burejensis*, *Czekanowskia* ex gr. *rigida*, *Phoenicopsis angustifolia*, *Pityophyllum* sp., *Pityocladus* sp., *Pityostrobus* sp. and *Carpolites* sp. Apart from this the upper strata of the Shandoronsk suite revealed *Cladophlebis sokolovii*, *C. pseudolobifolia*, *C. laxipinnata*, *Coniopteris depensis*, *C. nympharum*, *Onychiopsis tenuissima* and *Sphenobaiera angustiloba*. A number of species known from the Zeya and Bureya basins (*Raphaelia diamensis*, *R. stricta*, *Cladophlebis laxipinnata*, *Coniopteris depensis*) fail to continue into the Cretaceous system. *Cladophlebis lobifolia* is encountered in the Cretaceous while *Onychiopsis tenuissima* is familiar only from the Transbaikal area (the Gussinoye lake) although the genus *Onychiopsis* itself is typical for Eurasia in the Lower Cretaceous. The finds of these two species in the upper strata of the Shandoronsk suite possibly show that the host formations are transitional to the Lower Cretaceous.

The flora considered may be referred without any hesitation to the Amur basin though it is more impoverished than the flora of the Bureya basin. It is clearly dominated by ferns and Czekanowskiales and, in the midst of these, *Phoenicopsis* and ancient Pinaceae; cycads are very rare and uniform (*Butefia*, *Heilungia*) whereas Bennettitales and the genus *Ginkgo* are totally absent. On the whole, we do not have any doubts that the major portion of the Shandoronsk series, just as the flora enclosed in it, are of Late

Jurassic age. As was mentioned above, the fast accumulation of the deposits of the Shandoronsk series is signalled by the very similar composition of plant remains collected from both the lower and upper strata of the series.

It would seem that we are dealing here with the Undin–Dainsk flora stemming from the suite of the same name due to the extreme paucity of its composition (horsetails and ancient Pinaceae) and uncertainty of the correlation with the host deposits of the Shandoronsk series.

The coal-bearing formations containing plant remains are widely distributed in northern and north-eastern China. The upper portion belongs to the Lower Cretaceous possessing a flora characteristic of this age which will be described in the appropriate section. The lower half of the coal-bearing series exhibits plant remains of Late Jurassic age. Thus, the Shihetzi suite displays *Raphaelia diamensis*, *Coniopteris burejensis*, *Pterophyllum*, cf. *propinquum*, *Nilssonia sinensis*, *Ginkgoites orientalis*, *G. chiliensis*, *Baiera gracilis*, *Sphenobaiera gracilis*, *Sphenobaiera longifolia*, *Phoenicopsis manchuriea*, *Czekanowskia rigida*, *Elatocladus submanchurica*, etc. (Ye Meina and Li Bauxian, 1980).

Comparison of this list with the flora determined for the upper reaches of the Amur river and the basin of the river Zeya reveals many common forms. Their number would be still greater if it were possible to draw comparison between the plant remains from the USSR and China directly. *Raphaelia diamensis* is characteristically missing without rising higher than the boundary between the Jurassic and Cretaceous.

The floras of the Amur region considered which are situated on the territory of the USSR and the Amur-adjacent part of China show significant homogeneity in systematic composition. The flora of the Bureya basin appears to be the richest while the flora in the upper reaches of the Amur river is the least diverse. In comparison with the floras of the Lena province that of the Amur basin highlights a variety of ferns and cycads often represented by species unknown in the north, and the emergence of Bennettitales dominated by a species of *Pterophyllum*. *Pseudotorellia*, missing in the Lena province, also occurs.

In one of the more southerly habitats of the Late Jurassic flora situated on the river Tyrma, a tributary of the Bureya river, representatives of those genera which are inherent in the South Asia province adjoining to the south and making part of the Euro-Sinian region appear. These are *Phlebopteris*, *Dictyophyllum* and *Klukia*. Simultaneously, *Czekanowskia* are found abundantly in the floras of the Lena and the Amur provinces. The transition zone between the Siberian–Canadian and Euro-Sinian regions is surely located here.

The plant remains from both the Lena and Amur provinces are associated largely with the coal-bearing deposits without usually revealing any traces of a long transition.

2.6 Late Jurassic of the Euro-Sinian region

Initially this region was termed Indo-European and its extent embraced India. Later a great deal of palaeobotanic material accumulated coupled with palaeomagnetic research evidence which showed that India was situated in the Southern Hemisphere where it was close to the south of Africa. I therefore renamed the Indo-European region as the European-Sinian subregion of the Indo-European region (Vakhrameev, 1975) and subsequently gave it an independent status as the European–Sinian region (Vakhrameev, 1984). Originally within this region (Vakhrameev, 1964) I identified the European, Middle Asia and East Asia provinces. More detailed research by Doludenko (1984), who gave an in-depth analysis of the Late Jurassic floras of south-western Eurasia, supplied ground for splitting the European province into three: the Scottish, South-European and Caucasian province. The data of Doludenko are also an additional characteristic of the Middle Asia province whose Late Jurassic flora was studied by her in detail.

2.6.1 Scottish province

Several closely located habitats can be found in north-eastern Scotland (Sutherland peninsula). The age of the host deposits mostly exposed along the coasts of Culgower Bay is Kimmeridgian. Bivalves and ammonites (*Aulacostephanoides* cf. *mutabilis*, *Cardioceras alternans*) occur together with plant remains in dark-grey slaty clays. These deposits are supposed to have belonged to the submerged portion of the delta. The floras were first studied by Seward (1911) and much later by two other palaeobotanists, Van der Burgh and Van Konijnenburg-Van Cittert (1984); the latter examined the epidermis in many of the plants.

Below is the list of species determined by them. The majority of forms were quoted by Seward. Ferns are quite conspicuous and common in these lists and include *Phlebopteris dunkeri*, *Matonidium geoppertii*, *Hausmannia dichotoma*, *H. buchii*, *Gleichenites cycadina* and various *Cladophlebis*. Other species comprise *Pseudoctenis* cf. *eathiensis*, *Cycadopteris* sp., *Pachypteris* cf. *lanceolata*. *Ginkgo* (?) sp., *Czekanowskia* cf. *rigida*, *Phoenicopsis gunnii*, *Elatides curvifolia* and *Taxodiophyllum scoticum*.

The list Seward compiled on the basis of a study of plant remains from another locality of this area features many Bennettitales, i.e. *Ptilophyllum pecten*, *Pterophyllum nathorstii*, *Zamites buchianus* and *Z. carruthersii*, as well as *Nilssonia brevis*, *N. mediana* and *N. orientalis*. Apart from *Elatides curvifolia* the list also displays conifers such as *Brachyphyllum* sp., *Sphenolepidium* cf. *kurrianum* and *Araucarites* spp.

Unlike the other provinces of the Euro-Sinian region, the flora of the Scottish province exhibits a variety of ferns and abundance not only of Bennettitales but also cycads. Representatives of the drought-resistant conifers *Brachyphyllum* and *Pagiophyllum* were not encountered. On the other hand, the plants common in the Siberian–Canadian region, i.e. *Czekanowskia* and *Phoenicopsis*, missing in the rest of the provinces of the Euro-Sinian region, are recorded here.

Thus, the flora of the Scottish province presents a somewhat intermediate link between the floras of the Siberian and Euro-Sinian regions. However, the abundance of Bennettitales and thermophilic ferns compels us to refer this province to the latter.

The occurrence of *Pachypteris* is consistent with the position of this vegetation association near the coast of a marine basin which is confirmed by the finds of marine
faunal remains coexisting with plant vestiges. The overall composition of the flora
indicates a subtropical warm climate.

2.6.2 South European province

This province is characterized by plant remains detected in Portugal, France, West
Germany and Poland. The Late Jurassic floras of France are most fully covered in the
works of Barale (1970, 1981; Barale and Contini, 1976 – only summary works are indicated). He described the floras from the Callovian, Oxfordian and Kimmeridgian with
the two latter displaying half of the total genera.

The Lower Callovian in the environs of Chatillon (Etroche), specifically the marly
limestone series, exhibited pteridosperms (*Cycadopteris*), Bennettitales (*Otozamites* and
Sphenozamites), male fructifications of cycads (*Androstrobus*) and conifers (*Brachyphyllum*, *Palaeocyparis*). The presence of the woods with well-expressed rings points to
the seasonal nature of climate. Teeth and bones of marine crorodiles (*Machimosaurus*)
were discovered along with plant remains. The second habitat situated north of
Poitiers yielded *Ptilophyllum* sp., *Bucklandia* and *Araucarites* in addition.

The richest flora is from the Kimmeridgian. The largest habitat of this is to be found
in the southern part of the Jurassic mountains near the village Creys (Isère). The plant
remains here are confined to fine-grained limestones. Just as in the other localities of
the Jurassic mountains here the following finds were obtained: ferns *Stachypteris
spicans*, fragments of various *Sphenopteris*, pteridosperms (*Pachypteris*, *Raphidopteris*,
Cycadopteris), cycads (*Apoldia*, *Pseudoctenis*, *Paracycas*, *Cycadites*, *Cycadospadix*,
Bucklandia), Bennettitales (*Zamites*, 6 species; *Cycadolepis*, 7 species; *Ptilophyllum*,
Williamsonia), Ginkgoales (*Baiera verrucosa*), conifers (*Araucarites*, 4 species; *Brachyphyllum*, 6; *Masculostrobus*, 6; *Palaeocyparis*, *Elatocladus*, *Cupressinoxylon*).

The abundance of Bennettitales, pteridosperms and conifers is quite striking. The
former in the Oxfordian and especially the Kimmeridgian are dominated by *Zamites*,
while *Araucarites*, *Brachyphyllum* and the cones of *Masculostrobus* prevail among the
conifers. The lack of *Nilssonia* and rarity of *Ginkgo* are conspicuous. *Otozamites* was
encountered only in the Callovian. All the habitats of the Late Jurassic floras of France
are related to the coastal-to-marine deposits which are usually represented by marly
formations and limestones corroborating their shallow-water origin and a nearly
complete absence of displacement of the terrigenous material.

All these vegetation associations are likely to have grown on the coast including certain small islands of reefogenic origin. This is consistent with palaeogeographic maps.
Abundance of rigid-leaved Bennettitales, provided with a thick cuticle, and squamifoliate conifers, indicates a dry subtropical (or, perhaps even tropical) climate. The
presence in the Callovian of France of the remains of crocodiles suggests a hot climate.

Portugal provides two stratigraphic levels containing plant remains (Teixera and Pais, 1976). The Oxfordian flora (the capes of Mondego and Leiria) is similar in composition to the Late Jurassic floras of France whereas the flora of the Kimmeridgian–Portlandian (Montejunto ridge) proves richer in ferns (*Sphenopteris*, *Phlebopteris*) but is poor in Bennettitales (*Zamites* sp. and *Ptilophyllum*) and conifers (*Sphenolepis* and *Cupressinocladus*). Doludenko believed that these differences were brought about by dissimilar ecologic situations.

We may consider that the deposits hosting vegetation remains were of land origin and, most probably, alluvial or deltaic. This is supported by the presence of coals (Monte promontory) and also by the absence of pteridosperms. Possibly, the habitats of plants were more humid than in France because traces of horse-tails and a few shoots of conifers were found in them within the Kimmeridgian–Portlandian floras.

The flora from the province Lerida (Spain) earlier referred to the Late Jurassic, most probably belongs to the Berriasian since a *Frenelopsis* was recently detected there; this plant was unknown in the deposits of the Late Jurassic but became widely distributed in the Cretaceous sediments. *Weichselia reticulata* found in the same place is a widespread Early Cretaceous plant though it was occasionally encountered in Middle and Lower Jurassic rocks.

Vegetation remains of Late Jurassic age are also known from West Germany (Bavaria) and Poland. In Bavaria they are associated with platy limestones of the Lower Tithonian widely known as the lithographic stone of Solnhofen and traceable in the territory of adjoining Württemberg. Unfortunately, the plant fossils were in the main evaluated in the last century and have not been subjected to repeated studies. Doludenko, who critically reviewed the definitions and scrutinized their representations, characterizes the forms in this horizon as belonging to the genera *Sphenopteris*, *Furcifolium* (probably *Ginkgo*), *Brachyphyllum*, *Pagiophyllum*, *Palaecyparis*, *Arthrotaxites*, *Podozamites*, *Cycadopteris* which are especially common and which fully correspond to a coastal-to-marine origin of the host deposits.

In Poland, in the north-eastern part of the Mesozoic fringe of the Swentokshisk mountains (Vulka Baltovka), the following finds were made in the Oxfordian series: *Equisetum* sp., *Ctenozamites* sp., *Pachypteris* sp., *Pseudotorellia* sp., *Pagiophyllum connivens* and *Brachyphyllum* aff. *crucis*. Plant remains were detected in a formation of limestone rudites and limestones (Doludenko, 1984).

Thus, the Late Jurassic flora of the South European province is represented almost everywhere by littoral-to-marine associations, as confirmed by the joint finds of marine fauna and plant remains. This is in good harmony with the palaeogeography of south Europe in the Late Jurassic epoch characterized by a great number of islands including the reefs and by a rugged coastline.

The relative abundance of pteridosperms (*Cycadopteris* and *Pachypteris*) corroborates the view that they are restricted to marine coasts and they were, possibly, the Mesozoic analogues of the current mangrove vegetation. The abundance of pteridosperms, Bennettitales and conifers (particularly *Brachyphyllum*) with a thick cuticle, combined with almost total absence of Czekanowskiaceae, and rarity of *Nilssonia* and Ginkgoaceae, suggest a dry, probably, sunny subtropical climate. It can be visualized that forests composed of conifers wherein Bennettitales formed a lower level could be found beyond the narrow coastal strip inhabited by pteridosperms.

The vegetation of which remains were found in Portugal was somewhat different. These remains were discovered in carbonaceous deposits which suggests their growth in boggy areas in the upper part of the delta. The humid climate is confirmed by the appearance of horse-tails, missing in other localities, as well as by the abundance of ferns. Pteridosperms in this association are missing.

2.6.3   Caucasian province

Significant habitats belonging to this province are to be found in Georgia (Verkhyaya Racha, Tsesi village), Abkhasia (Bzybsk gorge of the river Kadgeripsh) and the south-east of Osetia (the environs of the lake Yertso, village Kemulta). The best studied and the richest is the first of these (Doludenko and Svanidze, 1969). The coastal deposits of the Callovian contain bivalves, Early Callovian ammonites (*Macrocephalites macrocephalus*) and numerous remains of plants which were perfectly preserved.

Single horse-tails were discovered here along with few ferns such as *Angiopteris* (Delle, Doludenko and Krassilov, 1986), *Cladophlebis*, *Sphenopteris*, the Caytoniales (*Sagenopteris*) and pteridosperms such as *Pachypteris*, *Cycadopteris* and *Ctenozamites*. Cycads are diverse including *Paracycas* (3 species) and especially Bennettitales such as *Nilssoniopteris* (6), *Otozamites*, *Pseudocycas*, *Pterophyllum* (11), *Ptilophyllum* (3) and *Cycadolepis*.

The composition of the Ginkgoales is impoverished; *Eretmophyllum*, *Sphenobaiera* and *Pseudotorellia* are found. The conifers are represented by *Brachyphyllum*, *Pagiophyllum*, *Elatocladus*, *Podozamites*, *Araucariodendron* and *Tomharrisia*. Czekanowskiales are totally lacking.

There is a conspicuous abundance and especially specific variety of the genera *Pterophyllum* and *Nilssoniopteris* established by Doludenko by examining the epidermis. At the same time a very rare occurrence of *Nilssonia* (1 species) is striking. It is noteworthy that remains of *Nilssonia* were not encountered in the South European province at all. Remains of certain squamifoliate conifers are abundant (*Brachyphyllum* and *Pagiophyllum*). Finds of *Podozamites* are exceptional.

The floras of the Caucasian province differ from those of the South European province by abundance of species of *Pterophyllum* (11), *Nilssoniopteris* (6) and *Sagenopteris* (3), as well as by the absence of the genus *Zamites* and shoots with small squamifoliate opposite scales (*Palaecyparis*, *Cupressinocladus*, *Cyparissidium*). The palynologic analysis reveals a high (50%) percentage of the pollen *Classopollis* (see Fig. 2.12). The floras of both the provinces display *Pachypteris* and *Cycadopteris* which is consistent with the habitat of these floras in a littoral strip of a marine basin. Prior to the monographic study of the Late Jurassic floras of Georgia undertaken by Doludenko and Svanidze (1969), their composition appeared to be similar to the coeval European flora. It was believed that the differences amounted to replacements of some Bennettitales with others. Therefore, earlier I (Vakhrameev, 1964) singled out only one European province.

2.6.4 Middle Asia province

This province is represented by two types of floras (Doludenko and Orlovskaya, 1976; Doludenko, 1984). A richer flora is associated with the littoral deposits of a large brackish-water lake situated in the southern part of the Karatau ridge. The second type of flora is confined to the littoral formations of the southern slope of the Hissar ridge.

In Karatau we know two stratigraphic horizons with vegetation remains. The more ancient of the two is related to the deposits of the Borolsait suite. The majority of plant remains were poorly preserved and many of them were evaluated only to a genus. The ferns *Coniopteris simplex* (= *C. angustiloba*) and the small-pinnule *Cladophlebis* sp. were discovered here. The Caytoniales are represented by *Sagenopteris phillipsi* and the cycads by individual fragments of *Nilssonia* sp. and *Pseudoctenis* sp. Bennettitales are more numerous including *Williamsoniella karataviensis*, *W. czochaiensis*, *Anomozamites* sp., *Pterophyllum* sp. and *Otozamites* (?) sp. *Ginkgoites* spp., *Eretmophyllum* sp., *Pseudotorellia* sp. and *Czekanowskia* sp. are observed. Especially numerous are the conifers such as *Elatocladus turutanovae*, *Pityophyllum* spp., *Pityospermum* spp., *Pityostrobus* sp., *Brachyphyllum* sp., *Pagiophyllum* sp. and *Pagiophyllum setosum*. Strobili and seeds of gymnosperms of uncertain systematic classification were also detected.

The flora confined to the Karabastausk suite seems to be richer. Its upper portion is composed of very characteristic thin-layer carbonate rocks widely known from literature as paper shales or 'fish' shales, the latter name being derived from the abundance of fish remains in these formations. Spore plants are represented here by the horse-tails, viz. *Equisetum laterale*, and ferns *Stachypteris turkestanica*, several species of *Coniopteris* (*C. simplex*, etc.) and *Hausmannia* sp., *Clathropteris* sp. and *Cladophlebis* sp. The ferns are mainly made up of small rarely encountered fragments. The most widespread and diverse representatives (Fig. 2.19) are those of Bennettitales (*Ptilophyllum*, *Otozamites* and *Sphenozamites*, *Zamiophyllum*, *Pterophyllum*, *Williamsoniella*, *Weltrichia* and *Cycadolepis*) and conifers.

The conifers include as their most diverse and numerous units *Brachyphyllum*, *Pagiophyllum* and *Elatocladus* (Figs. 2.20, 2.21) as well as the ancient Pinaceae, *Pityophyllum*, *Pityostrobus*, *Pityospermum*. *Araucarites* and *Podozamites* occur less often. The cycads present *Paracycas* (Fig. 2.22) and *Nilssonia* and the cycadophytes *Cycadites* and *Taeniopteris*. Rare Czekanowskiaceae (*Czekanowskia*, *Phoenicopsis*) were detected, too, in addition to *Ginkgo* (*Ginkgoites*, *Baiera*, *Sphenobaiera* and *Eretmophyllum*). Just as in the

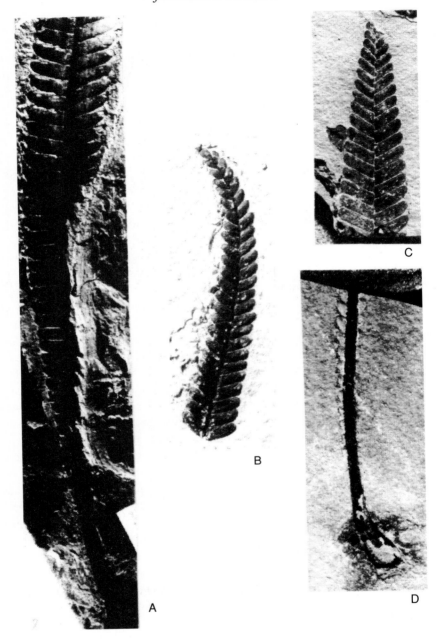

Fig. 2.19 Late Jurassic plant fossils from southern Kazakhstan; natural size.
A–D, *Ptilophyllum caucasicum* Doludenko and Svanidze.

Fig. 2.20 Late Jurassic plant fossils from southern Kazakhstan. A–B, *Pagiophyllum papillosum* Orlovskaya, natural size; C–D, *Elatocladus minutus* Doludenko, × 2.

Fig. 2.21 Late Jurassic plant fossil from southern Kazakhstan; natural size;
Brachyphyllum brickae Doludenko.

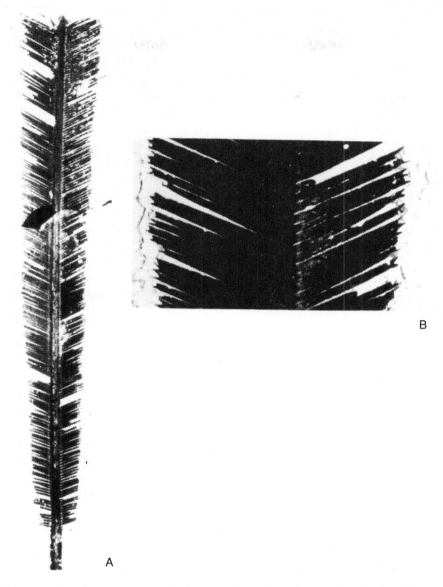

Fig. 2.22 Late Jurassic plant fossils from southern Kazakhstan; A–B, *Paracycas harrissii* Doludenko; A, × 0.25; B, natural size.

Borolsay suite seeds close to the unidentifiable gymnosperms are encountered (*Carpolithes, Platylepidium, Problematospermum*).

The floras of Karatau are different from those of the closest Caucasian province by the absence of *Pachypteris* which is associated with another ecologic situation. The Caucasus accommodates littoral vegetation of a coastal basin whereas in Karatau one can observe lacustrine plants. Certain Ginkgoaceae and Czekanowskiaceae, which are important elements of the floras of the Siberian region, are noticeably absent from the flora of Karatau. The Siberian region was separated from the Middle Asia province by the sea. Therefore, the plants could migrate from north to south and back. However, the principal impediment for such a migration was provided by climate. The climate of Karatau in the Late Jurassic epoch was dry and tropical, as supported by the nature of its plants (prevalence of Bennettitales and squamifoliate conifers). The same conclusion is borne out by the high content of the pollen *Classopollis* (95–100% in the Karabastausk suite (see Fig. 2.12)) of Oxfordian–Kimmeridgian age. The Borolsay suite seems to be of Bathonian age.

In the Late Jurassic epoch a wide belt of an arid subtropical climate stretched through south England (Purbeckian), France, and the countries of South Europe including the southern portion of the European part of the USSR, further extending across the Caucasus to middle and western China. It prevented a number of thermophilic elements from penetrating from the north. These were the elements of the Late Jurassic flora of the Siberian region, in the first place, ferns, Czekanowskiaceae, Ginkgoaceae and *Nilssonia*. Ancient Pinaceae were vanishing in the southward direction, too. The relatively northern position of the floras of Karatau belonging to the Middle Asia province situated close to the border of the Siberian region, as well as the absence of marine basin in between, account for the occurrence of relatively more thermophilic elements typical for the Siberian region.

It will be noted that in Southern Europe the Late Jurassic vegetation was largely insular, lacking in such elements typical for the Siberian region as Czekanowskiaceae and ancient Pinaceae while Ginkgoaceae are encountered very seldom and are represented mainly by *Ginkgo* or *Baiera*. The rich Late Jurassic floras of France did not yield a single *Nilssonia*.

Moving from the Karatau ridge to the southern slope of the Hissar ridge we observe certain alterations in the composition of coeval or virtually coeval floras. The few remnants of plants are confined here to the Baisum suite whose age is evaluated by some authors as Callovian, by others as Bathonian. An expected appearance is that of *Pachypteris lanceolata* indicating the habitats of vegetation in a coastal zone. In addition to *Pachypteris* the finds comprise *Gleichenia* sp., *Coniopteris* sp., *Ptilophyllum acutifolium*, *Otozamites* (4 species), *Ginkgo sibirica*, *Pagiophyllum* sp. and *Brachyphyllum mammilare*. The abundance of *Otozamites* brings this flora close to the Karabastau flora but, on the other hand, it is completely devoid of ferns and ancient Pinaceae whereas the content of the pollen *Classopollis* attains 60–70% and in the upper strata 95–96% (Kossenkova, 1975). This leads to the presumption that the lower part of the Baisun suite is of

Early–Middle Callovian age, while the upper portion belongs to the Late Callovian. Several small-sized and impoverished habitats of Callovian (or, possibly, Bathonian) flora, situated east along the southern slope of the Hissar ridge (Luchob, Khanaka) in deposits which are, probably, coeval with the Baisun suite, do not add anything new to the characterization.

2.6.5 East Asia province

This province occupies the central and southern areas of China, Japan, the Soviet Primorje area and south Mongolia. We know little about the composition of the Late Jurassic floras of this province because the continental deposits of this age in China and south Mongolia are mainly composed of red-coloured or variegated formations that developed under the conditions of a hot arid climate whereas the region of the Soviet Far East was invaded by the sea.

The Upper Jurassic in Mongolia is represented by the Ikhesnur, Tormkhon and Sharilin suites identified in various parts of this country and merged into the Sharilin horizon (Shuvalov, 1982). These suites mainly consist of red-coloured and variegated deposits and sometimes of effusive formations. So far researchers have failed to detect definable vegetation remains except for fragments of petrified trunks. However, going by the presence in these deposits of rarely encountered bones of dinosaurs, vegetation can be presumed to have been available here, probably growing on the slopes of highlands.

According to the date of palaeoentomofauna (personal communication by A. G. Ponnmarenko), the habitat Khoutin-Khamgar located some 250 km south of Ulan-Bator or 18 to 20 km from the village of Bayanzharalgan of the West Gobi aimak is presumed to belong to the Late Jurassic. Its host deposits were earlier considered as the Early Neocomian. Vegetation remains were found in thin-layer bluish-grey clayey (paper) shales enclosing intercalations of dolomites. From here were collected (Sodov, 1980) *Selaginella* sp., *Equisetites* sp., small fragments of ferns (*Sphenopteris* sp., *Cladophlebis* sp., *Raphaelia* ex gr. *diamensis*, *Heilungia houtensis*), Czekanowskiaceae (*Phoenicopsis, Czekanowskia, Leptostrobus*) and conifers (*Pseudolarix, Pityospermum, Schizolepis, Samaropsis*).

Conifers are prevalent according to frequency of occurrence being represented mainly by seeds with thistle-down. Plant remains indicate a long-distance transfer. Originally the age of this complex was thought most likely to be Early Neocomian (Sodov, 1980). In all probability this flora whose habitat is situated roughly in central Mongolia refers to the zone of ecotone although the availability of Czekanowskiaceae associates it with the floras of the Siberian region. *Coniopteris hymenophylloides, Ptilophyllum pecten, Nilssonia linearis, Pagiophyllum* cf. *expansum* and *Podozamites lanecolatus* were found in central China in the Dabashan area located north of the Yangtse river (Shantsi province) in the Dzyan-Foan series (Anon., 1960, 1963) composed of violet-red, brownish and green argillites and sandstones. The lower strata of the Dzyan-Foan series are deposited in the coal-bearing formations of the Syuidzya-Khe series.

Chinese geologists refer the series Dzyan-Foan to the Upper–Middle (?) Jurassic. Its most likely age is Late Jurassic. This supposition proceeds from the fact that the Middle Jurassic epoch in the central part of China witnessed deposition of carbonaceous formations whereas the onset of the Late Jurassic in this part of China was concomitant with cessation of coal formation due to the climatic change towards aridization accompanied by the beginning of sedimentation of red-coloured and variegated sequences.

The presence of the Late Jurassic floras in Japan is a very complicated issue. In works of the first half of the twentieth century the floras of Tetori were dated as Late Jurassic. Those floras were widespread in the internal part of Japan. The floras named Riosseki were assessed as Early Cretaceous being situated inland. However, later research undertaken by Kimura over many years (1980) revealed that these floras are coeval or almost coeval and belong to the Early Cretaceous. The correlation between these floras differing in composition is reviewed in the chapter on the Early Cretaceous.

The marine deposits of the Early Jurassic in the interior of Japan are represented by the Kudzyuru suite, at the base composed of sandstones and conglomerates succeeded upwards by alternating sandstones and clayey shales; its thickness attains 800 m. Ammonites, *Kepplerites* (*Seymourites*) *japonicus*, *K. acuticostatum* and *Reineckia yokoyamia* were found in the deposits of the Kudzyuru suite (subgroup) suggesting a Callovian age.

The Itosiro suite (subgroup) is located higher containing no marine fauna remnants and being laid on the eroded Kudzyuru suite and in places encroaching on the rocks of the ancient base. The plant remains grouped as the Tetori flora are related to the Itosiro suite. This flora so far remains intact rather than separated into stratified complexes and is regarded as a unified flora currently dated as the Early Cretaceous. It remains to be seen whether the break in the deposits should be classified as lasting for the greater period of the Late Jurassic or whether some part of the Itosiro suite with its inherent plant remains should be referred to the Late Jurassic. In the latter case it is necessary to determine that portion of the Tetori flora which corresponds to the Late Jurassic.

To resolve this task it is essential to gather stratum-wise samples in a number of sequences of the Itosiro suite and to draw a correlation between them. At present a review of the published lists of the Tetori flora from various habitats does not permit establishment of criteria for separating the late Jurassic part of the sequence. The lists of the Tetori flora available from literature were earlier interpreted as belonging to the late Jurassic floras whereas now they are referred to the Early Cretaceous.

In all probability the Late Jurassic and Early Cretaceous floras of Japan will prove very similar in composition since the climate of Japan did not change appreciably throughout the Jurassic, as confirmed by the absence of red-coloured or even multicoloured rocks in the Jurassic sequences. Such rocks are fairly developed in the Upper Jurassic deposits of south and central China, Mongolia and Middle Asia, suggesting a dry and hot climate in these regions in the Late Jurassic epoch. The humid, most probably monsoon, climate of Japan prevailing throughout the Jurassic and Cretaceous is due to its position on the edge of the continent of Asia washed by the Pacific.

2.7 Jurassic (undivided): North American region

The paucity of the remnants of the Jurassic deposits in North America and especially the USA and Canada does not at present permit identification of distinctly outlined phytochoria in this territory, nor the drawing of boundaries among them as was done for Eurasia where Jurassic floras are perhaps best studied. Therefore, this section will deal with a review of the factual material available followed by an attempt to establish likely analogues for phytochorial regions which were established for Eurasia.

The Toarcian floras are the most ancient ones in the Jurassic represented by macro-remains. One of the habitats described by Knowlton (1917) can be found in the basin of the upper reaches of the river Matanuska (south Alaska). The deposits referred to the Toarcian contain Early Jurassic invertebrates as well as plant remains, while overlying beds represent the marine Middle Jurassic. *Cladophlebis* sp., *Dictyophyllum nilssoni*, *Sagenopteris* sp., *Otozamites pterophylloides*, *Otozamites* sp., *Pterophyllum rajamahalense*, *Pterophyllum aequale*, *Nilssonia polymorpha* and *Pagiophyllum falcatum* have been collected from the Toarcian here.

Another habitat of the Toarcian floras is situated on the western shore of Vancouver Island (British Columbia) and contains imprints of plants whose principal representatives are *Ptilophyllum*, *Pterophyllum*, *Otozamites*, *Matonidium*, *Dictyophyllum* and *Nilssonia*. The comparison of the generic composition reveals a close similarity to the flora of south Alaska. Unfortunately, the specific composition of the Vancouver Island flora was not given (Fry, 1964).

A third locality is related to the Upper Jurassic formations of the Monte de Oro suite (northern part of central California) dated as Late Oxfordian–Early Kimmeridgian. Here representatives of the genera *Macrotaeniopteris*, *Taeniopteris*, *Pterophyllum*, *Ctenis*, *Sagenopteris*, *Baiera*, *Podozamites* and *Pagiophyllum* were detected. The ferns *Cladophlebis* and *Coniopteris* are encountered more rarely.

The comparison between the Toarcian and Late Jurassic floras indicates that the latter comprise fewer ferns and experience a total lack of *Dictyophyllum* which, as a rule, does not continue beyond the Middle Jurassic. The presence in the composition of the Late Jurassic flora of the conifers *Pagiophyllum* is typical. Scarcity of ferns in the Late Jurassic compared with the lower flora is characteristic of the Jurassic floras of the Euro-Sinian region, too.

The relative poverty of the Jurassic deposits of the USA results from the dry climate reigning over the greater part of this continent in Jurassic time. While western Europe

witnessed the onset of major climatic drying in the Late Jurassic, the USA was dominated by dry climate in the Lower and Middle Jurassic, as is supported by the development of obliquely laminated sandstones of aeolian origin in the Lower Jurassic and
the appearance of evaporites in the west of the USA in the Middle Jurassic. The
Jurassic rocks of the USA's western states include widespread red-coloured and
terrigenous rocks of continental origin, devoid of coals.

An example of this type of deposit is the lacustrine formation of the Morrison suite
distributed in the states of Utah, Colorado, Arizona and New Mexico. This vast
internal basin arose at the beginning of the Late Jurassic. The displacement of material
was directed from west to east. Therefore, the western part of the Morrison suite is
composed of variegated, more often obliquely laminated sandstones and conglomerates giving way to argillites. In the extreme east of its extent (Colorado) sandstones
emerge. The variegated colouring of the rocks and occurrence of sandstones of sub-
saline origin point to a dry climate.

Multitudinous burials of the bones of dinosaurs are associated with the deposits of
the Morrison suite. The plant remains are represented by petrified barrel-shaped
trunks of Cycadeoideans, although they are encountered very seldom. It is likely that
the dry climate inhibited growth of closed forests and formation of swamps because
beds of coal are missing. Wide open expanses are presumed to have been covered with
herbaceous vegetation occasionally replaced by rare forests where dinosaurs lived.

The data obtained show that during the Mesozoic history large land dinosaurs preferred to inhabit open spaces like contemporary savannah while closed and quite often
boggy forests were avoided because it was difficult to travel there. This regularity persists, as we shall see, for the Cretaceous period, too. Thus, in Mongolia the striking
abundance of the remnants of large dinosaurs is associated with the red-coloured
deposits of the Upper Cretaceous developing under the conditions of semi–arid or arid
climate and containing no coal-bearing members at all. These deposits, represented
by lacustrine and alluvial sediments, just as the Morrison suite, exhibit extremely rare
plant remains. Dinosaurs fed on herbaceous vegetation which densely covered lake
coasts.

Comparing the Jurassic floras of North America with those of Europe we so far find
no analogues to the Siberian region. The northernmost habitat of the Jurassic floras is
situated in south Alaska (the upper reaches of the river Matanuska) and contains the
forms typical for the Euro-Sinian region. However, we already know that in the
Toarcian certain thermophilic elements advanced far north to Siberia due to a wave
of warming. It is not unlikely that the floristic composition of the above-mentioned
locality can be identified through this warming whereas the pre-Toarcian and Middle
Jurassic times witnessed a more moderate vegetation typical for the Siberian region.

This possibility is indicated by the nature of the shallow-water marine deposits
evolved within Canada in the Jurassic time and especially the wide distribution of the
bivalvian *Buchia* of the Late Jurassic sediments which are also frequently encountered
in the coeval terrigenous sediments of northern Siberia and the Far East. These

disappear farther south thus suggesting a moderately warm climate habitat. In the south of western Canada the boundary of the Jurassic and Cretaceous houses a thick coal-bearing formation (Kootenai suite) containing remains of hygrophilic plants which are very similar in their composition to the coeval flora of the Siberian region. Therefore, we have good reason to believe that the Canadian Cretaceous and Jurassic saw the growth of seasonal leaf- and branch-dropping forests characteristic of the belt of moderate warm climate. The invasion to the north of the representatives of sub-tropical vegetation (the locality of the Toarcian flora in south Alaska) might have taken place during the warming phase as this happened in the Toarcian.

2.8 Jurassic (undivided): Equatorial region

While the data on the Early Jurassic floras are scarce and do not permit formation of a sound and distinct judgement, the composition of the Late Jurassic and particularly Middle Jurassic floras has lately been assessed adequately. The Jurassic floras of the Equatorial region are deemed to comprise the vegetation of south Mexico, Cuba, Colombia, Brazil, northern Africa (Tunisia, Libya) and Israel. The Late Jurassic flora of Madagascar stands somewhat apart. Joint consideration of the Jurassic floras of all three epochs is necessitated by the similarity of their systematic composition and still insufficient coverage. Their differences result not so much from age as from geographical position. The boundary between the floras of the Euro-Sinian and Equatorial regions is drawn rather generally in most places.

We shall start our overview with the rich floras of south Mexico whose main localities are to be found in the provinces of Oaxaca and Pueblo south and south-east of Mexico City. The Lower Jurassic deposits projecting in the provinces of Vera Cruz and Tamaulipis are impoverished in plant remains. Of these we can only mention a few forms not rising to the Middle Jurassic. They include *Pterophyllum propinquum*, *Sphenozamites* sp. and *Taeniopteris orvillensis*. The remaining species belong to the genera *Cladophlebis*, *Otozamites*, *Sagenopteris* and *Zamites*, and are encountered high in the sequence. The Rosario suite is richer in plant remains, and is developed in the province of Oaxaca (El Consuello area). This flora was first described by Weyland in a series of works published from 1909 to 1929. Recently it was subjected to a further investigation by Silva-Pineda (1984). Below is the list of the forms revised by her: *Piazopteris branneri*, *Coniopteris arguta*, *C.* cf. *hymenophylloides*, *Gonatosorus nathorstii*, *Cladophlebis browniana*, *Zamites lucerensis*, *Z. oaxacensis*, *Z. tribulosus*, *Otozamites mandelslohi*, *O. hespera*, *Ptilophyllum acutifolium*, *P. cutchense*, *Pterophyllum* cf. *munsteri*, *Anomozamites* sp., *Taeniopteris oaxacensis*, *Cycadolepis mexicana*, *Williamsonia cuauhtemoci*, *W. huitzilopochtlii*, *W. netzahualcoyotlii* and *Weltrichia mexicana*.

The age of the Rosario suite, according to current data, encompasses the uppermost strata of the Lower Jurassic (Toarcian) and the Middle Jurassic. The Middle Jurassic age is also suggested by occurrence of two species of *Coniopteris*. The works of Silva-

Pineda (1970, 1978) cite the species collected from other localities in the provinces of Oaxaca and Pueblo. Plant remains were also gathered from higher horizons of the Middle Jurassic separated from the Rosario suite by the conglomerates of the Kualak suite. This part of the Middle Jurassic is represented by the Zorillo suite and sequentially deposited suites overlying it, i.e. Taberai, Simon and Otatero. Apart from the forms attributed to the Rosario suite, the upper deposits of the Middle Jurassic contain *Pseudoctenis lanei*, *Otozamites graphicus*, *Zamites feneonis*, *Williamsonia diquiyni*, *W. oaxacensis* and *Podozamites* cf. *lanceolatus*.

We should point out two more localities with *Piazopteris branneri*. One of them is situated in the east of Brazil (Bahia), and the other is to be found in the western part of Cuba (the upper reaches of the river Nombre de Dias). This fern from the latter habitat was first described by myself (Vakhrameev, 1965) as a new species *Phlebopteris cubaensis* but was then re-evaluated and referred to the earlier known *Piazopteris branneri* which is quite acceptable. The genus *Piazopteris* differs from the well-known genus *Phlebopteris* by the double pinna-like frond. It is curious that in the locality in western Cuba the only species of this kind was encountered, being represented by a large number of individual specimens. This indicates the development of fern thickets here which were probably situated on the coastal plain.

Giving a general characterization to the Jurassic floras of south Mexico belonging largely to the Middle Jurassic we should note in the first place the prevalence of Bennettitales represented by the genera *Zamites*, *Otozamites*, *Ptilophyllum*, *Pterophyllum* and *Williamsonia*. Ferns are rather rare but *Piazopteris* should be singled out as belonging to Matoniaceae. Cycads are obviously very rare being represented by *Pseodoctenis*. A striking feature is the total absence of Ginkgoaceae and Czekanowskiaceae, as well as the rarity of conifers represented by *Podozamites* and other elongated leaves with parallel venation of unknown systematic classification and wrongly referred to the Late Palaeozoic species *Neoggerathiopsis hislopi*.

In the northern part of South America constituting part of the Equatorial region the Early Jurassic flora has been established in Colombia and Brazil. The former was found to accommodate *Sagenopteris* cf. *nilssonianum*, *Otozamites* sp., *Zamites* sp., *Cycadolepis* sp., *Podozamites* sp., *Brachyphyllum* sp. and *Pagiophyllum* sp. (Langenheim, 1961). *Otozamites* sp., *Pterophyllum* sp., *Nilssonia* sp. (?), *Sagenopteris* sp. and *Thinnfeldia* sp. are quoted for the strata which are transitional from the Upper Triassic to the Lower Jurassic of Brazil in addition to the *Piazopteris* mentioned above.

Localities with plant remains connected with the 'Valle Alto' suite are situated in Colombia approximately 120 km south of the town of Medellin. The above suite is composed of shallow-water terrigenous sediments deposited on the eroded surface of diorites. Lemoigne (1984), who determined representatives of the genera *Gleichenites*, *Cladophlebis*, *Sphenopteris*, *Pachypteris*, *Sagenopteris*, *Nilssoniopteris*, *Anomozamites*, *Otozamites*, *Zamites*, *Ctenozamites* and *Desmiophyllum*, referred the host deposits to the Upper Jurassic although before that they had been looked upon as Lower Jurassic. From our point of view this decision looks somewhat dubious because the list of plants described by Lemoigne includes such a species as *Cladophlebis exiliformis* so far known only from the Lower Cretaceous of Japan and the Soviet Far East Primorje area. Therefore we shall review this form in detail when characterizing the Early Cretaceous phytochoria.

In northern Africa plant remains are known from the Middle and Upper Jurassic. Thus, the Middle Jurassic of Tunisia underlying the Bathonian deposits with *Trigonia pullus* (Koeniguer, 1980) furnishes fragments of the stems of *Paradoxopteris* belonging, as was shown by Alvin (1971), to the fern *Weichselia reticulata*. In Algeria south-south-east of Oran the interior of carbonate deposits of the Upper Jurassic is permeated by shallow-water sandstones dated as Upper Oxfordian (Lusitanian). From these *Weichselia* sp., in conjunction with *Zamites* sp., *Otozamites* sp. and *Pterophyllum* sp., is reported. These sandstones are covered with reef limestones of Kimmeridgian age in a number of sequences (Nicol-Lejal, 1971); the most enriched localities are familiar from Libya. Its north-eastern part (Jabal-Nafusakh) is famous for the Chameae suite composed of gypsum clays with *Piazopteris branneri*, *Pagiophyllum* sp., *Brachyphyllum* sp., *Otozamites* sp. and *Samaropsis* sp., with underlying limestones of Callovian age (Salem, 1980) and overlain by the continental sublittoral sandstones with the bones of dinosaurs of Wealden age. Another habitat (the basin Hammada Al Hamra) reveals *Equisetites reticulata*, *Piazopteris branneri*, *Onychiopsis tenuiloba*, *Pagiophyllum* sp. and *Cupressinocladus* sp. in variegated and sometimes red-coloured compact clays.

Lorch (1967) discovered in the Jurassic deposits of the Negev desert (south of Israel) quite a rich flora including *Equisetum columnare*, *Piazopteris branneri*, *Sellingia microloba*, *Onychiopsis tenuiloba*, *Aspidites beckeri*, *Cladophlebis* cf. *stricta*, *Otozamites ramonensis*, *O. feistmantelii*, *O.* cf. *mimetes*, *Ptilophyllum* sp., *P.* cf. *cutchense*, *P.* cf. *acutifolium*, *Williamsonia atractylis*, *Elatocladus ramonensis*, *Brachyphyllum* cf. *mamillare* and *Podozamites* sp. Judging by the depth of the rocks with plant remains beneath the marine deposits of the Middle Bajocian, this flora may be dated as belonging to the lower strata of the Middle Jurassic. Lorch described *Dactyletrophyllum ramonensis*, *Brachyphyllum negevensis*, *B. pulcher*, *B. porrigente* and *Masculostrobus harrisianus* from the same area from the deposits referred to the Lower Jurassic. He also collected *Marattia curvinervis*, *Todites williamsonis*, *Phlebopteris polypodioides*, *Piazopteris* cf. *branneri* and *Stachypteris* cf. *spicans* from the Jurassic of the Sinai peninsula.

E. Buro determined *Brachyphyllum karpofii* and *B. multistenimium* from the Lower Jurassic deposits uncovered through a borehole in Saudi Arabia. Findings of Jurassic plants so far remain unknown for the Equatorial belt of south-east Asia.

The rich flora of Madagascar originating from its south-western part (the area of the Manama massif) is a separate issue. The member with plant remains attains a thickness of 150 m and the flora-bearing strata are devoid of coals. Their age is ascertained from the Oxfordian deposits characterized by ammonites. Appert (1973) described only horse-tails and ferns as follows: *Equisetites ferganensis*, *Mohriopsis plastica*, *Ruffordia goeppertii*, *Gleichenites nordenskioldii*, *Piazopteris lorchii* (= *Piazopteris branneri*), *Phlebopteris muensteri*, *Matonidium goeppertii*, *Matonia mesozoica*, *Culcitites madagascar-*

iensis, Coniopteris manamanensis, Eboracia lobifolia, Haydenia thyrsopteroides, cf. *Dictyophyllum,* cf. *Thaumatopteris, Onychiopsis psilotoides, Cladophlebis ankazoaboensis, Cladophlebis* sp. and *Sphenopteris* sp. Two fresh genera have been established, i.e. *Mohriopsis* and *Culcitites,* as well as five new species. Besides ferns, the remains of conifers were also found.

We see that the Late Jurassic flora of Madagascar is quite special. Appert supposes (personal communication) that the canopies of coniferous forests which grew here concealed localities of ferns; no cycadophytes have as yet been found.

Comparing the Jurassic floras of the Equatorial belt known to us from south Mexico, Colombia, Brazil, north Africa and Israel, we may characterize them as follows. These floras are dominated by the Bennettitales out of which the most frequently occurring are the representatives of the genera *Zamites, Ptilophyllum, Pterophyllum* and *Williamsonia.* Ferns are rather numerous with *Piazopteris branneri* standing out as being encountered in almost every locality of Mexico, north Africa and Israel. It should be noted that in some habitats it may prevail in the number of impressions or even constitute a monotopic association (Cuba). Outside the Equatorial belt this genus of ferns does not occur, being replaced by a similar genus, i.e. *Phlebopteris,* possessing a simple-pinna frond. The flora of Madagascar stands apart because it is dominated by ferns and contains no cycadophytes. But this flora has another fern, too, to which Appert gave a new specific name, i.e. *Piazopteris lorchii.* Appert believed that the typical species *Piazopteris branneri* from Brazil was based on inadequately complete and distinct imprints. Therefore, he thought it proper to identify a new species (*P. lorchii*) naming it in honour of Lorch (1967) who gave a far more detailed description of the impressions of this fern than the first writer White (1913).

The presence of *Weichselia reticulata* is observed for the Late and, possibly, Middle Jurassic of Northern Africa. This genus became quite widespread in the Cretaceous in the Euro-Sinian region. This fern monographically described by Alvin (1971) possessed xerophilic characters. *Coniopteris* and *Cladophlebis* are mentioned for the Middle Jurassic of Mexico. Conifers are represented throughout by the squamifoliate forms of *Brachyphyllum, Pagiophyllum* and *Cupressinocladus. Podozamites* was additionally detected in Mexico.

Ancient Pinaceae, Czekanowskiales and even Ginkgoales are totally missing including the genus *Ginkgo* which is widely distributed in the Euro-Sinian region. Cycads are very rare being represented in Mexico by the genus *Pseudoctenis.* Indications of the finds of leaf impressions of *Nilssonia* in the Jurassic of Mexico should be considered non-authentic since review of the photographs raises some doubt about their association with this genus.

Differences in composition between the established Middle Jurassic (Mexico) and Late Jurassic (northern Africa) floras are so far very obscure. The common characteristic species is *Piazopteris branneri.* To evaluate the differences between the Middle and

Late Jurassic floras on the basis of the specific composition of Bennettitales is very difficult because their systematic classification on the specific level has not been adequately elaborated; species described under different names are often identical. Another original fern *Weichselia reticulata* was found only in northern Africa where its remains were detected mainly in the Upper Jurassic deposits. Stems of *Paradoxopteris* possibly belonging to this fern were recorded only in the Middle Jurassic of Tunisia.

Going by the composition of the Middle Jurassic floras of south Mexico, the climate of this epoch within Central America was hot (as suggested by the abundance of Bennettitales) and, perhaps, semi-arid. The latter feature is corroborated by the absence of Ginkgoaceae and rarity of *Nilssonia*. The absence of plant remains in the Late Jurassic of South America deprives us of the palaeobotanical criteria for forming a judgement about climate but the formation of the Late Jurassic salts in the Bay of Mexico testifies to its enhanced aridity. The climate of the Late Jurassic in northern Africa was doubtless arid, as is signalled by the composition of its flora, viz. presence of *Weichselia reticulata* and squamifoliate conifers. Another supporting character is the nature of the Upper Jurassic deposits (variegated and red-coloured rocks accompanied by gypsum).

Given the abundance of ferns in the Middle Jurassic of Israel which co-occur with squamifoliate conifers, the climate of the Middle Jurassic was less dry in comparison with the Late Jurassic. The richness of the Late Jurassic flora of Madagascar in ferns speaks of a relatively more humid climate. However, the formation of coals did not take place here. Madagascar was situated on the edge of the Indian Ocean, probably within the monsoon zone.

2.9 Jurassic (undivided): Austral (Notal) region

This section is devoted to a review of the Jurassic floras of the Austral (Notal[1]) region encompassing Argentina and Chile in South America, the southern tip of Africa, India, Australia and New Zealand, as well as Antarctica. Shortage of data prevents us from identifying distinct provinces within this region although with time they are certain to be delineated since already certain differences are apparent largely referring to the specific level between the Jurassic floras of Australia, South America and, particularly, India. Therefore, following characterization of certain habitats in the regions mentioned above, we shall give a general description of the development of the floras of the Austral region.

---

[1] The term 'Notal' is to be preferred since 'Austral' causes difficulty in translation into other languages being cophonic with the adjective 'Australian' – Ed. (M.A.A.).

2.9.1 Argentina and Chile

The Jurassic floras of Argentina have been best studied. A small flora from the province of Neuken (Frenguelli, 1937) was dated as Early Jurassic. Plant remains were detected in the littoral faunistically characterized deposits of the Upper Sinemurian, and the following ferns were evaluated: *Cladophlebis oblonga*, *Dictyophyllum rothü*, *Thaumatopteris* sp., *Clathropteris* sp., *Sagenopteris nilssoniana*, *Otozamites* sp., *Ptilophyllum* sp. and *Araucarites* (?) sp.

Much later (Arrondo and Petriella, 1980), the same flora yielded for description the following: *Neocalamites carrerei*, *Marattia muensteri*, *Cladophlebis* spp., *Scleropteris vincei*, *Kurtziana brandmayri*, *K. cacheutensis*, etc. (4 species), *Otozamites*, *Ptilophyllum acutifolium*, *Taeniopteris* sp., *Araucarites philipsi* and *Elatocladus conferta*. The age of the deposits is estimated as Hettangian. The genus *Kurtziana* with the typical species *K. cacheutensis* was established in the Triassic of Argentina.

Middle Jurassic floras are known from the provinces of Neuken and Chubut. Below we give their general characterization based on the summary work (Menendez, 1969) and subsequent articles (Arrondo and Petriella, 1980; Baldoni, 1980 and others). The ferns are represented by *Thaumatopteris* sp., *Clathropteris* cf. *kurtsii*, *Todites williamsonii*, *Scleropteris furcata* and *S. lotenaense*, as well as by several species belonging to the form-genera *Cladophlebis* and *Sphenopteris*. *Sagenopteris nilssoniana* was found, and Bennettitales yielded *Otozamites sanctaecrucis*, *O. traversoi*, *Dictyozamites* sp., *Ptilophyllum hislopii* and *Williamsonia* cf. *gigas*. Conifers included *Brachyphyllum ramosum*, *B. lotenaensis*, *Elatocladus conferta*, *E. heterophylla*, *Pagiophyllum feistmanteli* and *Araucarites* sp. (Fig. 2.23).

It is important to note that in the boundary strata between the Middle and Upper Jurassic numerous petrified cones and trunks of *Araucarites* were found in the province of Santa Cruz in the south of South America. The same strata in the area of Grand Bayo de San Julian furnished for description *Cladophlebis* spp., *Gleichenites juliensis*, *Hausmannia ferrarisii*, *Ruffordia* sp., *Osmundites patagonica*, *Sphenopteris* sp., *Ptilophyllum* sp., *Otozamites* sp., *Athrotaxis* sp. and *Araucaria* sp. The floras from the strata which are transitional from the Jurassic to the Cretaceous also developed in the province Santa Cruz are characterized in the section dealing with the floras of the Early Cretaceous.

2.9.2 Antarctic

The largest habitat of the Jurassic flora is located on Graham Land (Antarctic peninsula). Its age was estimated by Halle (1913) who described this flora as Late Jurassic. Of late some Argentinian researchers have been inclined to date it as Late Jurassic while others date it as Early Jurassic.

The ferns are represented in this flora by *Dictyophyllum* sp., *Coniopteris hymenophylloides*, *C. lobata* and many species of *Cladophlebis* and *Sphenopteris*. *Pachypteris* and

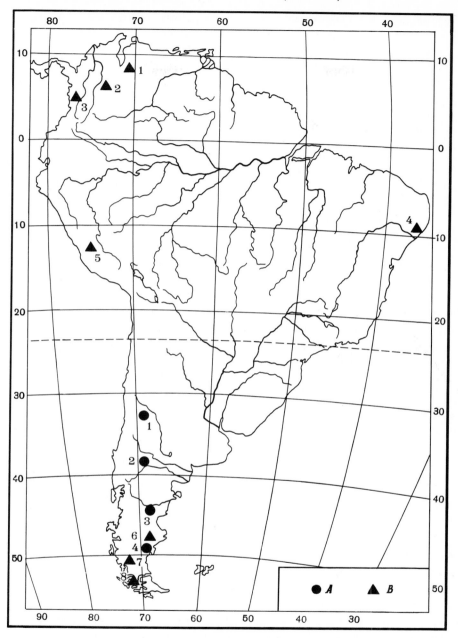

Fig. 2.23 Localities of Jurassic and Early Cretaceous floras in South America.
Jurassic floras of Argentina (A): 1, Mendosa; 2, Neuken; 3, Chubut; 4, Grand
Bayo de San Julian. Early Cretaceous floras (B): 1, Venezuela; 2–3, Colombia
(2, river Lebrilla; 3, south of the city of Medelin); 4, Brazil; 5, Peru; 6–7, Argentina (6, Tiko; 7, lake San Martin); 8, Chile (Spring-Hill).

Thinnfeldia have been encountered. Infrequent cycads are represented by *Nilssonia taeniopteroides* and *Pseudoctenis medlicottiana*. Bennettitales are more numerous including *Zamites* (4 species), *Otozamites* (4), *Ptilophyllum* (1), *Williamsonia* (1) and *Cycadolepis* (1). There are many *Araucarites cutchensis*, *Pagiophyllum* spp., *Brachyphyllum* spp. and *Elatocladus* spp. among conifers. The majority of these conifers must have belonged to the Araucariaceae.

For the coastline of the Ross Sea finds of *Otozamites antarcticus*, *Brachyphyllum* cf. *expansum*, *Pagiophyllum* cf. *peregrinum* and *Elatocladus* cf. *heterophylla* are recorded; the age of this flora may be estimated only within the Jurassic.

A locality of this middle flora is to be found on Snow Island which is one of the South Shetland islands. An effusive-sedimentary formation is developed here which supplied the finds as follows: *Dictyozamites falcatus*, *Ptilophyllum* aff. *cutchense*, *Otozamites* sp., *Pachypteris hallei*, *P. lanceolata*, *Scleropteris* sp., *Stenopteris* sp. and *Elatocladus* sp. This formation enters into the composition of a thick series whose upper strata on Livingstone Island provided a collection of remains of the Early Cretaceous (Funzalida, Araya and Herve, 1972; Gee, 1984).

2.9.3 South Africa

Here the Jurassic deposits are represented by red-coloured rocks (Red beds and Cave sandstone) whose deposition was followed by lava ejections contributing to the covering of the Drakensberg mountains. Impressions of *Otozamites* and shoots of conifers were found in the sandstones forming in the lower part of these lava beds in addition to petrified trunks.

2.9.4 Australia

This mainland houses Early Cretaceous floras that are represented only by two species of *Equisetites*. The richest floras are those of the Middle Jurassic (Hill, Playford and Woods, 1966) arising mainly from coal-bearing deposits of the Waloon basin, Queensland. Horse-tails and ferns, such as *Equisetites rotiferum*, *E.* cf. *rajmahalensis*, *Coniopteris delicatula*, *Osmundites dunlopi* and *Cladophlebis australis* were detected. The gymnosperms consist of *Pachypteris crassa*, *Sagenopteris rhoifolia*, *Ptilophyllum pecten*, *Otozamites feistmantelli*, *Taeniopteris spatulata*, *Williamsonia* sp., *Araucarites* sp., *Elatocladus planus*, *Pagiophyllum* sp. and *Brachyphyllum crassum*. *Otozamites* sp., *Brachyphyllum* sp. and *Elatocladus* sp. have been reported from the Middle Jurassic of north-western Africa. Plant macro-remains from the Upper Jurassic are unknown (Hill *et al.*, 1966).

2.9.5 New Zealand

Rather rich habitats of the Jurassic floras can be found on the southern tip of the northern island of New Zealand (Mildenhall, 1970). The age of the deposits is estimated as Late Jurassic. *Equisetites* spp., *Coniopteris hymenophylloides*, *Ruffordia goeppertii*, *Dictyophyllum rugosum*, several species of *Cladophlebis*, *Osmundites* and *Sphenopteris* were found here. Cycadophytes are represented by *Ptilophyllum acutifolium*, *Pterophyllum* sp., *Otozamites* sp. and *Nilssonia* (?) sp. *Araucarites* spp., *Araucarioxylon* sp., *Pagiophyllum* sp., *Elatocladus* sp. and *Podozamites gracilis* represented conifers. An important discovery was made by Harris (1962) who described *Carnoconites cranwelli* (female fructification of Pentoxylales) from the the upper strata of the Jurassic (Tithonian).

2.9.6 India

The Jurassic floras known on the Indian peninsula have been studied relatively extensively both morphologically and anatomically. However the issue of their age still remains obscure, and this is attributable to the fact that these floras are quite unique. On the other hand, the stratigraphy of the continental Mesozoic of this region has been adequately described.

The most important area comprising many localities is the Rajmahal hills situated in the state of Bihar in the north-eastern corner of India. At least a hundred articles are devoted to the description of plant remains from these hills; most were published in the annual issues of *Paleobotanist* edited by the Institute of Paleobotany in Lucknow (India).

The strata of sedimentary origin containing plant remains were deposited between the flows of basalt ejected on the surface of land and generally termed intertrappean strata. Earlier all this thick formation composed of the interspersed and predominant basalt layers with thinner intertrappean sedimentary formations was assessed as Jurassic, just as were the plant remains contained therein. However, lately the finds of the pollen *Cicatricosisporites* (Srivastava, 1983) in the intertrappean deposits raised the issue of re-estimating these formations as Early Cretaceous since pollen of this kind appears usually from Berriasian time onwards. Only in rare instances is it encountered in the uppermost strata of the Late Jurassic (the Lower Purbeckian of south England, etc.). However, the comparison of the flora composition from Rajmahal with the typical Early Cretaceous flora in other localities of India (Umia series on Kach peninsula, Jabbalpur series in the vicinity of Bansa, and Sehora) shows these floras to be not coeval. Without raising doubt about the evaluation of the spores of *Cicatricosisporites* it can be presumed that they were collected in the upper portion of a basalt series devoid of macrofossils of plant origin which are currently referred to the lower Cretaceous by Indian geologists.

Not so long ago Shah (1977) made a review of the Jurassic and Early Cretaceous floras of India in which he showed the distribution of plant taxa across certain sequences of the deposits of this age including the sequences of the Rajmahal hills. The same review quotes an extensive list of works dedicated to the fossil floras of India of this age.

According to these data and the courteous communication of A. K. Chatterji, the lower part of the sequence from the Rajmahal hills is constituted by the Dubrajpur suite. This is as thick as 120 m and is composed of terrigeneous formations which are coarsely clastic in the lower portion. In the higher and middle parts they contain remnants of *Ptilophyllum*. The age of these is probably Jurassic, although this estimate has not been updated.

The Rajmahal suite proper continues upwards and is as thick as 600 m. Its lower part is about 40 m thick and is composed of five basalt flows separated by sedimentary members containing a very rich flora whose complete list is cited in the work of Shah (1977). Rare remains of lycopods and horse-tails were discovered in the lower and middle parts of this 40-m-thick formation embracing four basalt layers and three intermediate strata of sedimentary origin. Ferns are represented by isolated finds of *Danaeopsis* and certain species of *Marattiopsis*, *Osmundites* (petrified rhizomes), *Klukia*, *Gleichenites*, *Todites*, *Coniopteris*, *Dicksonia*, *Hausmannia*, *Gonatosorus* and numerous and diverse *Cladophlebis* and *Sphenopteris* (Surange, 1966). Among pteridosperms *Thinnfeldia* were found, possibly belonging to the genus *Pachypteris*. Caytoniales are present (*Sagenopteris*). Bennettitales are richly represented by several species of *Ptilophyllum*, *Dictyozamites* and certain species of *Pterophyllum*, *Otozamites*, *Cycadolepis*, *Taeniopteris*, *Macrotaeniopteris*, various *Williamsonia* and two species of *Bucklandia*. Several species of *Nilssonia* represent the cycads. However, Sharma (1974) noted that the Indian species of *Nilssonia* differ from those known in other countries in incomplete coverage of the rachis leaf with the lamella and frequent dichotomy of veins. These characters are typical of Bennettitales (*Nilssoniopteris* and *Pterophyllum*). Therefore, the diversity of *Nilssonia* in India and even their presence may be doubted. *Ginkgoales* are represented by the genera *Ginkgo* and *Baiera*.

Araucarites, *Elatocladus*, *Brachyphyllum*, *Conites* and *Dadoxylon* wood were the conifers found.

The upper portion of the strata with plant remains embraces intertrappean layers laid between the fourth and fifth basalt units. They bear the name of layers with *Nipania*. They yielded remains of Pentoxylales: stalks, *Pentoxylon sahnii*; megastrobili, *Carnoconites compactum*; microstrobili, *Sahnia nipaniensis*; and leaves, *Nipaniophyllum*. The complex of plant remains from these deposits is quite specific. Apart from Pentoxylales they display rare ferns (*Protocyathea*, *Cladophlebis*) belonging to the genera mainly known from the underlying strata of the Rajmahal sequence as well as petrified stems, *Dictyostelopteris* and *Solenosteopteris*, found nowhere else except for the layers with *Nipania*. An almost complete absence of Bennettitales is an amazing feature; only one impression of *Ptilophyllum nipanica* was discovered. Conifers are more numerous (*Elatocladus sahnii, Brachyphyllum expansum, B. florinii, B. mamillare, B. spiroxylum* and *Pagiophyllum peregrinum*). *Nipanioruha* (3 species), *Indophyllum* (3), *Nipaniostrobus* (3), *Mehataia* (3) and *Sitholeaya* (1) were found only in these strata. As is seen from the list many genera of conifers are endemic.

It is probable that so endemic a composition of the strata with *Nipania* was induced primarily by some ecologic causes that were conducive to such an unusual community consisting of Pentoxylales and conifers composed mainly of the taxa encountered solely in these strata. Only a few species are known which are common with those of the underlying layers. A formation of basalts can be found upward in the sequence reaching a thickness of 500 m and enclosing members of sedimentary origin (intertrappean strata) which, however, revealed no plant microfossils. This thick formation of basalts is of Early Cretaceous age according to the Indian geologists. It is not excluded that it is in this formation, in one of the interbasaltic members, that the pollen of *Cicatricosisporites* was discovered (Srivastava, 1983). It is likely that its uppermost strata yielded a sample whose age, determined by the K–Ar technique, was established as Albian (100 to 105 million years).

The age of the strata with *Nipania* where the remains of Pentoxylales are concentrated is estimated as transitional from the Late Jurassic to the Cretaceous due to the fact that the fossils of Pentoxylales have so far been discovered only in this interval. Thus, Harris (1962) described *Carnoconites* from the very uppermost strata of the Jurassic–Lower Cretaceous (Tithonian–Neocomian?) of New Zealand. Recently, remains of Pentoxylales (Drinnan and Chambers, 1985) were found in the lower strata of the Lower Cretaceous of Victoria (Australia).

The discoveries of the remains of Pentoxylales are restricted to India, New Zealand and Australia which indirectly corroborates the statement that India was situated in the Southern Hemisphere not only in the Late Palaeozoic and Triassic, as is borne out by the occurrence of *Glossopteris* and *Dicroidium*, but also in the Jurassic and Early Cretaceous. Disagreement concerned only the issue of its association with an epoch within the Jurassic. Absence of dipterid ferns in this flora (except for *Hausmannia* having a wide range of stratigraphic distribution) dismisses conjectures about an Early Jurassic age. It is common knowledge that the remains of dipterid ferns are encountered not only in the lower Jurassic of the Northern but also in the Southern Hemisphere (South America). The abundance of *Ptilophyllum* and absence of *Gleichenites* also run counter to an Early Jurassic age.

Since the strata with the flora under consideration are situated directly under the layers with *Nipania* its age may prove to be Late Jurassic or else Late–Middle Jurassic. The habitats earlier referred to the Jurassic which are located on the eastern coast of India near the cities of Ellore, Ongole, Madras, Trichinopoli and Ramnad are currently dated as Early Cretaceous on the strength of palynologic evidence and on the basis of correlation with marine deposits characterized by fauna (Singh, 1974;

Shah, 1977; Venkatachala, 1977). There are several other localities dated as Jurassic but they do not contain any characteristic floras restricted in distribution to this period. Therefore their Jurassic age appears to be certain.

The main specific feature of the Jurassic (Middle–Late Jurassic) floras of India is the presence of the representative of the order Pentoxylales. A complete absence of Czekanowskiaceae is characteristic (earlier claims of their discovery were not confirmed), as also is the absence of ancient Pinaceae and the majority of Ginkgoaceae except for *Ginkgo* itself and, possibly, *Baiera*. The latter are represented by a small number of species (one or two) and exceptionally rare impressions. As was mentioned above, absence of dipterid ferns (except for *Hausmannia*) is conspicuous. Matoniaceae are represented by the rare finds of *Phlebopteris*. The genus *Coniopteris* is represented scantily, although it is so diverse in Middle Asia, Afghanistan and Iran.

The great variety of *Ptilophyllum* and *Williamsonia*, and the abundance of *Dictyozamites* and *Taeniopteris* is typical, while *Pterophyllum* is represented only by one species, i.e. *P. distans*. *Otozamites* (*O. bengalensis*) are also rare. Among the conifers several species are listed including *Elatocladus* (*E. conferta*, *E. sahnii*, *E. tennerrima*, etc.), *Brachyphyllum* (*B. expansum*, *B. feistmantelli*, etc.), *Araucarites* (*A. bindrabnunensis*, *A. cutchensis*) and the genera mentioned above, i.e. *Nipanioruha* and *Indophyllum*.

According to the maps compiled on the basis of palaeomagnetic data, India in the Jurassic period was located in the Southern Hemisphere within the Austral region near the border with the Equatorial region. The common features uniting it with the other provinces of the Austral region include, as was mentioned earlier, the presence of Pentoxylales and the abundance and variety of *Ptilophyllum* and, among the conifers, *Araucarites*. Mention should be made of the paucity of *Nilssonia* and *Ginkgo* as well as of the lack of Czekanowskiaceae and ancient Pinaceae. These features certainly corroborate the classification of the Jurassic floras of India as belonging to the subtropical floras of the Southern rather than the Northern Hemisphere. As is known, the former are rich in various Ginkgoaceae and *Nilssonia* and also contain Czekanowskiaceae and ancient Pinaceae; Pentoxylales are lacking in them.

The composition of the Jurassic floras of India suggests a subtropical climate (abundance of Bennettitales – *Ptilophyllum*, presence of *Phlebopteris*) which was moderately humid and typical for the Austral region.

2.9.7 Summary

Passing to the general characterization of the Jurassic floras known to inhabit the mainlands of the Southern Hemisphere, the least studied are the Early Jurassic floras with moderately rich localities situated in Argentina (Neuken). In Australia only horse-tails are familiar. The peculiar feature is the presence in these floras of large numbers of *Neocalamites*, Dipteridaceae (*Dictyophyllum*, *Thaumatopteris*) and Marattiaceae ferns just as in the Northern Hemisphere. In the Middle and Late Jurassic diverse species of the form-genus *Sphenopteris* are widely circulated. Many of the remains referred to this genus probably belong to the sterile leaves of the genus *Coniopteris* which is especially extensively spread in the Middle Jurassic of the Northern Hemisphere (Siberia and Euro-Sinian regions). In the Late Jurassic Dipteridaceae (*Dictyophyllum*, *Thaumatopteris*, *Clathropteris*) disappear except for *Hausmannia*.

Sagenopteris, as well as *Araucarites* and squamifoliate conifers, are typical of the majority of floras of the Middle and Late Jurassic. An interesting character which will be dealt with later is an almost intangible difference between the Middle and Late Jurassic floras. Only in certain areas (Argentina, the southern part of Neuken province) does the content of the pollen *Classopollis* increase in the Late Jurassic (30–65%).

The floras of all the three epochs of the Jurassic period emphasize richness in Bennettitales, which are found in every large locality and represented in the main by various species of *Otozamites*, *Zamites*, *Ptilophyllum*, *Williamsonia* and rarer *Dictyozamites*, i.e. genera which, as we saw earlier, failed to disseminate north of the Euro-Sinian region. Cycads are very rare (*Ctenis*, *Pseudoctenis*, *Nilssonia*). It is necessary to note that the leaves of *Nilssonia* occasionally ascribed in older works to some localities call for a revision because they may prove to belong to the genus *Nilssoniopteris* (i.e. Bennettitales) which is similar in external morphology. The repeated studies of the Jurassic floras of India revealed that classification of the ribbon-shaped leaves of Bennettitales as belonging to the genus *Nilssonia* had already taken place. But even if these leaf remains do belong to *Nilssonia* their occurrence pales into insignificance as compared with the number and specific variety of *Nilssonia* which one encounters in the Euro-Sinian or even Siberian regions of the Northern Hemisphere.

The most important peculiarity of the Austral region for the Jurassic period is the total absence of Czekanowskiaceae, ancient Pinaceae and the rarity of Ginkgoaceae. Caytoniales (*Sagenopteris*) are widely distributed and this applies to a lesser degree to pteridosperms (*Pachypteris*, *Thinnfeldia*). *Araucarites* are common and represented by shoots, petrified cones and woods. They may comprise part of the shoots referred to the form-genera *Pagiophyllum* and *Elatocladus*. Cheirolepidiaceae are also present embracing *Brachyphyllum* and, perhaps, some shoots of *Pagiophyllum*. Fluctuations in the amount of the pollen *Classopollis* depend not so much on age as on the ecologic conditions (Vakhrameev, 1970a). Representatives of the order Pentoxylales are found only in the floras of the Southern Hemisphere. This order makes part of the Phyllospermidae. Initially and most comprehensively they were studied from the Jurassic of India permitting the reconstruction of these plants. Later the plant remains belonging to this order were reported in the Tithonian of New Zealand and the Lower Cretaceous in the south-east of Australia (the state of Victoria).

In South Australia the strata with numerous leaf remains of *Taeniopteris daintreei* dated as Early Cretaceous (Valanginan–Aptian) recently yielded female strobili of *Carnoconites cranwellii* and some collections of microsporangia referred to the new species *Sahnia laxiphora* (Drinnan and Chambers, 1985). *Taeniopteris spatulata* known from India is currently considered as a synonym of *T. daintreei*. These two species of the form-genus *Taeniopteris* co-occurring with the reproductive organs of Pentoxylales (*Carnoconites*) are presumed to belong to the same plants. Leaves with retained anatomic structure classified in India as belonging to the genus *Nipaniophyllum* (*N. raoi*) possess a similar morphology. But the fact that the structure of vascular clusters is known makes us refrain from identifying them with the leaves of *Taeniopteris daintreei* because this structure is obscure in the latter.

Carnoconites cranwellii has been described from New Zealand while the related species *C. compactum* and *C. laxum* are familiar from India. Collections of microsporangia of *Sahnia* were also found in India but there they are represented by another species. Fertile organs of Pentoxylales were detected in conjunction with *Taeniopteris* in the upper Liassic of New South Wales in the paper shales of Talbragar. The finds of these plants in India, New Zealand and South Australia render their presence a singular feature of the Jurassic–Early Cretaceous floras of Gondwana.

In the Jurassic and Early Cretaceous floras of the Southern Hemisphere it is impossible to identify floras of the warm–temperate belt widespread in the Northern Hemisphere (Siberian region). The longest chain of major habitats of Middle Jurassic plants is observed in South America where it extends from the Equatorial region to the Antarctic peninsula. However, moving along this chain we cannot discern any appreciable changes to draw a boundary between subtropical and moderately thermophilic plants. Such indicators of subtropical climate not passing into the moderately warm belt of Northern Hemisphere as Bennettitales (*Otozamites*, *Zamites*, *Ptilophyllum*) are encountered from the Equator (Colombia) as far as the Antarctic (Graham Land). Neither Patagonia nor the Antarctic peninsula witness the emergence of any other groups of plants, such as Ginkgoaceae and Czekanowskiaceae, characteristic of the Siberian region, nor any local genera disseminated only in the extreme south of South America. Even Ross Island (Antarctica) displayed *Otozamites* sp. which are a typical subtropical bennettitalean, at 70° southern latitude.

All this is indicative of the subtropics reaching or as good as reaching the South Pole in the Southern Hemisphere. It goes without saying that a temperature gradient existed in the Southern Hemisphere, as is suggested by the distribution of foraminifera (Krasheninnikov and Bassov, 1985), but the fall of temperature southwards was apparently so insignificant that it made no impact on the distribution of plants belonging to one and the same generic taxon. Identification of distinctly defined species will, probably, illustrate the influence of a temperature drop down to the South Pole but at the present stage of taxonomic investigations of the Jurassic floras in the Southern Hemisphere this does not appear possible.

In conclusion, it may be stated that the climatic zonality mirrored in the composition of vegetation points to different distribution of climatic belts within the Northern and Southern Hemispheres. In the former subtropical and moderately warm climatic belts are observable whereas in the latter only a very wide subtropical belt is identifiable.

❀ 3 ❀

Early Cretaceous floras

The following regions are designated for the Early Cretaceous epoch: Siberian–Canadian, Euro-Sinian, Equatorial and Austral (Notal). The first of these is situated in the belt of moderate warm climate and is divided into the Lena, Amur and Canadian provinces.

The Euro-Sinian region, embracing the subtropics of the Early Cretaceous epoch, is split up into the European, Potomac, Middle Asia, and East Asia provinces including China, southern Mongolia and Africa without its southern tip. The Early Cretaceous floras of these territories are very similar to each other which inhibits identification of the basic separate provinces. The remaining part of the Equatorial belt is taken up by the ocean. No insular Early Cretaceous floras are known within its boundaries. There is evidence that the composition of the Early Cretaceous floras of south-east Asia is somewhat different to those of Africa and South America, but these data are insufficient to use in identifying a separate province. We therefore give their characterization in the section on Early Cretaceous floras of south-east Asia (Section 3.3.4).

The southernmost region, named here Austral (or Notal), is located in the subtropical belt of the Southern Hemisphere encompassing the southern part of South America, the southern tip of Africa, the Antarctic peninsula, Australia, New Zealand and the Indian peninsula. So far it is deemed impossible to identify belts of moderately warm climate by plant remains within the Southern Hemisphere. The southernmost localities on the islands adjacent to the Antarctic peninsula contain subtropical flora similar to that of the subtropics of the Southern Hemisphere.

3.1 Siberian–Canadian region

Initially this region was named Siberian (Vakhrameev, 1984) but later, when the Early Cretaceous floras of Canada came to be known to be very similar to the coeval floras of Siberia, I renamed it for the Early Cretaceous period as Siberian–Canadian. It is noteworthy that it is still often referred to as Siberian in the works concerning only the territory of the USSR.

The Siberian–Canadian region (Vakhrameev, 1964, 1965) embraced Siberia, north-east and northern China, Alaska, Canada, Spitsbergen, Franz Joseph Land, the north-east of the Russian Platform and, possibly, the very north of Scandinavia. The region was divided into three provinces: the Lena, Amur and Canadian. In the Neocomian all these expanses were probably covered with dense forests consisting largely of Ginkgoales, Czekanowskiales, Podozamitales and ancient Pinaceae whose canopy accommodated localities with herbaceous ferns, horse-tails, certain cycads (*Nilssonia, Ctenis, Heilungia*, etc.) and rare Bennettitales (*Nilssoniopteris, Pterophyllum, Neozamites*). The wide distribution of ancient Pinaceae is confirmed both by the abundance of bisaccate pollen and the finds of cones (*Pityostrobus*) and seeds (*Pityospermum*). The negligible amount of the pollen *Classopollis* (0.5 to 5%), the rarity of the shoots of *Brachyphyllum* and *Pagiophyllum*, and the absence of *Frenelopais* and *Pseudofrenelopsis* indicate that Cheirolepidiaceae were not in any way an important part of the composition of the vegetation (Vakhrameev, 1980, 1984).

Ferns were represented primarily by Dicksoniaceae (fertile pinnae of *Coniopteris*, spores of *Cyathidites*), Osmundaceae (the majority of pinnae of *Cladophlebis*, spores of *Osmundacidites*) and Gleicheniaceae (the spores of the latter were especially abundant locally). Schizeaceae appearing at the outset of the Early Cretaceous were at first very rare but from the Middle of the Neocomian their number somewhat increased. They are represented by the spores of *Cicatricosisporites, Appendicisporites, Pilosisporites* and the rare fragments of the leaves of *Ruffordia*.

The climate here was humid, moderately warm and seasonal, as is supported by the well-expressed annual rings in the woods and the brachyblasts in Czekanowskiales (*Czekanowskia* and *Phoenicopsis*) sometimes littering the planes of beds of fine-grained sandstones and argillites making part of the carbonaceous deposits. The humidity of the climate and the abundance of swamps is evidenced by the wide distribution of coal-bearing deposits. According to the evidence of Yasamanov (1980) the average temperature of near-surface marine waters washing Siberia did not normally exceed 10 to 12°C in the Neocomian. Winter temperatures in the north-east of Asia seemed to fall below zero for a short time. Despite the significant extent of the Siberian–Canadian region latitudinally from Baikal and Amur to Franz Joseph Land and Spitsbergen (which is at 25–28°N) the flora of this region retains its integrity. Only in the south (south of Canada and the Amur province) is there a zone of transition enriched by the elements inherent in the Euro-Sinian region. It is likely that a snow cover might have been preserved for a short time within the Arctic, Taimyr and Chukotka (the latter was closer to the Pole at the time according to many geologists).

From the Aptian onwards the composition of the flora of the Siberian–Canadian region that persisted in the belt of moderately warm seasonal climate with a relatively warm winter underwent only slight alterations. The genera *Arctopteris, Adiantopteris, Birisia* and '*Asplenium*' were represented by a number of species. V. A. Krassilov believes that the imprints referred to the genus *Asplenium* actually belong to the genus *Anemia* which seems to be probable.

Conifers are represented by the first Taxodiaceae (*Sequoia* and '*Cephalotaxopsis*'). Bennettitales become somewhat more diverse (*Neozamites, Encephalartites*) as do cycads (the new species of *Heilungia* and *Nilssonia*) which, probably, suggests some climatic warming in this epoch. This is corroborated by the penetration of the fern *Onychiopsis psilotoides* from the Euro-Sinian region where it was very widespread.

The most important changes in the composition of vegetation expressed in the emergence of the first angiosperms occur with the onset of the Albian. In the Early and Middle Albian they are represented by small-leaf forms (Vakhrameev, 1964) possessing an irregular venation, but it is in the Upper Albian that one can find leaf imprints of angiosperms with quite regular venation resembling the leaves of living plants. Palynocomplexes highlight tricolpate pollen of these plants. The Late Albian witnesses the predominance of Taxodiaceae within the conifers of the eastern areas of the USSR ('*Cephalotaxopsis*' and rarer *Sequoia*). *Czekanowskia* become rare abruptly although *Phoenicopsis* continues to be encountered often persisting even in the first half of the Late Cretaceous in the North Pacific region. This time, marked by greater humidity and some cooling of climate (Vakhrameev, 1978), is associated with extensive development of coniferous forests which came to be especially disseminated in mountainous areas adjacent to the Pacific Ocean (Okhotsk–Chukotsk volcanogenic belt). The Siberian–Canadian region comes to be divided into three provinces, i.e. Lena, Amur and Canadian.

3.1.1 Lena province

The flora of the Lena coal-bearing basin occupying the outer part of the Near Verkhoyansk foredeep and the Vilyui syneclise is the best studied and stratified flora of this province. This flora was studied over some years by Vassilevskaya (1959, 1966; Vassilevskaya and Pavlov, 1963; Vassilevskaya and Abramova, 1966), Vakhrameev (1958), Samylina (1963) and a little later by Kirichkova (1985) whose data coupled with the evidence of the earlier researchers are the fullest. A. I. Kirichkova singled out four stages in the evolution of the Early Cretaceous flora: Batalykh, Eksenyakh, Khatyrykh and Agrafenovo. These stages correspond to the horizons established on a palaeobotanic basis traceable from south to north over almost 2000 km from the lower reaches of the river Aldan to the delta of the river Lena. Here we shall consider only the floras of three horizons since that of the Agrafenovsk horizon whose age is estimated as Late Albian–Cenomanian, will be dealt with in the chapter devoted to the floras of the Late Cretaceous.

In the southern part of Priverkhoyanje area the Batalykh horizon comprises the Batalykh suite, in the middle part, the Ingyr, Dyangyn and Khos-Yuryakh suites, and in the northern part the Khalorgas (marine), Kigilyakh, Kyussyur and the lower portion of the Chonkogor suite. The flora of the Batalykh horizon, embracing Neocomian time, is split up into three floristic complexes (Ingir, Chongurgas and Sangar) replacing each other in time; at present it is known to comprise 160 species. For lack of space we shall give only a general characterization of the flora of the Batalykh horizon. Among spore plants ferns are represented most fully. The available genera include *Osmundopsis*, *Adiantopteris*, *Arctopteris*, *Gleichenites*, *Hausmannia*, *Birisia*, *Coniopteris*, *Gonatosorus*, *Eboracia*, *Jacutopteris*, *Cladophlebis*, *Raphaelia*, *Scleropteris* and *Sphenopteris*. The genus *Coniopteris* is the most numerous quantitatively and also specifically, with 16 species; the form-genus *Cladophlebis* has 14 species.

The species inherent in the lower, middle and upper complexes comprise, respectively, *Hausmannia leeana* and *Osmundopsis simplex* (lower), *Coniopteris columensis* and *C. samylinae* (middle) and *Birisia onychioides*, *Arctopteris heteropinnula* and *Gleichenites* sp. (upper). The entire Batalykh horizon is characterized by *Coniopteris burejensis*, *C. nympharum*, *C. setacea*, *Gonatorsorus ketovae*, *Cladophlebis argutula*, *C. ketovae*, *C. lenaensis* and *C. sangarensis*. Other spore plants reveal lycopods (*Lycopodites* sp.) and several species of horse-tails including *Equisetites rugosus* spread throughout the Batalykh horizon. Gymnosperms include numerous remnants of Ginkgoaceae, Czekanowskiaceae and conifers.

Ginkgoaceae are represented by the genera *Baiera*, *Ginkgo*, *Eretmophyllum*, *Sphenobaiera* and *Pseudotorellia*. Out of these the most widespread varieties comprise *Ginkgo papilionaceus*, *G.* ex gr. *sibirica* and *Baiera* ex gr. *czekanowskiana*. The upper strata of the Batalykh horizon yield *Ginkgo* ex gr. *adiantoides*. Czekanowskiaceae are represented by *Czekanowskia*, *Leptotoma* and *Phoenicopsis* (leaves) as well as by the reproductive female organs of *Leptostrobus*. The study of leaf epidermis undertaken by A. I. Kirichkova and V. A. Samylina paved the way for identifying many varieties of species according to differing epidermal structures. Publication of the descriptions of these species has begun and is certain to assist in distinguishing the Jurassic and Lower Cretaceous deposits.

The remains of conifers belong mainly to *Podozamites* (6 species) possessing great variability of leaves, and also to the ancient Pinaceae. The latter are represented by the offshoots of *Pityocladus* and *Pseudolarix*, isolated leaves (*Pityostrobus* and *Pityolepis*), and by various seeds (*Pityospermum*, *Schizolepis*). Apart from this shoots ('*Cephalotaxopsis*', *Parataxodium*, *Florinia*, *Elatocladus*, *Rhipidiocladus*) and fructifications (*Sorosaccus*, *Ixostrobus*) were found, although their systematic classification is still under discussion and some are still unidentifiable.

This work cites the generic name '*Cephalotaxopsis*', placing it in inverted commas. The examination of the cuticle removed from the type species *Cephalotaxopsis magnifolia* Fontaine 1889 from the upper strata of the Lower Cretaceous (Potomac series) of the USA Atlantic coast whence it was described, established its major difference from the cuticle of the shoots of conifers which are outwardly referred to *Cephalotaxopsis* widespread in the upper half of the lower Cretaceous and particularly in the upper Cretaceous of northern Asia and Alaska. The structure of the cuticle taken from the impressions of these shoots by I. N. Sveshnikova and studied by her, turned out to be similar to the cuticle of Taxodiaceae. She believes (personal communication) that the shoots termed in our literature *Cephalotaxopsis* belong to some extinct genus of the family Taxodiaceae. In future, before the matter is studied monographically and the genus is established, shoots of this kind ought to be classified as belonging to the form-genus *Taxites* Brongniart, 1828. This summary work based on generalization of a great deal of material describing and referring to these shoots under the generic name of *Cephalotaxopsis* retains this name, however placing it in inverted commas.

With the onset of the Early Cretaceous the flora of the Lena basin is markedly enriched with cycads and Bennettitales (Kirichkova, 1984). The greatest variety is attained by the representatives of these groups in the Neocomian (Batalykh horizon) especially in its lower and middle parts. Cycads feature *Nilssonia* (10 species), *Ctenis* (6), *Pseudoctenis* (1), *Jacutiella* (1), and *Doratophyllum* (1), as well as the endemic genus *Aldania* (2) close to *Heilungia* (3). Bennettitales, by and large inferior to cycads in diversity, display *Nilssoniopteris* (4) and *Pterophyllum* (5) to which we, in the wake of Krassilov (1973a), attribute the representatives of the genus *Tyrmia* (Kirichkova, 1984).

Relative rarity of finds of many forms makes it more difficult to elucidate their distribution within the overall sequence of the Batalykh horizon, hence the difficulty of using these finds with confidence for distinguishing parts of it. It will be noted, however, that the representatives of the genus *Aldania* were encountered only in the lower and middle parts of the Batalykh horizon, as were *Heilungia amurensis* and *H. auriculata* as well as almost all species of *Ctenis*. *Heilungia sangarensis*, *Ctenis jacutensis* and *Pseudoctenis zamiophylloides* are known only from the upper part of the Batalykh horizon. We should also note the presence of a representative of Caytoniales (*Sagenopteris lenaensis*) more characteristic of the Euro-Sinian province.

The younger Eksenyakhsk horizon is of Aptian age. In the south it encompasses the Eksenyakh suite and in the middle and northern parts of the trough, the upper portion of the Chonkogor, Bulun and Bakh suites. It produced plant remains belonging to more than 70 species whereas the Batalykh horizon exhibited about 160 species. It will be pointed out that the latter embraces a far greater time interval in comparison with the former corresponding approximately to the entire Neocomian. However, the correlation of the basic groups of plants established for the Batalykh horizon also holds for the Eksenyakhsk horizon.

Birisia onychoides which is widespread throughout is a characteristic fern of this horizon. '*Asplenium*' and *Acrosticopteris*, still not numerous, are also encountered. Finds of *Gleichenia* are more frequent than in the Batalykh horizon. On the other hand, the variety of *Cladophlebis* is much reduced, as, to a lesser extent, is that of *Coniopteris*. The latter gets deprived of the forms with finely dissected pinnules, i.e. *C. setacea* and *C. kolymensis*.

The Eksenyakh horizon witnesses a sharp reduction in the number of cycads and Bennettitales. Only *Ctenis burejensis*, *Nilssonia magnifolia* and *Nilssoniopteris rhitidorachis*

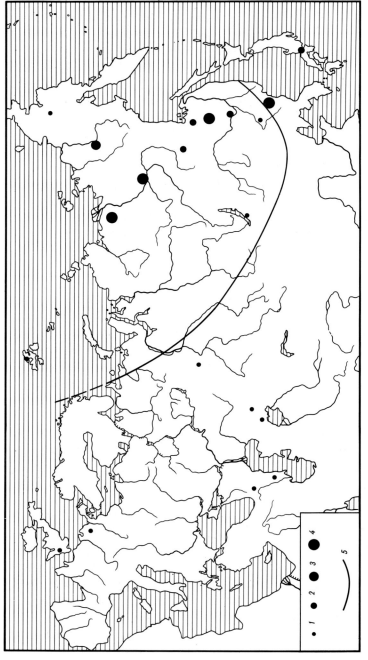

Fig. 3.1 Distribution of Early Cretaceous localities containing *Nilssonia*, 1, one species; 2, two to three species; 3, four to five species; 4, more than six species; 5, boundary between Siberian and Euro–Sinian regions.

rise here from the Neocomian. The new arrivals include *Anomozamites arcticus*, *Nilssonia gigantea*, *Neozamites verchojanensis* and *Pterophyllum bulunense*, as well as two new species of *Nilssoniopteris*. Representatives of the genera *Heilungia*, *Aldania* and *Ctenis* vanish. In the Ginkgoaceae the forms with a whole leaf, i.e. *Gingko* ex gr. *adiantoides* and *G. paraadiantoides*, are extensively developed. Conifers include representatives of the genera *Sequoia* (*S. ambigua*) and *Parataxodium* (*P. jacutensis*) already known from the Batalykh horizon floras. On the whole the floras of the Eksenyakh horizon appear to be a link between the floras of the Batalykh and Khatyryh horizons. Direct comparison of the floras of these formations separated by the Eksenyakh horizon reveals only a small number of common forms.

The age of the Khatyrykh horizon is estimated as early Middle Albian. This horizon is represented in the south by the suite of this name, in the middle part of the Near-Verkhoyansk area by the Djardjan suite and in the north by the Ogoner-Yurakh, Lukumay and Ukin suites. The number of species encountered here is 100.

Numerous ferns continue to indicate a great specific variety coupled with the wide distribution of the representatives of such genera as *Adiantopteris* (5 species), *Arctopteris* (2), *Birisia* (4), *Asplenium* (3) and *Scleropteris* (3). Some new species are so far known only from the Khatyryk horizon; these are *Adiantopteris gracilis*, *A. lepiskensis*, *A. minimus*, *A. polymorphus*, *A. sittensis*, *Birisia alata*, *B. vakhrameevii* and *Asplenium rigidum*. New species of *Coniopteris* evolve usually with small pinnules, i.e. *Coniopteris compressa*, *C. gleichenioides* and *C. minima*. Large-pinnuled *Coniopteris*, e.g. *C. burejensis* which were encountered in the Eksenyakh horizon, vanish; *Gleichenia lobata* and *Onychiopsis elongata* continue. Although the generic composition of cycads and Bennettitales does not change in comparison with the Eksenyakh stage, new species do arise, viz. *Nilssonia prynadai*, *N. orientalis* (altogether six species), *Neozamites lebedevii*, etc.

This overview shows that the maximal generic and species diversity of both cycads and Bennettitales is attained in the Neocomian. In the Aptian this group diminishes, becoming still rarer in the Albian. *Nilssonia* is most common and is known to number six species (Fig. 3.1), and is accompanied by one more species of *Encephalartites* (*E. borealis*). Bennettitales are represented by four species: *Anomozamites arcticus*, *Anomozamites* sp., *Neozamites verchojanensis* and *N. lebedevii*.

Ginkgos are dominated by the forms with an entire leaf which is sometimes small (*Ginkgo polaris*, *G. parvula*). The specific variety and frequency of occurrence of czekanowskias (particularly the representatives of the genus *Czekanowskia*) is reduced. Nonetheless, their composition is enriched by the characteristic species *Czekanowskia ninae*. Conifers continue to be dominated by the ancient Pinaceae already characterized in describing the floras of the Batalykh horizon, with new representatives of Taxodiaceae (*Parataxodium* and '*Cephalotaxopsis*') and Cupressaceae (*Cyparissidium*) emerging. The upper part of the Khatyrykh horizon (Lepis floristic complex) contains rare imprints of small leaves of angiosperms (*Prototrochondenroides jacutica*, *Morophyllum denticulata* and *Trochodendroides* sp.).

In comparison with the Late Jurassic the Early Cretaceous of the Lena province underwent rapid generic and especially specific development embracing almost all groups of plants. This is corroborated by the studies of epidermis (Samylina, 1970; Kirichkova and Samylina, 1979) in a number of plant remains particularly in the Ginkgoaceae and Czekanowskiaceae. A number of remains earlier classified as one and the same genus or even a specific taxon due to common outward appearance proved to belong to different genera or species. Thus, within *Nilssoniopteris* displaying hardly expressed morphology it was possible to identify eight species (out of which four were described for the first time). Within *Ginkgo* 22 species were discriminated (four new), and within *Sphenobaiera* 16 species. Certain earlier described species of this genus were transferred to the genus *Czekanowskia* (e.g. *C. ninae*). Re-definition concerned many individual plants on the basis of epidermis studies. These were earlier referred to *Sphenobaiera biloba*, *S. pulchella*, etc. A new genus *Leptotoma* was identified according to a sample earlier classified as *Baiera ahnertii*. One of the peculiarities of the Early Cretaceous floras of the Lena basin is their virtually uninterrupted evolution not impeded by sea transgressions or regional gaps encompassing considerable territories. This feature sometimes inhibits determination of distinct boundaries between different-age floristic complexes since changes in flora compositions occurred very gradually. This tendency is supported by the absence of abrupt changes in climate (up to the Albian) which can be seen from widespread coal development (up to the Albian) of the Lower Cretaceous series.

The Lower Cretaceous carbonaceous deposits are also widespread east of the Lena province in the north-east of the USSR. The plant remains contained therein were studied mainly by Samylina (1974, 1976) who monographically treated all floras from the Zyryansk and Omsukchansk basins. The sequence of the Zyryansk basin, divided into three suites – Ozhoginsk, Siyapsk and Buosukchansk, encloses almost all the lower Cretaceous (without the upper Albian). In the Omsukchansk basin there are deposits from only the upper half of the Lower Cretaceous divided into Omsukchansk, Toptansk and Zyryansk suites; the latter is very impoverished in plant remains.

The composition of the local Early Cretaceous floras is fairly similar to the coeval floras of the Lena basin. The same stages in the development in time are outlined as those established for the Lena basin. The variety of the fern *Arctopteris* (five species) should be mentioned as one of the peculiarities of the floras of the middle and upper subsuites within the Omsukchansk suite which is approximately coeval with the Khatytrykh horizon. But, perhaps the main feature is the abundance of mostly small-leaved angiosperms which are so plentiful in the Buorkemussk and Toptansk suites. The following were found here: *Cinnamomoides ievlevii*, *Ranunculicarpus quinque-carpellatus*, '*Cercidiphyllum*' *potomacense*, *Crataegites borealis*, *Sapindopsis* sp., *Celastrophyllum kolymensis*, *C. oppositiofolius*, *C. serrulatus*, *Zizyphoides* sp., *Carecopsis compacta*, *C. laxa*, *Ievlivia dorofeevii*, *Sugoia opposita*, *Araliaecarpum kolymensis*, *Kenella filatovii*,

K. harrisiana and *Rogersia denticulata*, as well as a number of leaves doubtless belonging to the dipterids but for the time being referred to *Dictyophyllum* sp.

The summary work by Kirichkova and Samylina (1978) gives a comparison between the upper and lower Cretaceous deposits and simultaneously between the enclosed stratofloras (according to the terminology of Samylina) of the Lena coal-bearing basin and the north-east of the USSR (Fig. 3.2).

In the South Yakutian basin the lower Cretaceous deposits which include the Kholodnikansk suite in the Aldan-Chulman area and the Undytykansk suite in the Tokinsk area contain a moderately rich flora (Vlassov and Markovich, 1979a, b). Its composition comprises *Equisetites rugosus*, *Coniopteris saportana*, *Lobiofolia lobifolia*, *L.* cf. *novopokrovskii* and several species of *Cladophlebis* currently encountered in the sequence in the Jurassic, as well as *Ctenis* sp., diverse Ginkgoaceae, Czekanowskiaceae, Podozamitaceae and ancient Pinaceae belonging to the species and genera widely disseminated in the lower Cretaceous in the Lena basin. It is revealing that *Pagiophyllum* and *Brachyphyllum* sp. were found here and were noted for the Upper Jurassic of the South Yakutian basin, too. The presence of these thermophilic plants is likely to be associated with the position of this basin in the south of the Lena basin.

The southernmost localities of the Early Cretaceous floras belonging to the Lena province were detected in the basin of the river Sutam, one of the tributaries of the upper reaches of the river Aldan (Vakhrameev and Blinova, 1971). Characteristic representatives of the Lena province, such as *Equisetites rugosus*, *Coniopteris saportana*, *Cladophlebis argutula*, *Phoenicopsis* ex gr. *angustifolia* and *Leptostrobus* sp. are found here.

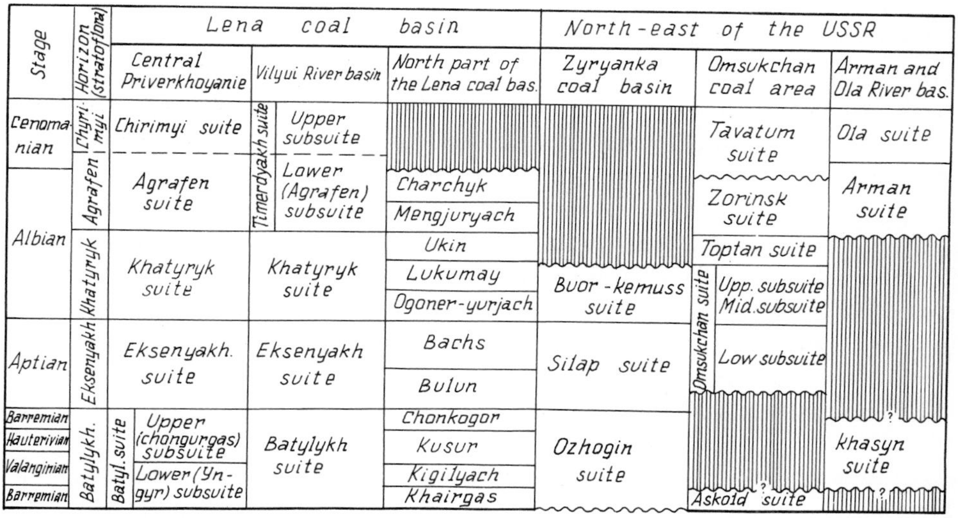

Stage	Horizon (stratofloru)	Lena coal basin			North-east of the USSR		
		Central Priverkhoyanie	Vilyui River basin	North part of the Lena coal bas.	Zyryanka coal basin	Omsukchan coal area	Arman and Ola River bas.
Cenomanian	Agrafen	Churimyi	mui	Chirimyi suite	Upper subsuite			Tavatum suite	Ola suite
Albian			Agrafen suite	Lower (Agrafen) subsuite	Charchyk		Zorinsk suite	Arman suite
					Mengjuryach			
	Khatyryk		Khatyryk suite	Khatyryk suite	Ukin		Toptan suite	
					Lukumay	Buor-kemuss suite	Upp. subsuite	
					Ogoner-yurjach		Mid.subsuite	
Aptian	Eksenyakh		Eksenyakh. suite	Eksenyakh suite	Bachs	Silap suite	Low subsuite	
					Bulun			
Barremian	Batylykh	Batyl. suite	Upper (chongurgas) subsuite	Batylykh suite	Chonkogor	Ozhogin suite		khasyn suite
Hauterivian					Kusur			
Valanginian			Lower(Yn-gyr) subsuite		Kigilyach		Askold suite	
Barremian					Khairgas			

Fig. 3.2 Comparison of Cretaceous deposits of the Lena basin and the north-east of the USSR.

The presence of *'Cephalotaxopsis' intermedia* and *Ginkgo* ex gr. *adiantoides* suggests that this flora belonged to the second half of the Early Cretaceous (Aptian–Early Albian).

It is interesting to point out the occurrence of two species of *Ctenis* (*C. stanovensis* and *C. harrisii*) featuring toothed edges; a similar species (*C. exilis*) is known only from the Middle Jurassic of Yorkshire, England. Mention should be made of the find of *Florinia borealis*; this species is known in the Aptian–Albian of Franz Joseph Land and from the coeval deposits of the river Vilyui.

It will be noted that the two above-mentioned habitats are situated on the northern slope of the Stanovy ridge which becomes an important water divide and the area of displacement of two basins of continental sedimentation, i.e. the Lena and Amur, in the Cretaceous epoch and later. It is along this ridge that the boundary between the Lena and Amur provinces passed.

The Lena province also comprises the Early Cretaceous flora of the island Kotyelny (Novosibirsk isles) studied by Vassilevskaya (1977). The presence of *Asplenium rigidum*, *Anomozamites arcticus* and *Sphenobaiera flabellata* and the simultaneous lack of angiosperms indicates that the host deposits belong to the lower part of the Khatyrykh horizon.

The Early Cretaceous floras of Spitsbergen and Franz Joseph Land are somewhat distinct. Spitsbergen (Vassilevskaya, 1980) accommodates continental deposits containing plant remnants enclosed between the faunistically characterised Hauterivian (*Simbirskites* ex gr. *decheni*) and the Late Aptian (*Tropaeum arcticum*). Thus, the age of these deposits corresponds to the Barremian–Early Aptian. According to its composition this flora is very close to that of the Lena province. Initially it was presumed (Vakhrameev *et al.*, 1970) that the floras of Spitsbergen and Franz Joseph Land probably belonged to a separate province featuring a paucity of ferns. However, the fresh samples scrutinized by Vassilevskaya revealed ferns in appreciable numbers including the representatives of the genera typical for the Siberian province (*Arctopteris*, *Birisia*).

Franz Joseph Land exhibited Early Cretaceous floras of two different ages. The more ancient is, in all probability, of Neocomian age, but its composition is not rich. It is dominated by Ginkgoaceae and Czekanowskiaceae as well as by various *Pityophyllum*. They are approximately coeval with the flora of Spitsbergen. It is with this flora that are associated the plant remains which guided Florin (1936) in identifying the genera *Stephenophyllum*, *Windwardia* and *Culgoweria* on the basis of their cuticle structure. These genera did not differ in their outward morphology from the genus *Phoenicopsis*. The younger flora comprises Bennettitales such as *Nilssoniopteris polymorpha* and *Tyrmia solsberiensis*. Later the genus *Tyrmia* was recognized as invalid (Krassilov, 1973a) and was included mainly in the genus *Pterophyllum*. The composition of this flora highlights a new genus of conifers, i.e. *Florinia*, which was subsequently discovered in the Khatyrykh horizon of the Lena basin as well as in the basin of the upper reaches of the Aldan river (Stanovy ridge). Another conifer, *Parataxodium* cf. *wigginsi* also points to the second half of the Lower Cretaceous. The absence of Czekanowskiales is also likely to be associated with the younger age. It is interesting that although the imprints of ferns were not recorded within the composition of this flora their occurrence in the vegetation of this period is confirmed by the find of spores of Schizeaceae, Gleicheniaceae and Cyatheaceae.

The northern part of Western Siberia, northern Urals and the Pechora depression so far have not exhibited localities of macro-remains of Early Cretaceous plants but the quite voluminous palynologic materials permit their reference to the Siberian–Canadian region. The transition from the Siberian–Canadian region to the Euro-Sinian is rather gradual contributing to a broad ecotone zone.

For palynologic characterization we shall quote here the data obtained by Gryazeva (1980) for the Pechora depression. The Neocomian is characterized by a significant content of the spores of Gleicheniaceae reaching a maximum of over 40% in the Hauterivian–Barremian. Spores of *Osmundacidites* are disseminated throughout the sequence. The rib-like pollen of Schizeaceae (*Cicatricosisporites* and *Appendicisporites*) emerges as individual grains in the Berriasian and Valanginian, while higher in the sequence it becomes more various and numerous. The content of the pollen *Classopollis* in the Berriasian still amounts to 10% but in the upper strata of the Neocomian it disappears. Bisaccate pollen is numerous.

The Barremian brings in the pollen of *Kuylisporites*, *Pilosisporites* and *Aequitriradites*, the latter being typical for the Aptian. The Aptian witnesses the greatest diversity of the spores of Schizeaceae and many spores of sphagnum mosses and the bisaccate pollen of ancient Pinaceae. The spores of Polypodiaceae appear as well as the pollen of the Taxodiaceae. The Albian complex resembles the Aptian but it regularly exhibits the pollen of angiosperms (*Tricolpites*).

In moving from the Pechora depression eastwards along the northern coast of Asia the composition of palynocomplexes is impoverished. In particular, the pollen of *Classopollis* disappears along with *Pilosisporites* which is accompanied by a rapid reduction of the spores of the Gleicheniaceae. The causes of this, associated with the position of the north-eastern edge of Eurasia in the higher latitudes in the vicinity of the North Pole, have already been covered in the section devoted to Late Jurassic floras.

Let us touch upon the issue of the boundary between the Siberian–Canadian and Euro-Sinian regions in these areas. Just as in the instance of the Late Jurassic I am inclined to draw it roughly along the sublatitudinal flow of the river Ob, recognizing a fairly wide transition zone (ecotone) between the floras of these regions. As one of the criteria for drawing the boundary I selected a find of the fern *Weichselia reticulata*, quite typical for the Euro-Sinian region, in the core of a borehole situated near the settlement Leushi south-west of the confluence of the Irtysh and Ob rivers. The age of the deposits is Valanginian (Maryanovsk suite). Vassilevskaya (personal communication) also found *Onychiopsis psilotoides* in the vicinity of Kolpashev; this is a typical representative of the Euro-Sinian region.

South of the sublatitudinal section of the river Ob the deposits of the Neocomian display an appreciably enhanced quantity of *Classopollis* pollen (15 to 20%) whereas north of this line it does not usually exceed 10% (Vakhrameev, 1978). South-east of Western Siberia this boundary is directed towards the southern tip of Baikal where there are a number of habitats with the floras intermediate between the Siberian–Canadian and Euro-Sinian ones (see the description of the Amur province below).

North-west of the Urals this boundary is drawn rather generally south of the island of Kolguyev whose Barremian and Aptian palynocomplexes have a composition closely resembling that of the basin of the river Pechora. North of Moscow (the environs of Klin) the Aptian sandstones contain the typical flora of the Euro-Sinian region with *Weichselia reticulata*. In Scandinavia the Lower Cretaceous deposits enclosing the Early Cretaceous complexes are known but it seems to me that the boundary under discussion should pass along Scandinavia's northern peripheral line or even slightly further north. Comparison of this boundary with the border drawn by Khlonova (Herngreen and Khlonova, 1981) far more to the north on the basis of disappearance of the spores of *Pilosisporites* north of this line (for the Neocomian) along with the vanishing of the spores of *Aequitriradites* and *Stenozonotriletes radiatus* (for the Aptian–Albian) shows that the border of Khlonova almost coincides with the boundary between the Lena and Amur provinces of the Siberian–Canadian region.

3.1.2 Amur province

The Amur province embraces the Early Cretaceous floras of the Bureya basin, the Tyl-Torom and Udsk troughs, the Transbaikal area, and north-eastern and northern China. The Stanovy ridge is the borderline between the Lena and Amur provinces. The southern boundary of the Amur province cuts off the southern part of the Far East Primorje area and then steeply descends to the south-west passing through the base of the Korean peninsula. Somewhat north of Beijing it sweeps to the north-west and, passing along the boundary of Mongolia, enters Kazakhstan north of lake Zaisan. The fullest articles and monographs devoted to the floras and stratigraphy of the Early Cretaceous deposits of the Amur province within the USSR include the works of Vakhrameev and Doludenko (1961), Vakhrameev and Lebedev (1967), Koshman (1969, 1970) and Krassilov (1972a, 1973a). References to earlier works are given in these publications.

The flora of the Bureya basin is the best studied and represented. The Early Cretaceous deposits of this basin are divided into four suites, i.e. Solon, Chagdamyn, Chemchuk and Kyndal. The Soloni suite corresponds to the upper portion of the earlier established Urgal suite which is still used by some geologists as a mapping unit. However, according to the composition of flora the suite is distinctly split into two parts. The lower part of this suite contains the characteristic Late Jurassic species of plants (*Raphaelia diamensis* etc.). The history of the Early Cretaceous floras of the Amur province highlights three stages: The Soloni, Chagdamyn-Chemchuk and Kyndal; the earliest of these corresponds to the Berriasian–Hauterivian.

The floras from the Chagdamyn and Chemchuk suites are regarded as a single entity due to the close similarity of their compositions. The age of this flora is not determined precisely. Vakhrameev and Lebedev (1967) refer it to the upper strata of the Hauterivian–Barremian while Krassilov (1973a) refers it to the Barremian–Aptian. The first point of view is based on the similarity of the composition of the Chagdamyn–Chemchuk suite to that of the upper portion of the Batalykh horizon of the Lena province as compared with the flora of the Eksenyakh horizon dated as Aptian. In addition, there is probably a vast gap corresponding to the Aptian at the base of the overlying Kyndal suite which is dealt with in detail below. The third flora is the Kyndal suite enclosing the remains of angiosperms and dated as Albian.

The Soloni complex (suite) is characterized by the ferns (Figs. 3.3–3.6) *Cyathea tyrmica*, *Dicksonia nympharum*, *Coniopteris burejensis*, *C. saportana*, *Eboracia lobifolia*, *Disorus nimakanensis*, *Dictyophyllum* cf. *nathorstii*, *Hausmannia leeiana* and the representatives of the genus *Cladophlebis* in whose midst *C. novopokrovskii*, *C. serrulata*, etc. are noteworthy. Bennettitales are composed of several species, viz. *Pterophyllum* (*P. burejense*, *P. sensinovianum*, *P. pterophylloides*) as well as *Anomozamites acutiloba*, *Pseudocycas polynovii* and *Jakutiella amurensis*. Cycads yielded *Nilssonia acutiloba*, *N. schmidtii*, *N. prynadii* (see Fig. 3.1), *Ctenis kaneharai* and *Heilungia amurensis*. Ginkgoaceae are represented by several species, i.e. *Ginkgo* (*Ginkgoites*), *Baiera* and *Sphenobaiera*. *Phoenicopsis* is the only Czekanowskiaceae retained whereas *Czekanowskia* and *Leptostrobus*, so abundant in the upper Jurassic of the Bureya basin, were not detected in the Soloni suite. Conifers are manifested mainly in ancient Pinaceae (*Pityophyllum*, *Pityospermum* and *Podozamites*).

In the basin of the river Tyrma situated south of the Bureya basin proper there are projected deposits which are most likely coeval with the Dublikan (the upper strata of the upper Jurassic) and Soloni suites. That the lower part of the sequence of the continental deposits belongs to the upper strata of the Upper Jurassic (perhaps the analogues of the Dublinkan suite) is corroborated by a find of *Raphaelia diamensis* (Krassilov, 1973a). The upper section of the sequence which we are inclined to compare with the Soloni suite supplied *Klukia tyganensis*, *Anemia asiatica* (transferred by Krassilov to the genus *Blechnum*) as well as *Phlebopteris* sp., *Dictyophyllum* sp. and *Jacutopteris lenaensis*.

The presence of southern elements inherent in the Euro-Sinian region and primarily in its East Asia province embracing the Far East Primorje area, constitutes the peculiarity of the flora of the Soloni suite as against that of the Batylykh horizon of the Lena province having roughly the same age. The above elements comprise certain representatives of the genera *Gleichenites*, *Eboracia*, *Klukia*, *Phlebopteris*, *Dictyophyllum*, *Anemia* and large-leaved *Hausmannia leeiana* missing from the Lena province as well as a relative abundance of diverse Bennettitales. The specific composition of all basic groups of plants is also quite different. A striking feature is absence in the Bureya basin of the remains of *Czekanowskia* ex gr. *rigida* encountered in fairly large numbers in the Lena province up to the lower boundary of the Upper Cretaceous.

The typical species of the Chagdamyn–Chemchuk complex include *Dicksonia*

Fig. 3.3 Plant fossils of Neocomian age from the Bureya river, Amur basin; natural size. A, *Birisia onychioides* (Vasilievskaya and Kara–Murza) Samylina, forma *gracilis* Vakhrameev; B, *Bisorus nimakanensis* Vakhrameev; C, *Sphenopteris interstifolia* Prynada.

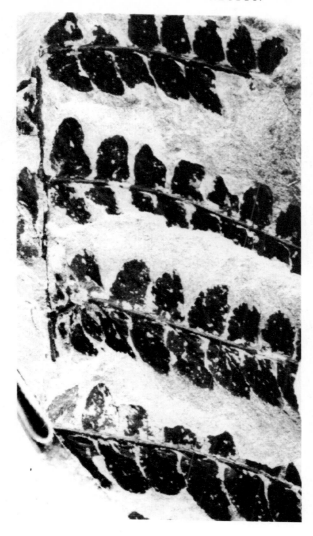

Fig. 3.4 Plant fossil of Neocomian age from the Bureya river, Amur basin; natural size; *Eboracia lobifolia* (Phillips) Thomas.

Fig. 3.5 Plant fossil of Neocomian age from the Bureya river, Amur basin; *Sphenopteris lepisensis*, × 3.

Fig. 3.6 Plant fossil of Neocomian age from the Bureya river, Amur basin; *Hausmannia leeiana* Sze, × 0.5.

arctica, Cladophlebis novopokrovskii, C. tschegdamensis, Cladophlebidium interstifolium, Sagenopteris sp., *Neozamites denticulatus* and *Ctenis formosa*. None of these save for *C. tschagdamensis* was detected lower in the sequence. The occurrence of *Hartzia angusta* is to be noted; this genus is very similar to *Czekanowskia* and some researchers do not differentiate between these two. Another feature is a find of *Leptostrobus* ex gr. *crassipes*. Ginkgoaceae are represented by the genera *Ginkgo* (*Ginkgoites*) and *Sphenobaiera* just as in the Soloni suite but the species described by Krassilov with due regard to their epidermal structure are different from the Soloni ones. In the midst of conifers the representatives of *Athrotaxopsis* (*A. expansa*) and *Florinia* sp. are encountered. Various *Podozamites* continue to be numerous.

Transition from the Soloni suite to the Chagdamyn–Chemchuk one is concomitant with the disappearance of a number of forms including *Heilungia amurensis, Pterophyllum pterophylloides* and *Pseudocycas polynovii* as well as such a Jurassic relict as *Dictyophyllum* cf. *nathorstii*. The specific composition of *Nilssonia* becomes somewhat altered but their variety remains intact. It is noteworthy that in comparison with the Soloni suite the number of forms in the composition of the Chagdamyn–Chemchuk relating it to the floras of the East Asia province is drastically diminished.

The Kyndal flora originates from the suite of that name whose basal part is tantamount, probably, to a thick formation of conglomerates (300 to 350 m thick) initially identified as an independent Iorek suite. The composition of the flora is quite different from the Chadgamyn–Chemchuk one (Koshman, 1973). The Kyndal flora is characterized by the ferns *Ruffordia goeppertii, Asplenium dicksonianum, A. rigidum, Polypodites polysorus, Birisia alata* and *Acrostichopteris* sp., as well as rare Ginkgoaceae including *Ginkgo adiantoides, Elatocladus manchurica* and *Sequoia* sp. The totally new elements of this flora comprise angiosperms represented by the imprints of leaves of *Araliaephyllum* sp. 1, *Araliaephyllum* sp. 2 and *Cinnamomoides elongata*. Later (Kapitsa and Ablayev, 1984) *Lindera jarmoljukii, Celastrophyllum kolymensis, Dalbergites* sp., *Cissites* sp. and the fern *Ruffordia* ex gr. *goeppertii*, etc. were detected. The overall appearance of the flora testifies to its Albian age, most probably the Early or Middle Albian. The latter supposition is borne out by the rare occurrence of the vestiges of angiosperms and the small size of their leaves.

Virtual absence of forms common both for the Chagdamyn–Chemchuk and Kyndal suites coupled with availability of thick conglomerates at the foundation of the Kyndal suite and, possibly, of stratigraphic incongruity between the Chemchuk and Kyndal suites make us think that the Bureya basin may be devoid of Aptian floral deposits. It is likely that the Aptian corresponds in part to the gap and partially to the thick basal formation of conglomerates of the Kyndal suite.

Another area which we incorporate into the Amur province is the Torom trough located within the West Near-Okhotsk Sea district. Two different-age groups are known from here. The older one associated with the Illinurek suite is dated by Lebedev (1974) as Berriasian owing to its position between marine deposits of Lower Valanginian age and the Volga stage characterised by *Buchia*. It contains the forms known both in the Soloni member (Lower Neocomian) and, to a lesser degree, in the Dublikan suite (upper strata of the Upper Jurassic) of the Bureya basin (*Ctenis burejensis*). But the absence of such characteristic species of the Upper Jurassic as *Raphaelia diamensis*, *R. stricta*, *Cladophlebis aldansis*, *C. orientalis* and some others induces us to regard the Illinurek flora as roughly coeval with that of the Soloni. However, the latter, probably, encompasses a wider range of the Early Cretaceous. The Neocomian age of the Illinurek member flora is signalled by the presence of a representative of the genus *Aldania* (*A. umanskii*) encountered in the Lower Cretaceous of the Lena province, and also plants known in the Late Jurassic, i.e. *Pterophyllum burejense* (which is widespread in the Neocomian of the Bureya basin) and *Sagenopteris* sp. The latter genus is not familiar in the Upper Jurassic either in the Lena or Amur province but is found in the Lower Cretaceous of both these provinces.

A more northerly position of the Torom trough in comparison with the Bureya basin was likely to lead to the occurrence in the former of *Czekanowskia* ex gr. *rigida* and its reproductive organs *Leptostrobus laxiflora*. No analogues of the Chagdamyn–Chemchuk flora have as yet been found in the Torom trough but the Tyl member yielded for extensive examination an Albian flora roughly coeval with that of the Kyndal.

According to the composition (around 80 species) it is far richer than the Kyndal. It is characterized by such forms as ferns *Osmunda denticulata* (Fig. 3.7), *Ruffordia* ex gr. *goeppertii*, *Onychiopsis psilotoides*, *Arctopteris tschumikanensis*, *Asplenium dicksonianum*. *A. rigidum*, *Acrostichopteris vakhrameevii* (Fig. 3.8) and *Ochtopteris ochotensis*. Several species of *Taeniopteris* (Fig. 3.9) probably belong to Bennettitales; cycads are represented by *Nilssonia* (Fig. 3.9). Ginkgoaceae belong primarily to the genera *Ginkgo* (*G. adiantoides*) and *Sphenobaiera* (Fig. 3.10). Czekanowskiaceae are composed of *Phoenicopsis* ex gr. *angustifolia* (Lebedev, 1974).

Conifers include a number of genera typical for the upper strata of the Lower Cretaceous, i.e. *Parataxodium*, *Athrotaxites*, '*Cephalotaxopsis*' and *Sequoia*. More ancient forms comprise *Pseudolarix*, *Pityophyllum* and *Podozamites* (Fig. 3.11). The Albian age predetermines the presence of angiosperms such as *Lindera jarmoljukii* (Fig. 3.12), *Celastrophyllum* aff. *kolymensis*, *Cissites* cf. *parvifolius* and *Kenella harisiana*. The list of angiosperms covers more than half the forms found in the Kyndal member.

The fourth region of dissemination of the Amur province flora is situated in the basin of the river Uda (Vakhrameev and Lebedev, 1967). A flora close in composition to the Soloni one was collected on the right bank of the river Uda downstream of the settlement of Udsky. Over and above some common forms of ferns the local finds comprised *Pterophyllum burejense* whose distribution in the Bureya basin fails to cross the boundary of the Soloni member. Younger elements were also found in the vicinity of lake Bokon, i.e. *Nilssonia* cf. *sinensis* and *Cephalotaxopsis* sp. *Czekanowskia* is present here just as in the river Tyl basin.

This review shows that three main floras are identifiable within the limits of the Amur province (stratofloras according to V. A. Samylina). The most ancient Soloni

Fig. 3.7 Plant fossils of Albian age from the Tyl river, West Priokotje; natural
size. A, *Lobifolia tenuifolia* E. Lebedev; B, C, *Sphenopteris achmeteevii* E. Lebedev;
D, *Osmunda denticulata* Samylina.

Fig. 3.8 Plant fossils of Albian age from the Tyl river, West Priokotje; natural size. A, B, *Acrostichopteris vakhrameevii* E. Lebedev.

Fig. 3.9   Plant fossils of Albian age from the Tyl river, West Priokotje; natural size. A, *Taeniopteris aborigena* E. Lebedev; B, *Nilssonia menneri* E. Lebedev.

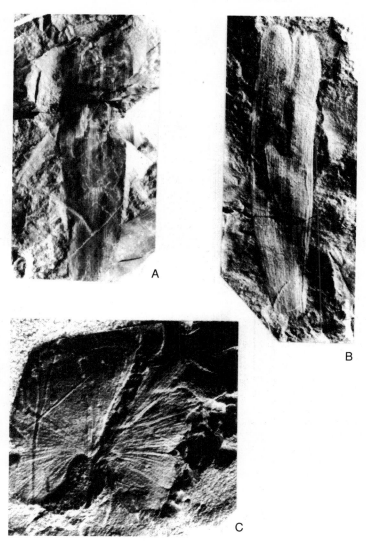

Fig. 3.10 Plant fossils of Albian age from the Tyl river, West Priokotje; natural size. A, B, *Arctobaiera florinii* E. Lebedev; C, *Ginkgo harrisii* E. Lebedev.

Fig. 3.11 Plant fossils of Albian age from the Tyl river, West Priokotje; natural size. A, *Podozamites eichwaldii* Schimper; B, *Elatocladus manchurica* (Yokoyama) Yabe; C, *Elatocladus platyphyllus* E. Lebedev.

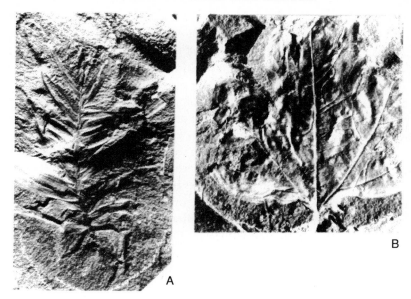

Fig. 3.12 Plant fossils of Albian age from the Tyl river, West Priokotje; natural size. A, *Cephalotaxopsis* aff. *acuminata* Kryshtofovich and Prynada; B, *Lindera jarmoljukii* E. Lebedev.

flora includes the flora of the Soloni member of the Bureya basin and that of the Illinurek member of the Torom trough. Its age is the lower half of the Neocomian (Berriasian–Valanginian and, probably, part of Hauterivian). The age of this flora in the Torom trough is confined to the Berriasian. The Chagdamyn–Chemchuk flora known from the Bureya basin has the majority of genera and species in common with the Soloniysk member despite a substantial difference in specific composition. I am inclined to estimate its age as the upper strata of the Neocomian (upper Hauterivian–Barremian). A flora which on the whole may correspond to the Soloniysk and Chagdamyn–Chemchuk taken together, was found in the basin of the river Uda and so far has not been associated with any separate part of the sequence.

The youngest flora is that from the Tyl river basin combining the Tyl flora proper and the Kyndal one. The specific feature of these floras is the emergence of angiosperms. However, this is not the only difference of the Tyl flora from the more ancient floras of the Amur province although it is fairly important. In comparing the Tyl flora with the older floras one fails to discover even one common species. The generic composition also proves fairly different (appearance of *Onychiopsis*, *Arctopteris*, *Asplenium*, *Acrostichopteris*, '*Cephalotaxopsis*', *Parataxodium* and *Sequoia*).

This indicates that the epochs of the prevalence of known floras of the Amur province did not follow each other continuously as was the case for the Lena province wherein the flora of the Eksenyakh horizon contains the elements of the older Batalykh one. This is conducive to the supposition that the Early Cretaceous floras of the Lena basin developed gradually. In contrast, the Bureya basin, the Torom basin and, north of this area, the Priokotje area witnessed a break in sedimentation between the deposits of the Neocomian and Albian, i.e. roughly in the Aptian. This gap is corroborated not only by missing intermediate floras between the Neocomian and Albian but also by the presence of thick conglomerates both at the base of the Kyndal and the Tyl members.

The Early Cretaceous floras of the Amur province, as was mentioned above, highlight the presence of the elements of the Early Cretaceous flora of the East Asia province from the Euro-Sinian region. It is important to note that the greatest number of these elements were in the southernmost habitat in the river Tyrma; *Klukia tygaensis* was discovered here. It is common knowledge that the representatives of this genus were distributed, as a rule, in the more southerly Euro-Sinian region. *Anemia asiatica* was found in the Tyrma river basin, too. It had also been detected in the Primorje region (lake Khanka). These are almost unknown north of this locality, for example, in the Torom basin.

New data coupled with revision of the composition of the floras earlier known from the Primorje and Transbaikal regions make it possible to change slightly the configuration of the southern boundary of the Amur province incorporating the Primorje and Transbaikal regions, northern and north-eastern China and, generally, the northern part of Mongolia.

The localities of the Early Cretaceous floras in the Sutar ridge and in the vicinity of the railway station of Bira are known within Malyi Khingan mountain (Vakhrameev, 1964). The former of these refers to the Neocomian while the latter refers to the Aptian or Aptian—Albian. Its composition includes various *Coniopteris* and *Cladophlebis* as well as *Onychiopsis psilotoides* = *O. elongata*. Watson (1969) showed that *Onychiopsis elongata* known from East Asia was a synonym of *Hymenopteris psilotoides* described in 1824 from the Wealden of south England and later transferred to the genus *Onychiopsis*. Therefore, henceforward we shall keep on replacing the specific name *O. elongata* established in 1894 according to the imprints from Japan and often encountered in the lists of Japanese, Chinese and sometimes Soviet researchers, with the specific name *O. psilotoides*. The age of this flora is estimated as Neocomian.

Earlier the presence of this species was considered by me as one of the proofs that the host flora belongs to the Indo-European (currently Euro-Sinian) region. As it turned out later certain localities of this fern are found further north. Some finds of this species are recorded far into the basin of the Lena river (Lena province), while south, in the basin of the Amur river, they are more frequent (becoming common) in the south of the Amur province, i.e. in the territory of China. *Onychiopsis psilotoides* turns out to be an ecotone form more than anything else being widespread along phyto-chorial boundaries.

The composition of cycadophytes and conifers as well as the presence or lack of Czekanowskiales seem to be more weighty features for drawing boundaries between regions. Cycadophytes in the Small Hingan are very rare and represented by *Anomozamites* and *Nilssonia* which are more typical for the Siberian region. Among Czekanowskiaceae *Phoenicopsis* was found. In the vicinity of the railway station at Bira the younger flora yielded the fern '*Asplenium*' and conifers '*Cephalotaxopsis*' spp. indicating an Aptian—Albian age. *Elatocladus manchurica* was detected too, being very characteristic of the southern part of the Amur province.

Let us turn to the Transbaikal area along whose southern border we earlier drew the boundary between the Siberian and Indo-European (Euro-Sinian) regions (Vakhrameev, 1964). According to the data of Bugdayeva (1984), who studied both the literature sources and the evidence of her own long-term research, two large floras of different age in the Early Cretaceous are singled out here, i.e. the Turga and the younger Kuta floras confined to the corresponding members. The Turga suite is mainly composed of lacustrine and alluvial sediments with the so-called paper shales consisting of fine interlayers of argillites and aleurites. The Kuta member is made up of coal-bearing deposits. Certain researchers (Y. S. Sinitsa) refer the lower strata of the Turga member as far back as to the Upper Jurassic, but due to the sufficient homogeneity of the Turga flora complex its age should be estimated as Early Cretaceous.

The restriction of plant remains to the deposits of large lakes, burial in whose deposits requires in most cases long transportation, affected the systematic composition of the fossil plants discovered in the Turga member. This flora has few ferns which have usually been represented by fragments of *Coniopteris* and of *Cladophlebis* with a thin leaf blade (including *C. lenaensis*). Mosses, Sellaginellaceae and horse-tails are encountered. Bennettitales are rather diverse (*Neozamites, Nilssoniopteris, Otozamites, Vitimia, Baikalophyllum*) as are cycads (*Ctenis, Nilssonia*). However, the number of specimens is limited. They are met largely in the basin of the Vitim river. Ginkgoaceae are represented by *Ginkgo* (*Ginkgoites*) spp., *Sphenobaiera* spp. and *Pseudotorellia angustifolia*.

Czekanowskiaceae are manifested in *Czekanowskia* ex gr. *rigida, C. vakhrameevii, Leptostrobus laxiflora, Phoenicopsis* ex gr. *angustifolia* and *Windwardia burjatica*. Remnants of conifers are especially plentiful and are dominated by winged seeds capable of long-distance travel. Conifers are represented mainly by the remains of ancient Pinaceae. These include *Pityocladus, Pityolepis, Pityophyllum, Pityospermum, Pseudolarix, Pityanthus* and *Schizolepis. Samaropsis, Podozamites, Brachyphyllum, Elatocladus* and *Nageiopsis* are also encountered.

A find of very small angiosperm leaves is quite interesting. These are *Dicotylophyllum pusillum* (Vakhrameev and Kotova, 1977) and *Baisia hirsuta* (Krassilov and Bugdaeva, 1982). The latter species is represented by reproductive organs resembling the dry fruits of Cyperaceae or the strobili of Bennettitales with long diverging numerous hairs; the latter supposition is more probable.

In the overlying Kuta member (Aptian–Lower Albian) the nature of the flora is dramatically changed but this is mainly due to the peculiarities of deposition. Largely lacustrine Turga deposits are ousted by carbonaceous ones. Distances of transportation of plant remains to the site of their burial are sharply reduced especially in the instance of coal beds; this was the main cause of a change in the systematic composition of the Kuta flora as against the Turga flora. In particular, this was expressed in an abruptly diminishing role of cycadophytes whose only remaining species is *Nilssoniopteris prynadae*. Ginkgoaceae prevail among plant remains found in the carbonaceous deposits of the Kuta member (*Ginkgo, Sphenobaiera, Pseudotorellia*). Conifers are also common (*Pityophyllum*, more rarely *Pagiophyllum*). The maceration of rocks from a section of the roof of the coal-bearing bed in the deposit of Kharanor conducted by Bugdayeva showed the predominance of the leaves of *Pityophyllum* and *Pseudotorellia*. Intercalations with mono-dominant burials of the shoots of *Pagiophyllum* and leaves of *Nilssoniopteris* and *Ginkgo* are encountered; *Phoenicopsis* is common.

The flora of the Kuta member is close to that of the Gusinoozyorsk series developed in the Transbaikal area in the vicinity of Gussinoye lake whose upper portion contains coal beds (Vakhrameev, 1964). This part of the series comprises such ferns as *Coniopteris burejensis, Coniopteris* (*Birisia*) *onychioides, Onychiopsis psilotoides, Sphenopteris* (*Ruffordia*) *goeppertii* and certain *Cladophlebis*. There are many *Ginkgo, Podozamites* and *Pityophyllum*. Cycadophytes, including *Pterophyllum sensinovianum* and *Nilssonia* ex gr. *mediana*, are encountered rarely. The presence of *Onychiopsis psilotoides* is conspicuous; it is often encountered in the Amur province especially in northern and north-eastern China.

South of the Amur river, in the territory of north-eastern China, the composition of the floras of the Amur province remains substantially unchanged. In the lower reaches of the river Sungari there are washed-out deposits of coal-bearing formations of the Mishan series which are presumed to be overlain by another coal-bearing series of Khuashan (Anon, 1960, 1963).

The floras enclosed in this series are very similar although Chinese geologists dated the lower of these as Upper Jurassic and the upper as Lower Cretaceous. They may be coeval, though, because the Chinese geologists are not fully confident about the genuine interaction of these two.

It will be noted that the composition of both the series (Mishan and Khuashan) is devoid of any species which, being encountered in the deposits referred by the Chinese geologists to the Jurassic, would not also be known in the Lower Cretaceous.

The typical species found in both the series include *Onychiopsis psilotoides*, *Sphenopteris goeppertii*, *Ginkgo* ex gr. *sibirica*, *Czekanowskia* ex gr. *rigida*, *Elatocladus manchurica*, *E. submanchurica* and various *Pityophyllum*. The Khuashan series also displayed *Pseudocycas* sp. and *Brachyphyllum* sp., and the Mishan, series, *Nilssonia sinensis*.

The scarcity of these species in the floras in question (likely to be due to insufficient samples) does not permit in-depth analysis of their composition. An almost identical composition was identified in the flora collected from the coal-bearing series of Shikhetszy jutting out in the upper reaches of the river Sungari. Local *Phoenicopsis* is concurrent with *Onychiopsis psilotoides* and *Elatocladus manchurica* (Fig. 3.13).

The flora discovered near the river Lofokhe (upper reaches of the river Sungari) approximately 60 km south-east of the town of Girin (Li and Ye, 1980) proves to be most interesting and diverse. This habitat is confined to the coal-free member composed of the terrigenous continental deposits enclosing washed-out boundaries. The following finds are to be noted: *Equisetites*, *Lycopodites*, *Dryopteris*, *Onychiopsis*, *Coniopteris* spp., *Acanthropteris*, *Cladophlebis* spp., *Arctopteris*, *Ctenis*, *Nilssonia* sp., *Phoenicopsis*, *Podozamites*, *Brachyphyllum*, *Elatocladus*, *Pityospermum*, *Pityocladus*, *Rhipidiocladus*, *Cissites* and *Phyllites*.

Apart from these, two new genera of cycadophytes were described from this flora, i.e. *Chillinia* (probably Cycadales) whose segments resemble the segments of some species of *Ctenis* with serrated edges, and *Chiaohoella* (2 species) resembling *Neozamites*. A new species of *Rhipidiocladus* was found, too; this genus known solely from the Bureya basin whence it was established by V. D. Prynada underscores the similarity of the floras of north-eastern China and the Bureya basin. The age of the flora of this locality may be estimated as Albian, most probably the Lower or Middle Albian according to the presence of a fairly large angiosperm leaf established as *Cissites* sp.

The province of Lyaonin has lower Cretaceous deposits in the Fushun basin which also exhibited a similar complex of macro-remains with prevalent *Acanthopteris* (*Birisia*), *Coniopteris* spp., Ginkgoaceae (*Ginkgo*, *Baiera*, *Sphenobaiera*), *Phoenicopsis* ex gr. *angustifolium*, *Nilssonia sinensis* and *Nilssoniopteris* sp., as well as *Elatocladus* spp.,

Pityophyllum spp. and *Podozamites* spp. A similar complex containing *Coniopteris burejensis, Onychiopsis psilotoides, Acrostichopteris* sp., *Ginkgo sibirica* and *Podozamites reinii* was discovered in the vicinity of Beijing (Chen, Yang and Chow, 1981; Chen and Yang, 1982).

A typical feature of the southern part of the Amur province is the presence of the ferns *Onychiopsis elongata* and *Birisia onychioides* (this species is described by Chinese palaeobotanists as *Acanthopteris gothani*). Cycadophytes are extremely rare and represented mainly by *Nilssonia* (largely by *N. sinensis*). *Neozamites* or similar *Chiaohoella* are also found. Ginkgoaceae still continue to be diverse but Czekanowskiaceae (especially *Czekanowskia*) prove rare. Among conifers a combination of *Elatocladus* (mainly *E. manchurica* and *E. submanchurica*) with *Pityophyllum* is quite characteristic. The upper part of the Lower Cretaceous sees the emergence of *Acrostichopteris* and *Rhipidiocladus*.

Concluding the overview of the Amur province it is necessary to point out the mingling of the southern and northern elements typical of its southern part. The northern elements comprise *Czekanowskia* especially in the lower strata of the Lower Cretaceous and *Phoenicopsis* which is more widespread. *Ginkgo* and *Podozamites* are diverse. Rare *Nilssonia* and *Neozamites* or its close relative *Chiaohoella* are constantly present. The genus *Klukia* as well as occasionally numerous squamifoliate *Brachyphyllum* accompanied by *Elatocladus manchurica* and *E. submanchurica* (according to V. A. Krassilov both the species belong to *Elatides asiatica*) are the southern elements.

The Amur province is rich in ancient Pinaceae represented by bisaccate pollen, winged seeds, strobili, separate leaves and shoots. The abundance of the vestiges of ancient Pinaceae (particularly seeds) is noted in lacustrine deposits which include among other things, the paper shales which are widely disseminated in the Transbaikal area.

At the same time the Amur province is totally devoid of elements inherent in the Euro-Sinian region such as ferns (*Matonidium, Phlebopteris, Weichselia*), Bennettitales (*Zamites, Zamiophyllum, Otozamites, Dictyozamites, Ptilophyllum*) and conifers *Frenelopsis* and *Pseudofrenelopsis* (= *Manica*). It is on the basis of the disappearance of these genera towards the north that we draw the boundary between the Euro-Sinian and Siberian regions. The quantitative distribution of *Classopollis* pollen is also of probable importance for delineating the boundary. The content of this pollen in the Euro-Sinian region including its East Asia province attains 40% and more.

3.1.3 Canadian province

The Canadian province incorporates Alaska in addition to Canada. It will be pointed out that remains of the Early Cretaceous flora are familiar only from western Canada while those of the Late Cretaceous flora are known from both areas. The Early Cretaceous floras of western Canada (Alberta and British Columbia) were studied by

Bell comparatively long ago (Bell, 1956) and after that were never taken up for description again although the lists of the Early Cretaceous plants did appear in later geological works (e.g. Stott, 1961, 1963, 1968).

The habitats of older Early Cretaceous floras of western Canada are related to the middle and upper parts of the Kootenay member represented by coal-bearing deposits. The lower part of this member is composed of marine deposits which is suggested by a find of the ammonite *Titanites occidentalis* in the Fernic area. This ammonite is a form characteristic of the upper Portlandian (the middle part of the Volgian stage).

A rich flora was collected higher in the sequence, *Coniopteris brevifolia, C. yukonensis, Cladophlebis virginiensis, C. heterophylla, C. impressa, Sphenopteris acrodentata*

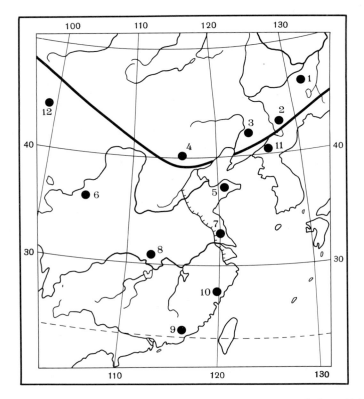

Fig. 3.13 Sites of the Early Cretaceous floras in eastern Asia including China. 1, lower reaches of river Sungari (Michan and Khuashan suites); 2, upper reaches of the river Sungari; 3, Fushun area; 4, Beijing Hills; 5, Shandung peninsula (Layan suite); 6, Khuankhe river bend; 7, Dzyansu province (Gegun suite); 8, Hupei province; 9, Futszydan province (Pantou suite); 10, Chzhe-Dzyan province (Mishishan suite); 11, Korean Peoples' Democratic Republic (river Yaludzyan); 12, Mongolia (find of *Cycadeoidea* trunk 300 km east of Ulan-Bator). The continuous line represents the boundary between the Amur and East Asia Provinces.

and *S. latiloba* being the most typical ferns. Such relics as *Dictyophyllum fuchsiforme*, *Phlebopteris elongata* (?) are also encountered. The presence of *Klukia canadensis* is to be noted. Cycadophytes are represented by '*Ptilophyllum*' *arcticum*, *Nilssonia nigracallensis*, *N. schaumburgensis*, *Ctenis borealis* etc. *Ginkgo pluripartita*, *G. nana*, *G.* cf. *lepida* and *Czekanowskia* cf. *rigida* were met among Ginkgoaceae. Conifers are represented by *Pityophyllum* cf. *nordenskioldii*, *Podozamites lanceolatus* and *P. corbinensis* coupled with rare *Pagiophyllum*.

Comparing the descriptions and appearances of certain species quoted in the work of Bell with the floras from Siberia we perceive that *Coniopteris brevifolia* proves very close to *C. burejensis*, *Coniopteris yukonensis* may turn out to be identical with *C. nympharum*, while *Cladophlebis virginiensis* is similar to *C. argutula* and *Cladophlebis heterophylla* to *C. pseudolobifolia*. The meticulous scrutiny of the photographs supplied in the work of Bell revealed that the imprints referred by him to the genus *Ptilophyllum* belong to the genus *Pterophyllum* (Vakhrameev and Doludenko, 1961).

The flora from the Kootenay member proves most like that of the Soloni suite of the Amur province which is in good harmony with the geographical position of the two habitats. Their nearness is supported by the presence in both floras of such thermophilic elements as the representatives of the genera *Dictyophyllum* and *Klukia* (the river Tyrma) inherent in the Euro–Sinian region situated in the subtropical belt.

The plants of the lower part of the Cretaceous except for the Kootenay member are encountered in the form of its age analogues, i.e. Nikanassin and Tantalus members and Haselton series.

The Kootenay member is overlain by the member, or rather series, Blairmore which is differentiated from the former by a thick bed of conglomerate testifying to the existence of a gap (Vakhrameev, 1974). Within this Bell identified two floras. One conformed to the complex of the middle portion of the Lower Cretaceous, the other to its upper strata. The middle complex of the Lower Cretaceous displays *Gleichenites* (*Gleichenia*), *Onychiopsis psilotoides* and *Ruffordia goeppertii*. *Sagenopteris* are common though they are absent from the lower complex. The disappearance of *Czekanowskia* is concurrent with the persistence of *Phoenicopsis*. Conifers exhibit an abundance of *Elatides* and *Elatocladus* lacking in the lower complex. Bell points out the emergence of an angiosperm, i.e. *Sapindopsis angusta*. The fullest representation is provided by the middle complex in the lower half of Blairmore series as well as in its stratigraphic analogues, i.e. the Luscar and Bullhead members. The vanishing of *Czekanowskia* upwards in the sequence is consistent with the disappearance of this genus in the sequence in the Bureya basin (Amur province).

The upper complex dated by Bell as Albian sees the disappearance of *Coniopteris* and *Klukia* and reduction of the specific variety of *Cladophlebis* and *Sphenopteris* out of which *Cladophlebis virginiensis* and *Sphenopteris mcclearnii* continue to be encountered in abundance. *Gleichenites* and *Onychiopsis* persist while the composition of cycadophytes is drastically diminished. The only remaining species comprise *Pterophyllum* (1 species), *Pseudocycas* (2), *Zamites* (1) and *Nilssonia* (1). Ginkgoaceae are totally missing.

Conifers exhibit numerous *Sequoia condita* and *Cyparassidium gracile*. The latter are coupled with *Elatocladus* and less numerous *Pityophyllum, Pagiophyllum, Brachyphyllum* and *Podozamites*.

A diversity of angiosperms proves to be the most salient feature of the upper complex. These are *Salix, Populites, Ficus, Trochodendroides, Menispermites, Nelumbites, Magnolia, Cinnamomoides, Platanus, Celastrophyllum, Rhamnites, Mytrophyllum, Sapindopsis, Fontainea* and *Araliaephyllum*. The Mill Creek locality, situated in southern Alberta near the tributary of the river Castie, yielded plant remains described by Bell (1956). These remains were found in the deposits of the upper part of the Blairmore series. They include *Cladophlebis alberta, Sphenopteris mcclearnii, Zamites tenuinervis* and *Brachyphyllum crassicaule* along with *Trochodendroides potomacense, Celastrophyllum acutidens, Araliaephyllum westonii, Sapindopsis angusta*, etc.

After Bell's monograph was published (1956) the stratigraphic research by Stott and Singh (Stott, 1963, 1968; Singh, 1975) updated the age of the floras of the middle and upper complexes. The following sequence was established in the basin of the Pine river, the right tributary of the river Mirnoy. The Cadomin member comparable with the Kootenay suite is overlain by the Hetting member containing the flora of the middle complex (lower flora of Blairmore series). It is covered by the Moosebar member of marine origin enclosing ammonites of the middle Albian (*Lemuroceras*). The Commotion suite is situated above being represented by alternating marine and continental deposits. The subsuite Gates, making up the lower part of the Commotion suite which houses the first angiosperms (*Sapindopsis*), is located inside marine sediments of the Middle Albian. Its foundation accommodates the above-mentioned Moosebar suite while the roof houses the subsuite Hulcross with *Gastropolites*.

The upper strata of the Commotion suite (subsuite Boulder) exhibit *Cladophlebis* cf. *parva*, *C.* cf. *rigida*, *Menispermites reniformis*, *Cercidiphyllum* sp. (probably *Trochodendroides* sp.), *Platanus* sp., *Magnolia* sp. and *Rhamnites* sp. The presence of these angiosperms and the position of the Boulder subsuite above the Hulcross subsuite containing the Middle Albian ammonites *Gastropolites* testifies to the Albian, most probably Late Albian, age of this flora. The Hasler suite follows up in the sequence enclosing no definable organism remains and still higher lies the Goodrich suite with the Late Albian ammonites *Neogastropolites*. This is followed by the Cruiser suite overlain by the Dunvegan suite with a Cenomanian flora and fauna. Stott (1963) combined the deposits from the foundations of the Moosebar suite to the base of the Dunvegan member into the series (group) Fort Saint John, but Singh (1975), whose comparison table we make use of (Fig. 3.14) regards the unit of this name as a member confining it to the roof of Commotion suite and to the base of the Dunvegan suite.

Comparison of the floras of western Canada and the Amur province shows their main differences to be associated with diverging specific compositions. However, we already pointed out above that certain forms described under different specific names in the USSR and Canada bear a close resemblance to each other and, possibly, should be referred to the same species. The review of the specific composition establishes that

Canada is deprived of representatives of the genera *Hausmannia* and *Heilungia*, and the Early Cretaceous flora of Canada is also far more impoverished in Ginkgoaceae. For instance, the latter are not represented by the genus *Sphenobaiera*. The genus *Czekanowskia*, just as in the Bureya basin, rapidly disappears upwards in the sequence. The presence of such conifers as several species of *Pagiophyllum* and *Brachyphyllum crassicaule*, as well as ferns *Klukia*, *Dictyophyllum* and *Phlebopteris*, are indicative of the proximity of these localities to the boundary of the Canadian province which brought about the penetration of elements inherent in the subtropical belt climate.

However, we do not agree with Samylina (1976) who suggested that this border should be drawn further to the north and referred the localities already described to the subtropical belt (Potomac region). The Canadian floras are best compared with the coeval floras of the Amur province, in the first place with that of the river Tyrma which also directly borders on the East Province located within the subtropical belt.

The Early Cretaceous floras from the currently known localities of western Canada, like the floras of the Amur province, belong to a transition zone (ecotone) between the subtropics and the belt of moderately warm climate. But they are distinctly dominated by plants from the latter. However, to combine the Amur and Canadian provinces into one is not appropriate in my opinion because there is every reason to believe that a more careful investigation of the floras of western Canada will reveal new features of their systematic composition setting them aside from the floras of the Amur province. It will be noted that after the works of Bell no further systematic research into the Cretaceous floras of Canada has been undertaken.

Canadian floras have so far not yielded any analogues of the floras of the Lena province since its northern part has well-developed marine deposits which have so far not furnished any fossil plant remains. Cretaceous deposits are totally missing in the east of Canada, too, which is occupied by the Canadian shield.

3.2 Euro-Sinian region

As was already noted in describing the Jurassic floras, originally the Euro-Sinian region was regarded as a subregion of the Indo-European region (Vakhrameev, 1970a). However, accumulated data showed that India, which during the Cretaceous period was situated far in the south close to the Antarctic, should not be associated with the rest of Eurasia. It appeared more correct to refer it to the Austral region embracing the gradually diverging parts of the Gondwana continent. Therefore, the Euro-Sinian subregion was transferred to the rank of an independent region (Vakhrameev, 1984) occupying the whole of Europe, the Caucasus and, possibly, part of the Middle East as well as the southern half of Western Siberia, Kazakhstan and Middle Asia. The territory of the USA is referred to the same region although the Early Cretaceous of this territory is still little known except for the Albian floras.

The flora of the Euro-Sinian region is almost twice as rich as that of the Siberian–

Canadian one according to the systematic composition. Disappearance of Czekanowskiales in the Euro-Sinian region and presence in significant numbers of various Bennettitales and Cheirolepidiaceae proves to be one of the most striking differences between these two floras in the Early Cretaceous epoch. Cheirolepidiaceae are represented by the shoots of *Brachyphyllum*, *Frenelopsis* and *Pseudofrenelopsis* as well as by major proportions of the pollen of *Classopollis* (Vakhrameev, 1978).

At the southern boundary the climate of the Euro-Sinian region was subtropical nearing tropical, but its humidity decreased in the central and southern parts of Eurasia (Kazakhstan, Middle Asia, western and central areas of China) where the climate continued to be arid with the onset of the Late Jurassic. In the west of the region the climate became more humid as compared to the Late Jurassic (western and central Europe). This is corroborated by the wide dissemination of ferns within which the most diverse and numerous units were Gleicheniaceae and Schizeaceae. The presence of these families is borne out by both macro-remains (*Ruffordia*, *Pelletiera*, *Gleichenia*) and spores (*Cicatricosisporites*, *Appendicisporites*, *Pilosisporites*, *Gleicheniidites*, etc.). *Weichselia* (close to the family of Matoniaceae) and *Onychiopsis* were characteristic forms. Finds of Matoniaceae (*Matonidium*, *Phlebopteris*) are frequently reported. On the boundary of the Jurassic and Cretaceous, representatives of the genus *Coniopteris* disappear but they continue in the Siberian–Canadian region. *Hausmannia* appears

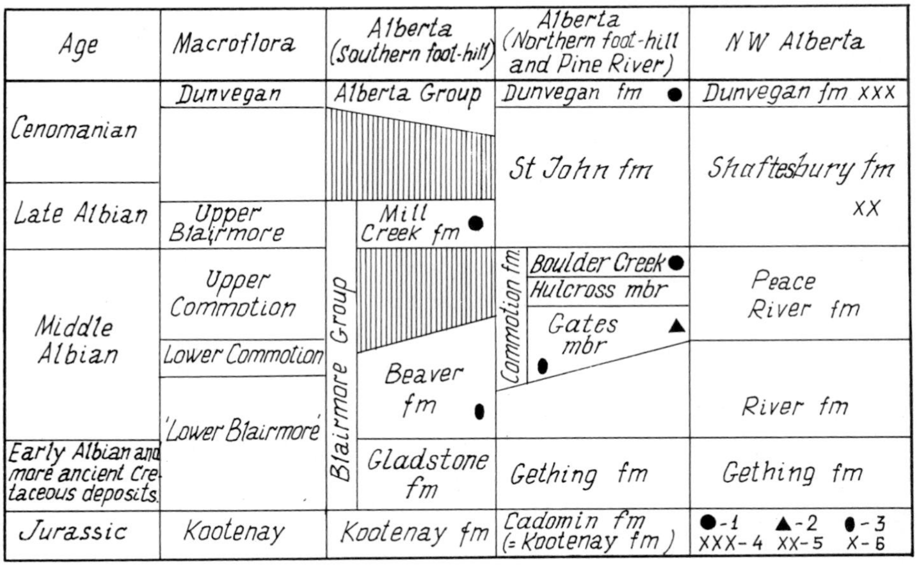

Age	Macroflora	Alberta (Southern foot-hill)	Alberta (Northern foot-hill and Pine River)	N W Alberta
Cenomanian	Dunvegan	Alberta Group	Dunvegan fm ●	Dunvegan fm xxx
			St John fm	Shaftesbury fm
Late Albian	Upper Blairmore	Mill Creek fm ●		xx
Middle Albian	Upper Commotion	Commotion Group / Blairmore Group — Beaver fm ●	Commotion fm — Boulder Creek ● / Hulcross mbr / Gates mbr ▲ ●	Peace River fm
	Lower Commotion			
	'Lower Blairmore'			River fm
Early Albian and more ancient Cretaceous deposits.		Gladstone fm	Gething fm	Gething fm
Jurassic	Kootenay	Kootenay fm	Cadomin fm (= Kootenay fm)	●-1 ▲-2 ●-3 XXX-4 XX-5 X-6

Fig. 3.14 Correlation of Lower Cretaceous and Cenomanian deposits of western Canada containing plant remains (according to Singh, 1975). Mega-fossils: 1, domination of angiosperms; 2, frequent occurrence of angiosperms; 3, rare finds of angiosperms. Pollen: 4, domination of angiosperm pollen; 5, moderate content of angiosperm pollen; 6, rare angiosperm pollen.

to be the only surviving representative of the Dipteridaceae. In comparison with the moderately warm belt occupied by the Siberian–Canadian region various Bennettitales are common here, such as *Anomozamites*, *Ptilophyllum*, *Zamites*, *Otozamites* and *Cycadites*, as well as cycads including *Pseudoctenis*, etc. Numerous finds of the trunks of *Cycadeoidea* (USA, western Europe, Mongolia, Japan) that possessed manoxylic structure suggest a frost-free climate (Fig. 3.15).

As against the Siberian–Canadian region the Euro-Sinian region can claim of far less numerous and diverse Ginkgoaceae. They are particularly rare in the European province from where only the genus *Ginkgo* is known. In the East Asia province (the Far East Primorje area, USSR) they are more varied being, however, not too numerous. Apart from Cheirolepidiaceae, ancient Pinaceae and Podocarpaceae, the ancient vegetation of western Europe included *Araucarites*, though to a lesser extent. The presence of the latter is ascertained mainly from macro-remains.

The composition of the Neocomian floras of eastern Europe, the Caucasus, Kazakhstan and Middle Asia is known virtually only from palynologic analysis data. These data indicate the increased content of the spores of *Classopollis* (50 to 70%) for the Valanginian (in the Berriasian it is still higher) in the eastern and south-eastern direction (Middle Asia). This is naturally accompanied by a drastic reduction in the number of spores of ferns and the amount of bisaccate pollen. As was shown by numerous researches (Vakhrameev, 1970a, 1978) the pollen *Classopollis* was produced by gymnosperms of wood appearance whose shoots were covered with squamifoliate or awl-shaped leaves (*Brachyphyllum*, *Pagiophyllum*, *Frenelopsis*, etc.) combined into the family of Cheirolepidiaceae (see Fig. 2.12, p. 53). The latter were a success in with-standing drought conditions that led to the extinction of the majority of such mesophilic plants as ferns, *Ginkgo* and *Nilssonia*.

Eastwards the climate within the subtropical belt of the Euro-Sinian region changed from humid in western Europe to semi-arid (south of Western Siberia, Kazakhstan). Still farther to the south-east it became arid in Middle and Central Asia and humid again only on the coast of the Pacific. This is corroborated by the composition of the Neocomian floras of the Far East and Japan which are very rich in varied ferns (Krassilov, 1967; Kimura, 1975a, b).

Recent finds from eastern China and Tibet (Lhasa) represent typical Early Cretaceous plants of the Euro-Sinian region, i.e. *Weichselia reticulata* and varied Cheirolepidiaceae (*Frenelopsis*, *Pseudofrenelopsis*) as well as *Brachyphyllum*. An earlier flora with *Otozamites*, *Ptilophyllum* and *Brachyphyllum* typical for the Euro-Sinian region was discovered from the western part of Futszyan province.

Throughout the Aptian and especially the Albian the flora of the Euro-Sinian region undergoes major changes due to the increasing humidity of climate reaching its peak in the Albian (Vakhrameev, 1978). This is especially pronounced in the eastern parts of the Euro-Sinian region (Kazakhstan, Middle Asia) where the content of the pollen *Classopollis* is reduced in the Albian to 2 to 4% (see Fig. 2.12, p. 53). Thus, Cheirolepidiaceae cease to be the dominant forms giving place to other conifers, in the first place to ancient Pinaceae possessing bisaccate pollen. The role of ferns is greatly increased. Such changes record greater humidity of climate, which reaches a maximum in the Albian and is accompanied by some cooling.

These changes are borne out by consideration of the composition of large-sized remains of plants. Localities of Neocomian age within the European part of the USSR, Kazakhstan and Middle Asia have not so far been found. The localities of Aptian age (Moscow basin, Dniepr-Donetsk depression) are dominated by remains of ferns (varied *Gleichenia*, *Phlebopteris*, *Ruffordia* and rarer finds of *Hausmannia* and *Weichselia*) testifying to the advance of a more humid climate. The Albian is characterized by the wide distribution of aborescent ferns *Tempskya* encountered in England, Mugodjary, Kyzylkum, Mangyshlak and in a number of localities in the western states of the USA. It is characteristic that the Early Cretaceous deposits of the USA pre-Albian have as yet yielded no localities of leaf floras; this is likely to be associated with the dry climate of this time replaced by the greater humidity only in the Albian.

The Albian epoch is also connected with the emergence in a·number of localities of the leaves of angiosperms (the Atlantic coast of the USA, western Kazakhstan). While in the Middle Albian angiosperms are represented mainly by small-sized forms

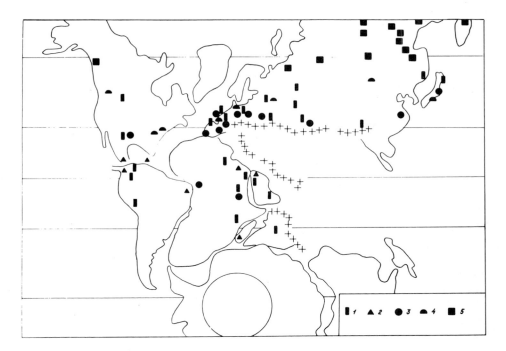

Fig. 3.15 Localities of plant remains used as indicators of climate (J_2 = Late Jurassic; K_1 = Early Cretaceous). 1, *Weichselia* (J_2–K_1); 2, *Piazopteris* (J_2–K_1); 3, *Frenelopsis* (K_1); 4, *Cycadeoidea* (K_1); 5, *Czekanowskia* and *Phoenicopsis* (J_2–K_1).

with irregular venation, the Late Albian witnesses the appearance of broadleaved trees represented mostly by platanoids that made up deciduous forests throughout alluvial valleys spread in the Late Albian and Cenomanian along the northern edge of the Euro-Sinian region (Kazakhstan, East Germany, Czechoslovakia).

In comparison with the Late Jurassic the climate in Mongolia also became more humid. This is signalled by formation of large lakes with sediments producing remains of ground vegetation (Krassilov, 1982). The most intense humidity occurred at the end of the Early Cretaceous since this was the time of coal-bed formation (Vakhrameev, 1983). In the extreme Far East near the coast of the Pacific the climate was humid and subtropical throughout the Early Cretaceous. The Early Cretaceous floras of the South Far East Primorje area and Japan are rich in ferns, Podocarpaceae, Araucariaceae, ancient Pinaceae, Taxaceae and Taxodiaceae (Krassilov, 1967; Kimura, 1975a, b).

Three provinces are identified in the territory of Eurasia within the Euro-Sinian region, i.e. European, Middle Asia, and East Asia. In North America the Potomac province is regarded as embracing the USA which by nature of its flora is very similar to the European province. This similarity is easily explained by the narrow strait existing at that time in place of Northern Atlantic between Greenland and northern Europe. This affinity allows incorporation of the Potomac province into the Euro-Sinian region. It is not excluded that further research will show the Potomac province to be merely part of the European one. The European province also comprises Greenland with its only but major locality in the vicinity of Disco island situated on the western coast of this very large island. It will be noted that close inter-relation between the Greenlandian floras and the European ones had existed before, i.e. in the Triassic–Early Liassic.

3.2.1 European province

The most ancient Neocomian floras usually referred to as the Wealden are well-studied in the areas adjacent to the North Sea, i.e. in south-east England, northern France, Belgium, the Netherlands and north-west and East Germany. The lagoon and delta deposits hosting the vegetation remains in the above-mentioned areas are of slightly different age which, however, does not exceed the span Berriasian–Barremian, as is observed in south-east England. In West and East Germany as well as in Belgium adjoining the North Sea it embraces a far smaller time interval (Berrisian–Early Valanginian) since the Late Valanginian transgression ended the local deposition of the lagoon and delta sediments. Despite differences in age within the Wealden in the above areas the composition of the plant remains is largely unaltered. This is quite obvious from the floristic characterization of the various horizons of the Wealden from south England (Hughes, 1975) encompassing deposits ranging from the Berriasian to the Barremian. The composition of plants keeps on changing to a very small degree and, what is most important, irregularly upwards in the sequence.

The Wealden flora is most fully represented in the localities of south England. Its composition comprises liverworts (*Hepaticites*), Lycopods (*Lycopodites* including a very special *L. hannahensis* and *Selaginella*) and horse-tails (*Equisetum*). It is rich in ferns such as *Ruffordia* and *Pelletiera* (Schizeaceae), *Matonidium* and *Phlebopteris* (Matoniaceae). Certain palaeobotanists single out the Early Cretaceous *Phlebopteris* as an independent genus *Nathorstia*. Other typical genera include *Onychiopsis*, *Weichselia* and *Hausmannia*. There are many remains classified as belonging to the form-genera *Cladophlebis* (*C. albertsii* and *C. browniana* are quite common) and *Sphenopteris*.

Dictyophyllum is occasionally encountered as a relict form. Remains of aborescent ferns *Protopteris* and *Tempskya* are also present.

From pteridosperms rare *Cycadopteris dunkeri* and *Pachypteris lanceolata* are to be noted with Caytoniales being represented by the leaves of *Sagenopteris* of which the number is appreciably reduced as against the Jurassic.

Cycads are manifested only in rare *Nilssonia* appearing as imprints of leaves and as reproductive organs (macrosporophylls *Beania* and microsporophylls *Androstrobus*). Bennettitales are much more numerous featuring, in particular, a variety of *Anomozamites*, *Otozamites*, *Pseudocycas* (rarely), *Zamites* and *Williamsonia*. Of major interest are the petrified trunks of bennettitalean *Cycadeoidea*.

Ginkgoales occur only as rare impressions of *Ginkgo* itself and as *Pseudotorellia heterophylla* which is close to *Ginkgo*. This is probably the result of aridization of climate that occurred in the Late Jurassic epoch and led to a sharp reduction in this mesophilic group of plants.

Conifers are represented by numerous shoots of *Brachyphyllum*, *Cupressinocladus* and *Sphenolepis*, and cones of *Elatides*. Recently the stalks of *Pseudofrenelopsis parceramosum* were described from the Isle of Wight (Watson, 1977) found together with the male cone of *Classostrobus* containing the pollen *Classopollis*. The dispersed pollen of this genus is plentiful in the underlying Purbeckian layers (upper Tithonian–lower Berriasian) and the Wealden deposits. But its amount in the Wealden proper is not great (4–5%). Remains of ancient Pinaceae (*Abietites*, *Pityostrobus*) and bisaccate pollen are encountered. These remains, even if studied thoroughly, cannot be referred to any contemporary genus of Pinaceae since they usually possess mixed characters.

The vegetation associations of the Wealden are presumed (Batten, 1975) to have grown both within the delta proper and on the relatively higher lands directly adjacent to it. The delta itself was covered by thickets of ferns and lycopods under the canopies of rare conifers. Horse-tails inhabited small lagoons as well as semi-inundated areas of the delta. The bars and spits of the littoral delta accommodated the ferns *Weichselia*. Higher lands adjoining the delta were mostly covered by coniferous forests with ferns in the lower levels and in the open.

The marine Barremian of south England characterized by flora corresponding to the continental deposits of the upper Wealden (Hughes, Drewry and Laing, 1979) exhibited the most ancient pollen of angiosperms (*Clavatipollenites*, *Liliacidites*, *Stellatopollis*).

Plant remains detected in the pre-Aptian deposits are known from various parts of Europe, i.e. Spain (Lerida, Ortigosa), Portugal (vicinity of Sesimbra), Czechoslovakia (Stramberg), Yugoslavia (Tsrne Gore), Hungary, East Germany and Poland. The fern *Weichselia* and representatives of Cheirolepidiaceae (*Frenelopsis* and *Pesudofrenelopsis*) were found in the majority of localities represented by uncommon species of plants. Shoots of *Brachyphyllum* belong here, too, along with *Araucarites*, *Sphenolepidium* and *Elatides*, as well as remains of ancient Pinaceae (*Pityophyllum*). A number of localities (Buckeburg in West Germany, Quedlinburg in East Germany, Tsrne Gore in Yugoslavia) supplied fragments of pinnae of Gleicheniaceae often carrying sori.

A lycopod *Nathorstiana* was discovered in the Hauterivian of Quedlinburg (East Germany); it is a connecting link between the Triassic *Pleuromeia* and contemporary Isoetopsida. In addition to this a *Hausmannia* was recorded, and *Weichselia reticulata*, *Phlebopteris dunkeri* and *Matonidium goeppertii* were encountered upwards in the sequence (Barremian?). In Silesia (Poland) trunks of *Cycadeoidea*[1] are known, while Hungary and Austria can claim finds of the arborescent fern *Alsophilina*.

The Aptian floras are better known from the European part of the USSR. The more western parts of Europe possess the deposits of this age of marine origin almost universally. The localities of the Aptian flora are distributed within the Moscow area (the town of Klin, Tatarovo village), Kaluga (Karovo village) and Voronezh (Devitsa village, Kriushi village and Latnaya village) regions (Prynada, 1937; Peresvetov, 1947).

The ferns are dominated by Gleicheniaceae among which an original *Gleichenia rotula* is to be noted. Just as in the Neocomian of western Europe, *Ruffordia*, *Phlebopteris*, *Matonidium* and *Weichselia* continue to be found. Certain researchers single out the Early Cretaceous *Phlebopteris* as an independent genus *Nathorstia* but differences between them are very few. Only *Cycadites* represent cycadophytes. Remnants of conifers are not numerous and belong to *Pagiophyllum*, *Cyparisidium* and *Pityostrobus* which represent, respectively, Cheirolepidiaceae (or Araucariaceae), Cupressaceae and Pinaceae. No remains of Ginkgoaceae have so far been found.

The Albian leaf flora of the European province is familiar from France, the Iberian peninsula and the Ukraine. Its remains are ordinarily associated with littoral and marine deposits.

In recent years (1984 to 1986) several contributions have been published on the earlier known and newly discovered localities of the Albian of Ukraine. Vegetation remains were collected from the faunistically characterized deposits of the Upper Albian in the vicinity of the city of Simpheropol (Stanislavsky and Kiselevich, 1986), Upper Albian in the environs of the town of Kanev (Radkevich, 1895; Pimenova,

---

[1] A fragment of a *Cycadeoidea* trunk from the Valanginian of the Tula region (P. Gerasimov's collection) was recently discovered by M. P. Doludenko in the collection of the M. V. and A. P. Pavlovs' Geological-Paleontologic museum in MGRI. This find is marked in Figs. 3.8 and 5.4 – Ed. (M.A.A.).

1939; Barale and Doludenko, 1985; Shilkina and Doludenko, 1985; Doludenko and Pons, 1986), and Upper Albian (and Cenomanian) from the Bakhchisaray district of Crimea near the Prokhladnoye village (Krassilov, 1984).

The following were determined from the Middle Albian flora of Simpheropol: *Phlebopteris dunkeri*, *Sphenopteris* cf. *delicatissima*, *Protophyllum* sp., *Sphenopteris kurriana*, *Elatides curvifolia* and *Elatides* sp. Preservation of the material was not very good. The flora is dominated by the leafy shoots of the three latter species of conifers.

The dominant species of the Late Albian flora of Kanev (Figs. 3.16, 3.17) include the representatives of two families, i.e. Cheirolepidiaceae (*Frenelopsis kaneviensis*, *Brachyphyllum squammosum*) and Taxodiaceae (*Sequoia* sp., *Cryptomeria*? *pimenovae*). The find of a female cone attached to a shoot of *C.*? *pimenovae* permit subsequent reference to a new genus. Apart from conifers, rare *Sagenopteris*, *Dioonites*? (2 species) and *Dicotylophyllum* were discovered, being manifested in two fragments of small narrow leaves of angiosperms with unordered type of venation and several cones of poor preservation with unestablished systematic classification.

In Podolia according to the date of Pimenova (1939) the same conifers are prevalent but the flora of Kanev appears somewhat richer. In the Bakhchisaray area of the Crimea (Krassilov, 1984) the lower part of the upper Albian (strata with *Hysteroceras*) displayed remains of ferns, conifers, caytoniales and primitive angiosperms. Conifers prevail in the uppermost strata of the upper Albian. In the Albian of France shoots and cones of *Araucarites* and ancient Pinaceae were recorded in addition to the petrified trunk remnants of arborescent ferns (*Protopteris*) and Bennettitales (*Cycadeoidea*). The Albian floras of Portugal (strata Buarcos) and Spain (Asturia, Cuenca) contain remains of angiosperms in addition to the Early Cretaceous typical ferns (*Cladophlebis browniana*, etc.), rare Bennettitales (*Zamites* sp.), Cheirolepidiaceae (*Frenelopsis*, *Pseudofrenelopsis*) and other conifers (*Sphenolepis*, *Podozamites*, *Sequoia*). The angiosperm leaves possess a still unstable and asymmetrical venation; they are referred to the genera 'Aralia', *Magnioliaephyllum*, *Cissites*, *Menispermites* and *Proteophyllum*. They doubtless include leaves belonging to aqueous plants (*Nymphaeites*, *Potamophyllum*).

The localities of the Aptian and Albian floras situated near the border between the European and Middle Asia provinces are to be found in Georgia and Azerbaijan. According to the evidence of Loladze (1978) the south and south-eastern fringe of the Dzirul massif includes a number of localities of this age. The marine deposits dated faunistically exhibit diverse *Cladophlebis*, *Phlebopteris* sp. and *Sagenopteris*, as well as the shoots of *Araucarites* and *Pagiophyllum*.

The Albian flora proved to be more enriched. Besides the forms mentioned above it contains *Pterophyllum magnum*, *Nilssonia* sp. and *Sequoia* sp., I. V. Palibin and P. A. Mchedlishvili discovered from the Albian of the same areas (Vakhrameev, 1964) *Sagenopteris* sp., *Zamites buchianus*, *Nilssonia schaumburgensis* and *Sphenolepidium sternbergianum*. The flora under consideration is poor in cycads (only *Nilssonia*) and Ginkgoaceae have not been found at all. Representatives of the Taxodiaceae (*Sequoia sublata*) appear among the conifers.

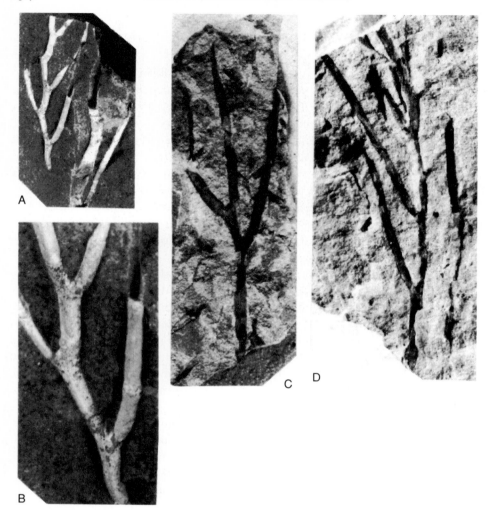

Fig. 3.16 Fossil plants of Albian age from the Ukraine; *Frenelopsis kaneviensis*
Barale and Doludenko; A, natural size; B, × 3. Fossil plants of Albian–
Cenomanian age from Tadjikistan; natural size. C, D, *Frenelopsis harrisii*
Doludenko.

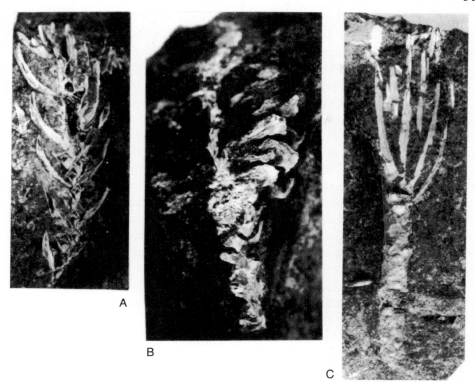

Fig. 3.17 Fossil plants of Albian age from the Ukraine. A, B, *Kanevia pimenovae* (Doludenko and Shilkina); A, natural size; B, female cone, × 2. C, *Kanevia teslenkoi* Doludenko, Kostina and Shilkina.

In Azerbaijan copal-bearing deposits are known from the Verkhny Agjakend village. They enclose plant remains (Vakhrameev, 1964) whose age was first estimated by V. P. Rengarten as Aptian. However, the research by Aliyev (1977) who collected marine gastropods and bivalves of the Albian from basal conglomerate led him to the conclusion that the overlying copal-bearing stratum belonged to the Early Cenomanian. However, there are no data contradicting its classification as belonging to the Upper or Middle Albian which, in particular, is indicated by an almost total absence of angiosperms. Here I found fragments of very narrow pinnae of Gleicheniaceae as well as *Onychiopsis psilotoides*, *Matonidium* sp., *Cladophlebis* sp., *Sagenopteris* sp., *Zamites* sp., *Nilssonia* sp., *Brachyphyllum* sp. and *Elatocladus* sp. I. V. Palibin managed to discover an impression of an angiosperm evaluated by him as *Diospyros* sp., although careful further examination yielded no more remains of angiosperms.

The flora of Agjakend is similar in composition to the Albian flora of the peripheral part of the Dzirul massif, especially if we take into account that the fern evaluated by Y. M. Loladze as *Gonatosorus* sp. now belongs to the genus *Gleichenia* (*Gleichenites*) judging from the pattern.

According to palynologic data Gleicheniaceae were widespread in the Early Cretaceous of the European province which is consistent with the composition of the leaf flora. Bolkhovitina (1968) distinguished four genera according to the morphological characters i.e. *Gleicheniidites*, *Clavifera*, *Plicifera* and *Ornamentifera*. These are most abundant in the Aptian and Albian. Schizeaceae constituted another widely disseminated group comprising the genera *Cicatricosisporites*, *Appendicisporites* and *Pilosisporites* identified according to the peculiarities of spore morphology. As was shown above, this family includes the pinnae of the ferns *Ruffordia* and *Pelletiera* exhibiting the typical spore-cases wherefrom spores were isolated. It is possible to draw the boundary between the Upper Jurassic and Lower Cretaceous going by the appearance of the costate spores of *Cicatricosisporites*. But the greatest variety of the spores of Schizeaceae along with their most frequent occurrence is witnessed in the second half of the Lower Cretaceous (Aptian–Albian). Bisaccate pollen mostly belonging to the ancient Pinaceae is abundant in the Lower Cretaceous of more northern areas of the European province.

Towards the more southerly areas (southern France, Spain, Portugal, Ukraine, Crimea, the Caucasus) it begins to be replaced by the pollen of Cheirolepidiaceae (*Classopollis*). Thus, in the Wealden of England the content of *Classopollis* pollen does not exceed 2 to 3% whereas for the Barremian–Lower Aptian of southern France it fluctuates from 14 to 25%. Still more contrasting data are furnished by the comparison of the *Classopollis* content in the deposits of the Moscow basin where it keeps within 2 to 3% of its proportions in the formations of the same age in the Crimea and particularly the Caucasus where it surpasses 50%. In the second half of the Aptian and Albian the content of *Classopollis* pollen in southern areas is dramatically curtailed due to the climate becoming more humid.

3.2.2 Potomac province

The Potomac province occupies the current territory of the USA (minus Alaska) situated in the Early Cretaceous epoch in a belt of subtropical climate. We know very little about its Jurassic vegetation which preceded the Early Cretaceous. From a find in the Morrison suite dated as the Late Jurassic (trunks of *Cycadeoidea*) one may suppose that the Jurassic period witnessed a subtropical and, probably, semi-arid climate. The fossil plant remains of the first half of the Early Cretaceous epoch (up to the Aptian) are represented in the USA by isolated localities. More frequently they are encountered in the deposits of the Aptian and, especially, the Albian.

We shall start our review with the western states of the USA. The oldest finds are from the upper part of the Neocomian of California whence a fruit of the angiosperm *Onoana californica* was described (Chandler and Axelrod, 1961). The same deposits yielded the leaf imprints of *Cladophlebis*, *Ctenopteris*, *Sagenopteris*, *Ctenophyllum*, 'Cephalotaxopsis', *Sphenolepidium* and *Acaciaephyllum*. Two species of cones belonging to Pinaceae (*Pityostrobus*) and a shoot with an attached cone referred to Taxodiaceae were described (Miller, 1976) in California from a formation dated as Barremian–Lower Albian.

The borderline region between the states of Colorado and Utah supplied for study two floras (Brown, 1950; Tidwell, 1966; Thayne and Tidwell, 1983). The older one is related to the Burro-Canyon suite presumably of Aptian age. The coeval deposits within Utah, west of the Colorado river, are recognized as the Sider-Mountain suite. *Brachyphyllum crassicaule*, *Frenelopsis varians*, *Sphenolepis kurriana* and a trunk fragment of *Cycadeoidea* were established from the Burro-Canyon suite. A wood referred to the genus *Paraphyllanthoxylon* (*P. utahense*) was detected in the Sider-Mountain suite lying between the Morrison suite of Late Jurassic age and the Dakota suite whose lower strata are dated as upper (?) Albian. Anatomical studies showed this genus should be related to the dicotyledons.

Upwards in the sequence is situated the Dakota suite (Lesquereux, 1892) being composed mainly of sandstones and widespread in the central part of the USA. Its formation within Colorado started as early as the Albian whereas in Kansas this happened later. In the latter the lower boundary of this suite is drawn at the base of the Cenomanian. *Asplenium* sp., *Matonidium americanum* and *Astralopteris coloradica*, as well as angiosperms of the genera *Juglans*, *Ficus*, *Nelumbium*, *Mahonia*, *Sassafras*, *Platanus*, *Celastrophyllum*, *Sterculia* and *Capsulocarpus* were described from the lower part of the Dakota suite. The fertile pinnae of the fern *Astralopteris* were initially referred by Brown (1950) to the contemporary genus *Bolbis* but later a new genus *Astralopteris* was established (Tidwell, Rushforth and Reveal, 1967) related to the Matoniaceae.

The flora from the lower strata of the Dakota suite which is dominated by broad-leaved angiosperms with well-formed venation cannot be older than the Late Albian.

A similar flora can be found from the Upper Albian of western Kazakhstan (Vakhrameev, 1952).

On the border of Wyoming and South Dakota (Black Hills) *Acrostichopteris*, *Cladophlebis*, *Onychiopsis*, *Nilssonia*, *Zamites* and many petrified trunks of *Cycadeoidea* (Berry, 1911) were found in the Dakota suite (Barremian–Aptian). Later a *Sphenobaiera ikorfatensis* f. *papillata* was also described from here. In the south of the USA in Texas the carbonate member of Glen Rose supplied for description a new genus of conifers, i.e. *Glenrosa* (Watson and Fisher, 1984) with an original structure of stomatal apparatus suggesting adaptability to arid conditions. *Frenelopsis alata* and *Pseudofrenelopsis varians* were found in conjunction with the latter.

Of special interest are the finds of petrified trunks of the aborescent ferns of the genus *Tempskya* whose localities are confined to a meridional strip, around 2000 km long, corresponding to the outcrops of the Albian deposits stretching from Montana in the north all across the USA to the border of Mexico (Ash and Read, 1976). All in all there are 41 localities. They are especially numerous in the frontier strip between Wyoming and Idaho. The above-mentioned work shows the distribution of these localities in sequences, although several are regarded as re-deposited.

The most instructive are the sequences of the second half of the Lower Cretaceous made up of continental sediments laid close to the coastline. They are recorded in the Potomac series outcropping along the Atlantic coast of the USA north and south of the mouth of the Potomac river in Maryland and Virginia. The Potomac series is divided into three suites (upwards), i.e. Patuxent, Arundel and Patapsco, whose ages are estimated as Aptian (Patuxent) and Albian (Arundel and Patapsco), respectively. Cenomanian deposits can be found still higher.

The Potomac series is deposited on an irregular eroded surface of an ancient foundation, filling the hollows. Recently (Doyle, 1983) it was discovered that in one of these hollows the Potomac series foundation is composed of the Berriasian deposits with *Cicatricosisporites* and *Pilosisporites*. The amount of the pollen *Classopollis* fluctuates from 8.5 to 25.3%.

The plant remains from the Aptian–Albian were studied by many researchers (Fontaine, 1889; Berry, 1911; Doyle and Hickey, 1976; Hickey and Doyle, 1977; Watson, 1977; Upchurch and Doyle, 1981; Upchurch, 1984). The richest floras are, probably, those established in the Patuxent (lower) and Patapsco (upper) suites. The paucity of the flora from the Arundel suite (middle) and the similarity of its angiosperms to those from the Patuxent suite makes one consider them jointly (Doyle and Hickey, 1976) as the first complex.

This complex displays rare horse-tails (*Equisetites*) and numerous ferns including those of the genus *Schizaeopsis* carrying the sporangia on the tips of ribbon-shaped pinnae drawn into a cluster. The sporangia contain the spores of *Cicatricosisporites*. Other finds comprise *Ruffordia*, varied *Acrostichopteris*, *Cladophlebis* and *Onychiopsis*. Several species of *Onychiopsis* described by Berry (1911) are likely to belong to *Onychiopsis psilotoides* which is a very characteristic species of the Lower Cretaceous of

Europe. Petrified trunks of the arborescent ferns *Tempskya* were found as fragments deposited on the surface of the Patuxent and Patapsco suite rocks.

Petrified barrel-like trunks of *Cycadeoidea* and imprints of the leaves of *Sagenopteris*, *Thinnfeldia*, *Taeniopteris*, *Ctenopteris*, *Zamiopsis*, *Zamites* and *Nilssonia* were discovered. Gingkoaceae are represented by a single species (*Baiera*). Conifers are quite abundant, including varied *Podozamites*, *Nageiopsis* and *Cephalotaxopsis*. The latter is marked by a difference in cuticle structure from '*Cephalotaxopsis*' widespread in the Upper Cretaceous of Alaska and in the north-east of the USSR. *Brachyphyllum* and *Frenelopsis* are encountered (Watson, 1977; Upchurch and Doyle, 1981) as well as *Abietites*, *Sphenolepis*, *Arthrotaxopsis* and *Sequoia*.

In the Patapsco suite (second complex) the specific variety of ferns is diminished although their generic composition remains little changed. *Schizaeopsis* vanishes while *Knowltonella* appears; Berry approximated the latter with Matoniaceae. *Sagenopteris* is not detected, cycadophytes become less diverse with *Ctenopteris*, *Zamiopsis* and *Nilssonia* being missing altogether though known from the first complex. *Cycadeoidea* were found only in the re-deposited state. *Athrotaxopsis* and '*Cephalotaxopsis*' are absent among conifers. On the other hand, *Araucarites* (cone scales) and *Pinus* (cones, scales and winged seeds) were encountered.

The angiosperms of the first complex are represented by the genera *Acaciaephyllum*, *Proteaephyllum*, *Quercophyllum*, *Rogersia*, *Ficophyllum*, *Plantaginopsis*, *Vitiphyllum* and *Celastrophyllum* identified by Fontaine (1889) and Berry (1911) according to leaf morphology. At a later stage the nature of their venation and leaf-form features were thoroughly investigated by Doyle and particularly by Hickey (1973). They found the leaves of the first complex to possess weakly differentiated veins of several orders as well as indistinct orientation of leaves. In some forms the petiole is hardly separated from the leaf lamella descending along the petiole. In the genus *Plantaginopsis* several parallel veins imperfectly dichotomizing in the leaf blade enter the leaf from the narrowed petiole-like foundation. In *Acaciaephyllum* the venation is asymmetrical and the middle vein is not manifested. The same applies to *Proteaephyllum*.

The second complex (Patapsco suite) angiosperms are represented by *Alismophyllum*, '*Populus*' *potomacense*, *Populophyllum reniforme*, *Nelumbites*, *Celastrophyllum*, *Sapindopsis*, *Araliaephyllum*, *Menispermites potomacensis* and *Sapindopsis* spp. In the leaves of the second complex venation becomes more orderly with the middle vein being well-expressed as are the veins of the second and third order. These feature leaves with both feather-like and digital-feather-like or plain digital venation.

The age of the first complex is estimated as Aptian–Early Albian and the age of the second as Middle–Late Albian (Doyle and Hickey, 1976). It is noteworthy that the lower strata of the Patapsco suite (IIA zone) are devoid of leaf remains probably due to this suite being partly of Middle Albian age. Stratigraphically highest are the complicated pinna-like leaves of *Sapindopsis* (IIB sone) and three-lamina leaves (*Sassafras*) taking up the highest position in the Albian of the Atlantic coast of the USA.

The nature of alterations in morphology of the leaves of angiosperms throughout the Albian on the Atlantic coast of the USA proves close to that in western Kazakhstan (Vakhrameev, 1952) and in the north-east of the USSR (Samylina, 1960).

Passing on to the estimation of the USA climate in the Early Cretaceous epoch it is necessary to pay attention to the rarity of plant discoveries in the Neocomian as well as to their increased number in the Albian. This is likely to be associated with increased humidity of climate toward the end of the Early Cretaceous. In pre-Albian time (at least in the southern part of the USA) the climate was dry which is signalled by the finds of Cheirolepidiaceae in Colorado (Aptian) and Texas (Glen Rose suite).

From the second half of the Aptian the number of localities with remains of ferns (Atlantic coast of the USA) increases. The Albian brings in the arborescent fern genus *Tempskya*. The fragments of their petrified trunks are scattered in the littoral, largely continental, deposits which evolved along the coasts of the 'middle sea' of Northern America from the boundary with Canada to the frontier with Mexico (Fig. 3.18). It is hard to visualize as wide a distribution of *Tempskya* under arid climatic conditions.

As far as temperature was concerned the climate throughout the USA during the Early Cretaceous epoch was subtropical, as is supported by the extensive dissemination of the barrel-shaped trunks of *Cycadeoidea*. These are especially common on the border between South Dakota and Wyoming and on the Atlantic coast of the USA. The prevalence of subtropical climate is also corroborated by the abundance of localities with the bones of large-sized dinosaurs which did not survive low temperatures (Berry, 1911).

This review of the Potomac province elucidates a major similarity between its flora and that of the European province. The common features comprise the occurrence of Schizeaceae (*Ruffordia*) and Matoniaceae ferns as well as the arborescent fern *Tempskya*. The spores of *Cicatricosisporites* and *Pilosisporites* are present. Petrified trunks of *Cycadeoidea* were found. Ginkgoaceae are represented by rare imprints of *Ginkgo* and *Sphenobaiera*. Cheirolepidiaceae are noted for *Frenelopsis* and *Pseudofrenelopsis*. Many leaves of angiosperms detected in the Peruč strata (Lower Cenomanian) of Czechoslovakia show imperfect venation just as do the leaves from the Paotapsco suite.

However, the specific composition of the Early Cretaceous plants of the USA and Europe are markedly different from each other which, in my opinion, suggests their association with different provinces. There are also some differences in generic composition. A find of *Gleichenia* (*Gleichenites*) was noted in only one locality of the Potomac province while in the Lower Cretaceous of Europe diverse species of this genus are abundant. The presence of authentic *Weichselia* in the USA is still dubious. Matoniaceae are manifested in Europe in *Phlebopteris* (*Nathorstia*) and *Matonidium*, but in the USA by *Astralopteris*. This provides grounds to identify different provinces but at the same time indicates a close relationship between Northern America and Europe existing in Early Cretaceous time.

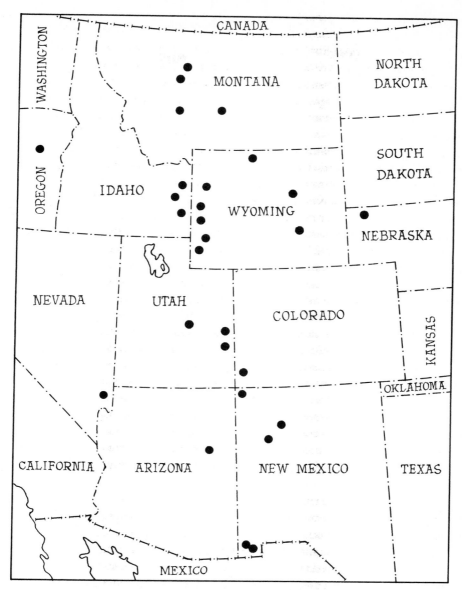

Fig. 3.18 Localities of fossils of the tree-fern *Tempskya* in the Albian deposits of
the western USA (according to Ash, 1976).

3.2.3 Middle Asia province

The nature of the flora that grew in the territory of the Middle Asia province embracing Kazakhstan and the republics of Middle Asia can be inferred only from spore–pollen analysis data. Plant impressions occur only in the Albian deposits. The pollen *Classopollis* is prevalent in the commonly undivided deposits of the Berriasian, Valanginian and Hauterivian (i.e. Neocomian of European stratigraphers) which extended over a vast area encompassing central and eastern Turkmenia, Uzbekistan and the Fergana basin. The content of the pollen here exceeds 60% while in some samples it reaches 75 to 90%. The bisaccate pollen of conifers is usually not more than 10%. According to all data available it belongs to the ancient *Pinus*. The pollen *Ephedripites* is encountered, with morphology very close to that of the contemporary Gnetales. Its content is very variable fluctuating from 1 to 2 up to 14%.

The number of spores of the ferns is insignificant (around 10%) but the generic composition is rather varied. From the very outset of the Lower Cretaceous costate spores of Schizeaceae belonging to the genus *Cicatricosisporites* make their appearance, and some way up *Appendicisporites* appears followed by *Pilosisporites*. The spores of *Lygodiumsporites* are also usually referred to Schizeaceae; they are morphologically similar to contemporary *Lygodium*.

The spores of *Lygodiumsporites* appear in the Jurassic and are relatively widespread in the Lower Cretaceous. However, no fossil leaves at least to some extent resembling the very special three-lobed leaves of contemporary *Lygodium* are known from the pre-Palaeogene deposits. It is likely that the spores of *Lygodiumsporites* might belong to some extinct genus of the family Schizeaceae which is similar to the genus *Lygodium* in spore structure. The costate spores of the genera enumerated above are virtually identical with those isolated from the sporangia of *Ruffordia* and *Pelletiera* frequently encountered in the Lower Cretaceous and morphologically attributed to the family Schizeaceae.

The spores of Gleicheniaceae are found more frequently than those of Schizeaceae; they are usually referred to the genera *Gleicheniidites*, *Clavifera*, *Plicifera* and *Ornamentifera*. *Gleicheniidites* is common for the lower half of the Cretaceous whereas the other three become most widespread in the Aptian–Albian. Smooth trilete spores described as *Cyathidites* are also found; in all probability they belong to the families of Cyatheaceae or Dicksoniaceae.

The composition of the deposits whence the above-mentioned spore–pollen complex was identified changes somewhat from west to east. In central Turkmenia and on the Ustyurt (Karakalpakia) they are represented by marine shallow-water sediments. Eastwards (eastern Turkmenia, Bukhara-Khiva region, Fergana basin) these marine deposits gradually turn into continental terrigenous formations (argillites, sandstones) of red, cherry and grey tints containing intercalations of anhydrite and gypsum. The combination of lithologic (red-coloured strata and gypsum) and palaeobotanic characters (marked predominance of *Classopollis* pollen) points to a dry, hot and, most probably, subtropical climate. Vast lowland surfaces and adjoining slopes of highlands were covered by forests likely to have been grouped into fairly small units. Cheirolepidiaceae were prevalent among trees and bushes (*Frenelopsis*, *Pseudofrenelopsis*, *Cupressinocladus*, *Brachyphyllum*, etc.) producing the pollen *Classopollis*. To a lesser extent ancient Pinaceae were also included in this group. Araucariaceae were also probably growing because their pollen is mentioned by some researchers. However, due to the shortage of clear-cut distinctive features the majority of palynologists do not think it possible to separate this pollen from the form-genus *Inaperturopollenites*. Ferns seem to have been represented by herbaceous forms growing under the forest canopy or else on the fringes. However, their fronds dried out in a dry climate with only the spores, covered with more durable exine, being capable of survival.

It will be noted that the abundance of the pollen *Classopollis* does not depend on whether the samples studied belong to shallow-water littoral deposits or to the continental deposits formed on lowlands more than 200 to 300 km inland (Vakhrameev, 1980). This indicates that Cheirolepidiaceae were not necessarily situated along the coastline as certain researchers think but could grow in abundance at great distances from the coastline and far inland. This is borne out by the significant amount of the pollen *Classopollis* in the Jurassic and Lower Cretaceous continental deposits in Kazakhstan and China that formed very far inland.

In the Barremian the quantitative spore–pollen ratio remains unchanged. The pollen *Classopollis* continues to prevail, its content in some samples of many areas reaching 80 to 90%. The amount of bisaccate pollen of ancient Pinaceae is less than 10%. In comparison with the preceding times both the number and diversity of the spores of Gleicheniaceae and especially Schizeaceae are increased.

The Aptian witnesses a major breakthrough in the composition of vegetation. This is expressed in the sharply reduced pollen of gymnosperms largely made up of conifers (including Cheirolepidiaceae) as well as in the corresponding increase in the number of fern spores. This process which we associate with climatic humidification appears to have proceeded irregularly in the territory of Middle Asia. Thus, in central Turkmenia the average content of spores attains 70% while the pollen of gymnosperms amounts to a mere 30%. The same ratios are observable for Ustyurt, central Kara-Kum and the Bukhara-Khiva regions. The content of the pollen *Classopollis* in places does not exceed 10%. The remaining part of the pollen of gymnosperms is largely *Disaccites* (bisaccate pollen of ancient Pinaceae), *Ephedripites*, *Araucarites*, *Inaperturopollenites*, etc. The predominant part of the Aptian palynologic complexes is represented by the spores of Gleicheniaceae and, to a lesser extent, by Schizeaceae. In the Aptian the former become most varied (4 genera, 12 species) judging from spore morphology.

So far we are unable to ascertain the differentiated presence of the pollen of such widespread Early Cretaceous groups of plants as Ginkgoaceae, cycads and Bennettitales whose macro-remains are known from the coeval deposits of the European province. The fact is that all these groups of plants possess a monosulcate structurally rather simple pollen generally described under the form-generic name of

Cycadopites or *Ginkgocycadophytus*. This pollen is usually encountered in all the stages of the Lower Cretaceous of the Middle Asia province, though in fairly small amounts. But its simple morphology prevents pollen grains from being associated with one of the plant groups.

In certain regions, for instance in eastern Turkmenia, *Classopollis* continues to prevail in palynologic complexes (over 50 to 60%). Spores predominate all along within the compositions of the spore–pollen spectra of the Albian of various areas of Middle Asia (60% and more). Gleicheniaceae are especially numerous but their specific variety is reduced as compared with the Aptian. Schizeaceae are represented most frequently by *Cicatricosisporites* and *Lygodiumsporites*. The content of the pollen *Classopollis* changes from area to area (1 to 10%) but on the whole it is inferior to that of the Aptian.

The bisaccate pollen (*Disaccites*) and the pollen of unestablished systematic classification, *Inaperturopollenites*, are the prevalent remaining pollen. The presence of the pollen of *Tricolporopollenites* belonging to the angiosperms (see also Figs. 3.19 and 3.20) appears to be typical. The minor content of the pollen *Classopollis* as well as the prevalence of fern spores indicates a rather high humidization of climate in the Albian.

Imprints of *Gleichenia zippei*, *Phlebopteris pectinata*, *Cladophlebis* cf. *browniana*, *Asplenium dicksonianum*, *Weichselia reticulata*, *Pityospermum* sp. and *Sphenolepis kurriana* were found at different times in the Albian deposits of Kyzyl-Kum. Comparing this list with that of the pollen found in the Albian deposits, certain coincidences can be seen. Thus, imprints of Gleicheniaceae and the seeds of ancient Pinaceae (*Pityospermum*) were encountered. It is not excluded that certain pinnae related to the form-genus *Cladophlebis* might belong to Schizeaceae ferns. The find of the shoots of *Frenelopsis harrisii* by Doludenko (1978; Doludenko and Reymanówna, 1978) from the Albian–Cenomanian deposits of Darwaz is to be noted. As is known the representatives of this genus produced the pollen *Classopollis*.

The Albian age is associated in Middle Asia with the development of the transgression that began as far back as the Aptian and gained momentum in eastern and northern directions. The sea invaded the Tadjik depression without penetrating to Fergana.

The basic difference between the Middle Asia province and the European in the Early Cretaceous was the uncommon abundance of Cheirolepidiaceae represented by the *Classopollis* pollen in the former. Within the European province the content of this pollen usually remains below 20%. It should be assumed that this peculiarity was due to the far drier continental climate of Middle Asia and Kazakhstan situated on the edge of the vast Asiatic mainland whereas the European province in the Early Cretaceous was an archipelago of more or less small islands whose climate was much less arid, and milder.

The evidence on the palaeogeographic situation prevalent at the time of Early Cretaceous runs counter to the views of those researchers who considered Cheirolepidiaceae to be mainly the inhabitants of marine coasts. In this case it would seem that

Fig. 3.19 Fossil plants from Kazakhstan; natural size. A, *Caspiocarpus paniculiger*
Krassilov and Vakhrameev; terminal paniculate inflorescences with leaves of
Cissites type, Middle Albian; B, *Celastrophyllum kazachstanense* Vakhrameev,
Middle Albian; C, *Otozamites jarmolenkoi* Vakhrameev, Upper Albian;
D, *Cissites uralensis* Krysht., Upper Albian; E, *Anacardites neuburgae* Vakhrameev,
Upper Cretaceous.

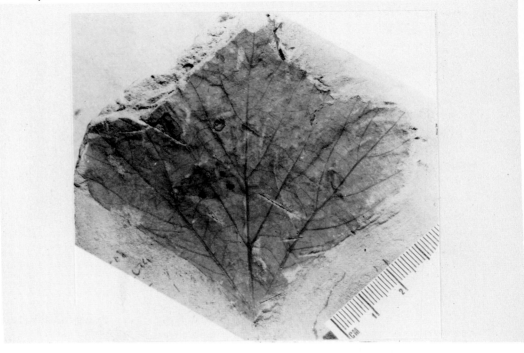

Fig. 3.20 Fossil plant from the Upper Cretaceous of Kazakhstan, natural size; *Platanus cuneifolia* Bronn emend. Vakhrameev.

in western Europe, made up of more or less small islands, the content of *Classopollis* pollen produced by Cheirolepidiaceae should be superior to that in Middle Asia whose eastern part was never inundated by the sea. However, it is precisely here in the deposits formed far eastwards from the coastline that the content of the *Classopollis* pollen in the first half of the Early Cretaceous on average exceeded 50 to 60%, running into 90% in some samples, whereas in the coeval littoral deposits of Europe it is usually no more than 10 to 20%.

Just as earlier, I believe that such a high proportion of this pollen in Middle Asia is attributable to the arid climate of this area suppressing the growth of other hygrophilic plants, in the first place the ferns. By and large Cheirolepidiaceae could grow both on the sea coasts and far inland, both on dry lowlands and slopes of highlands. Their adaptability to dry climate is revealed by the structure of shoots covered with clinging scaly leaves that possessed a thick cuticle and deeply seated stomata often covered by papillae of adjacent cells (Doludenko, 1978; Alvin, 1982).

3.2.4 East-Asia province

This province embraces the Southern Primorje area, Japan, the Korean peninsula, and a major part of China except for its northern and north-eastern portions, as well as the southern parts of Mongolia.

3.2.4.1 SOUTHERN PRIMORJE AREA

The localities of this area situated in the west in the basin of the river Razdolnaya and in the east in the basin of the river Partizanskaya constitute the northernmost group of localities of the Early Cretaceous floras of the East Asia province. Study results of these floras were published in many articles by A. N. Krishtofovich and V. D. Prynada. But the fullest coverage was achieved by Krassilov (1967) who employed cuticular analysis.

The littoral deposits of the Berriasian (Taukhe suite) and the continental deposits (Sibaigay suite) of the Berriasian and marine Valanginian (Kluchevskaya suite) were shown to host representatives of the following genera: *Equisetites*, *Ruffordia*, *Alsophilites*, *Onychiopsis*, *Cladophlebis*, *Ctenozamites*, *Sagenopteris*, *Zamiophyllum*, *Otozamites*, *Dictyophyllum*, *Nilssonia*, *Pseudotorellia*, *Podozamites*, *Ussuriocladus* and *Brachyphyllum*. In the younger sediments of the Barremian and Aptian in the Razdolnaya river basin (Ussuriysk and Lipovtsy suites) and in the Partizanskaya river basin (Starosuchan and Severosuchan suites) flora compositions are further enriched. The ferns and horse-tails are represented here by *Neocalamites*, *Equisetites*, *Osmunda* sp., *Ruffordia*, *Gleichenites*, *Gleichenia*, *Gleicheniopsis*, *Nathorstia*, *Matonidium*, *Alsophilites*, *Coniopteris*, *Onychiopsis*, *Adianopteris*, '*Asplenium*', *Polypodites*, *Weichselia* and *Cladophlebis*.

Gymnosperms yielded *Caytonia, Sagenopteris, Williamsonia, Zamiophyllum, Zamites, Neozamites, Dictyozamites, Pterophyllum, Ptilophyllum, Cycadites, Cycadolepis, Nilssoniopteris, Doratophyllum, Ctenis, Nilssonia, Ginkgo, Baiera* and *Pseudotorellia.* Conifers appear to be very diverse including *Podozamites, Araucariodendron, Ussuriocladus, Podocarpus, Cephalotaxus, Torreya, Tomharrisia, 'Cephalotaxopsis', Athrotaxites, Elatides, Sciadopitites, Nageiopsis, Elatocladus* and *Brachyphyllum.* There are also gymnosperms of uncertain affinity, i.e. *Aralia lucifera, Nyssidium orientale* and *Onoana nicanica.* In the deposits of the Albian in the Southern Primorje area (Galenki suite) in the west as well as in its age-equivalent Frentzev suite in the east, the following gymnosperms were recorded: *Sapindopsis* cf. *angusta, Arctocarpidium* sp., *Laurophyllum* sp., *Cissites* sp. and a newcomer to the Severosuchan suite, *Aralia lucifera.*

Floral compositions in the Far East Primorje area are markedly different from those of the Siberian–Canadian region including the flora of the Amur province containing certain southern elements. This discrepancy amounts in the floras of the Far East mainly to a complete lack of Czekanowskiales which are so inherent in the Siberian–Canadian region. Another point of difference is relative paucity of Ginkgoales. On the other hand, the Early Cretaceous floras of the Southern Primorje area are rich in such ferns as *Matonidium, Nathorstia, Ruffordia, Weichselia,* and Bennettitales (*Dictyozamites, Cycadites, Ptilophyllum, Zamites, Zamiophyllum* and *Williamsonia*) which are missing in the Siberian–Canadian region. The variety of conifers is strikingly represented by such thermophilic elements as *Araucariodendron, Podocarpus, Arthrotaxites* and *Brachyphyllum.* Early emergence of angiosperms is noteworthy; their impressions were found as far back as the Aptian (Severosuchan suite).

All this permits reference of the floras of the southern Far East to the East Asia province of the Euro-Sinian region without hesitation.

3.2.4.2 MONGOLIA

Within southern Mongolia Early Cretaceous deposits are well-developed having recently been divided into three horizons (Shuvalov, 1982). These are the Tsagantsab (Berriasian, Valanginian), Shinkhuduk (Hauterivian–Barremian) and Khukhtyk (Aptian–Albian) horizons. Each of the horizons is represented in various parts of Mongolia by coeval but lithologically different suites having names of their own.

The Early Cretaceous is composed in great measure of sandstones and argillites intercalated with marls and less often with limestones tinted grey, red-grey, brownish and yellowish. Members and lenses of conglomerates and basalt flows are encountered. Lacustrine deposits include a range of thin-layered 'paper shales' which are especially well-developed in the Shinkhuduk horizon. The higher Khukhtyk horizon typically contains coal-beds attaining major thicknesses. The predominant grey-coloured deposits in the lower and middle parts of the Lower Cretaceous are concomitant with occurrences of red formations. The latter, for instance, comprise the

Manlaysk suite (Lopatin, 1980) evolved in south-eastern Mongolia making part of the Tsagantsab horizon.

A number of localities with Early Cretaceous remains are associated with the Lower Cretaceous deposits. These are (westwards): Khuren-Dukh, Shin-Khuduk, Manlay, Anda-Khuduk, Bon Tsagan, Kholbo (Kholboto-Gol), Erdene-Ula and Gurvan-Eren.

Until recently the plant remains reported therefrom were not studied at all. Only the recent publication of the monograph of Krassilov (1982) fills the gap. The plant remains he examined show that we are concerned here with the vegetation related to numerous lakes existing at that time in Mongolia and, in part, with forest vegetation covering elevated land slopes.

Semi-aquatic plants include the remains assessed by Krassilov as *Cypercites* sp., *Typhaera* sp., *Sparganium* (?) and *Potamogeton* (?) having similarity to the present-day representatives of these genera. Whether they do belong to these contemporary monocotyledons will be established by further investigations. Semi-aquatic plants apparently include a new genus of lycopsids – *Limnonioba* (*L. insignis*) and, according to Krassilov, *Otozamites lacustris* displaying a well-developed aerenchyma.

Rarity and extreme fragmentariness of the fern fossils is conspicuous (*Onychiopsis* sp., *Cyathea* sp., *Osmunda* (*Raphaelia*) sp., *Cladophlebis* sp.). Just as in some other sites of the Transbaikal area this is likely to be due to the fact that lacustrine shores used to be forest-free while during transportation from forested highlands fern pinnae usually possessing a thin lamina rapidly turned into tiny fragments and detritus. Forest vegetation was represented by Ginkgoales (*Ginkgo*, *Baiera*, *Sphenobaiera*), Czekanowskiales (*Phoenicopsis* and *Hartzia*, the latter being assigned to *Czekanowskia*), and conifers (*Podozamites*, *Araucaria*, *Brachyphyllum*), as well as by the forms most probably belonging to ancient Pinaceae (shoots of *Pseudolarix* and *Pityophyllum*, winged seeds of *Schizolepis* and *Pityospermum*, bisaccate pollen).

Bennettitales are not numerous; apart from the above-mentioned *Otozamites lacustris* the recorded species included *Nilssoniopteris denticulata*, *Pterophyllum* cf. *burejense*, *P.* cf. *acutilobum*, *P.* cf. *sutschasense*, *Cycadolepis* sp. and, what is particularly exciting, *Neozamites verchojansis*. It will be noted that this species is recorded both in the Siberian–Canadian region and in the Euro-Sinian region (Southern Primorje area).

We should agree with Krassilov who indicated the ecotone, i.e. the mixed nature of the vegetation in Mongolia which comprises both southern (*Otozamites*, *Araucaria*, *Brachyphyllum*) and northern elements (*Phoenicopsis*, ancient Pinaceae). On the map of phytochoria we place south Mongolia (there are no data for its northern part) within the East Asia province included in the Euro-Sinian region due to the lack of the above-mentioned subtropical elements (*Otozamites*).

A most convincing proof in favour of referring Mongolia to the Euro-Sinian region was recently supplied by a find of a petrified barrel-shaped trunk of *Cycadeoidea* uncovered around 300 km south of Ulan Bator (Neiburg, 1932). Until recently, the trunks of this genus of Bennettitales had been encountered only within the Euro-

Sinian region. Expansion of *Cycadeoidea* is restricted by the latitudinal belt passing through the USA to western Europe (England–Poland). Southwards, within the USSR, not a single find of *Cycadeoidea* had been reported. Now it has been discovered in the Tula region and farther east is reported from Mongolia and Japan. Absence of finds of *Cycadeoidea* within the USSR is thought to be due to unfavourable conditions for their establishment and burial. The section of the subtropics in the European part of the USSR and the Caucasus housed marine basins, while eastwards in Middle Asia the Cretaceous consisted of well-developed red beds, and often clastic rocks virtually devoid of identifiable plant remains.

Krassilov (1982) identified flora-bearing complexes roughly corresponding to the three horizons of V. F. Shuvalov. However, one to two forms indicated by him for each horizon can hardly be regarded as the index fossils. This is inhibited by the dearth of material and the difficulty in comparing these new species confined in their distribution to Mongolia with the species known from other provinces. However, it is interesting to note that the previously described species *Neozamites verchojansis* from the second half of the Lower Cretaceous was found in the second horizon in Mongolia which is dated as Aptian; this can hardly be considered as accidental.

The survey of the Early Cretaceous floras of Mongolia should be concluded by looking at the issue of the presence of angiosperms. V. A. Krassilov attributes certain remains to the monocotyledonous *Graminophyllum primum*, *Potamogeton* (?) sp. and *Sparganium* (?) sp. Almost all of these are derived from the deposits of the fossil lake Manlay assigned to the Undurkhan suite; higher in the sequence they were not found. Their disappearance is probably connected with altered conditions of burial. In the same part of the section (Undurkhan suite enclosed within the Tsagantsab horizon) Krassilov described the fruits of dicotyledons *Girvanella* gen. nov. and *Erenia* gen. nov.

Appearance of the remains of angiosperms in the Neocomian brings to mind the find of a leaf imprint (Fig. 3.21) of an angiosperm (*Dicotylophyllum pusillum*) in the Zazin suite of the Vitim river in the Transbaikal region. Palaeoentomologists estimate the age of the Zazin suite as Neocomian while I believe it to be Barremian–Aptian (Vakhrameev and Kotova, 1977).

Palynologic research conducted by Bratseva and Novodvorskaya (1979) in the Anda-Khuduk locality in the region of the central part of Mongolia detected a very young complex incorporating, among other things, various tricolpate pollen of angiosperms. These were *Retitricolpites vulgaris*, *Retitricolpites* sp., *Bacutricolpites constrictus*, *Tricolpites* sp., *Tricolpopollenites micromunus* and *Tricolporopollenites* sp.; with these the pollen of *Ephedripites* and *Gnetaceaepollenites* was encountered. Unfortunately the black clays member whence this complex was collected was attributed to the Anda-Khuduk sequence according to the exposure and is undoubtedly younger than the locally outcropping Anda-Khuduk suite. The palynocomplex collected from this suite is dominated by the pollen of gymnosperms (Podocarpaceae, Pinaceae, Araucariaceae, etc.) with spores being also diverse. The angiosperms are represented solely by the pollen of *Asteropollis* and *Stephanocolpites*. The tricolpate pollen was not recorded here. This palynocomplex is certainly more ancient than the previous one which contained a variety of angiosperms. In addition to Anda-Khuduk the pollen of *Asteropollis* (Fig. 3.21) is peculiar to Mongolia in the habitats of Ulzit, Khobur and Khuren-Dukh. The first of these exhibited isolated pollen of tricolpate type and the pollen of *Clavatipollenites* in addition to *Asteropollis*. I. Z. Kotova (Vakhrameev and Kotova, 1977) found this pollen type from the Beklemishev, Chita–Ingoda, East-Urulyunguy, Kondin and Arbagar basins. It should be admitted that the palyno-complex with *Asteropollis* is inherent in the Anda-Khuduk suite making part of the Shikhunduk horizon. Its age is determined for Mongolia as Hauterivian–Barremian (Shuvalov, 1982) or Barremian–Aptian for the Transbaikal area (Vakhrameev and Kotova, 1977). The palynocomplex with a more diverse tricolpate pollen discovered in Anda-Khuduk is likely to be descended from the Khukhtyl horizon indicating Albian rather than Late Albian age.

One can presume that the central parts of Asia embracing Mongolia and Trans-baikalia were one of the centres of emergence of angiosperms because it is precisely here that the pollen of *Asteropollis* occurs earlier than anywhere else (usually from the Albian). An imprint of a leaf of one of the dicotyledons (*Dicotylophyllum pusillum*) was also found here. Judging from the alternation of red beds and variegated members with the grey shales the ecologic situation in Mongolia changed markedly with time.

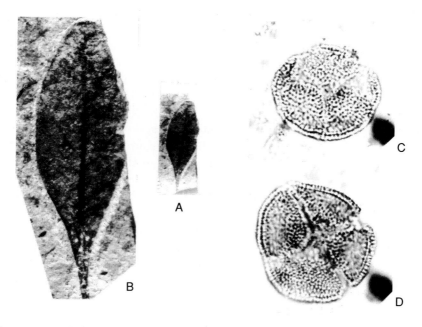

Fig. 3.21 Fossil plants of Early Cretaceous age, Lake Baikal area. A, B, *Dicotylo-phyllum pusillum* Vakhrameev; A, natural size; B, × 3; C, D, *Asteropollis asteroides*, × 1000.

Moister periods were replaced by dry spells and this factor must have influenced the physiognomy of the vegetation, and its composition which underwent continuous change. These conditions might have been conducive to the appearance of one of the first angiosperms enjoying a miscellany of advantages to adapt to the changing situation.

3.2.4.3 JAPAN

The Early Cretaceous floras of Japan in the first half of the twentieth century were mostly studied by Oysi, Yabe and Yakoyama. Larger-scale research was initiated in the second half of the century; Kimura and a number of other workers were especially prolific turning out numerous works. These include the following principal contributions: Kimura (1975a, b, 1976, 1979, 1980), Kimura and Hirata (1975), Kimura and Seikdo (1976, 1978).

Kimura came to the conclusion that the Early Cretaceous of Japan witnessed two types of Early Cretaceous floras (Fig. 3.22). Its outer eastern side opening into the Pacific Ocean provided the Riosseki flora of which the remains are confined to the series of that name, whereas the inner side looking on to the Japan Sea housed the Tethori flora (Kimura, 1979). In the south-east of Japan (Shikoku and Kyushu islands) in the stratotype area, the Riosseki series is divided into three suites (upwards) named Riosseki (Neocomian), Lower Monobegava (Upper Neocomian or, possibly, Barremian) and Upper Monobegava (Aptian–Albian). Here the Riosseki suite is deposited unconformably on the Palaeozoic or on the Torinosu suite composed of marine sediments containing an Upper Jurassic fauna (Kimura and Hirata, 1975). Age-equivalent suites of the Riosseki formation are traced all along the outer edge of Japan where they are known under different names (Omoto, Ayukava, Yuasa, Kavaguchi suites, etc.).

The Berriasian and Valanginian–Hauterivian age of the Omoto and Ayukava (Kimura and Sekido, 1976, 1978) is determined from the marine fauna including ammonites. The Arida, Kobosura and other suites are thought to be analogous to the Lower Monobegava suite while the Nisikhiro, Yatsusiro and Miyako suites are considered to be analogous to the Upper Monobegava. The composition of the floras of the Riosseki series alters little upwards permitting a general characterization founded primarily on the materials from Sikoku island (Kohl prefecture).

The Riosseki and Lower Monobegava suites (Kimura, 1975a, b, 1980) have *Neocalamites, Nathorstia, Klukia, Weichselia* and *Naktongia*, as well as *Adiantopteris, Gleichenites, Cladophlebis, Onychiopsis* and *Sphenopteris* including *S. goeppertii*. The numerous *Cladophlebis* are dominated by three-lobed leaves with tiny pinnules having a lobed margin. The only exception is *Cladophlebis takezaki* which probably belongs to the Osmundaceae. The genera *Otozamites, Pterophyllum, Ptilophyllum, Zamites, Cycadolepis* and *Nilssonia* are widespread, and are predominantly represented by the species *Zamiophyllum buchianus, Ptilophyllum* ex gr. *pecten* and *Nilssonia schaum-*

bergensis. Ginkgoales (*Baiera brauniana*) are very rare. The presence of *Pseudotorellia* is presumed. *Podozamites lanceolatus* is known. Horse-tails, *Brachyphyllum*, *Elatocladus*, *Frenelopsis* and *Nageiopsis* are encountered rather seldomly.

The floras of the Lower Monobegava suite differ but slightly from that of the Riosseki. They become devoid of *Naktongia*, with the specific composition of *Cladophlebis* being somewhat curtailed (*C. toyomensis* and *C. triangularis* are vanishing). *Pachypteris* sp. has been recorded testifying to the coastal habitat of the flora of Lower Monobegava, and *Baiera brauniana* was found.

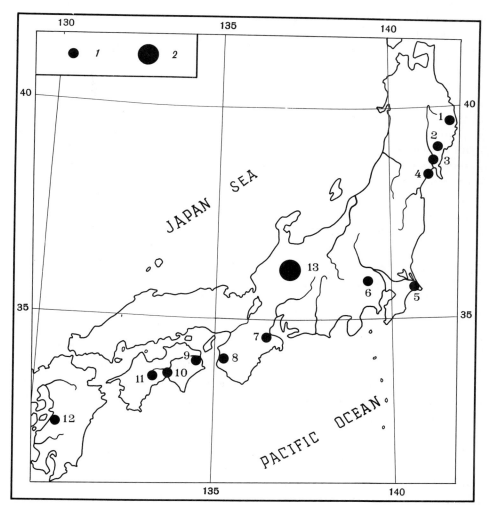

Fig. 3.22 Localities of Early Cretaceous floras in Japan. 1–12, localities situated within the outer zone (Riosseki flora); 13, locality within the inner zone (Tethori flora) (according to Kimura and Hirata, 1975).

The flora of the Upper Monobegava suite is more appreciably different. *Neocalamites*, *Klukia*, *Naktongia*, *Matonidium*, *Weichselia* and sphenopterid ferns completely disappear, just as do certain *Cladophlebis*; *Otozamites* are reduced in diversity. The Sebuyasi suite deposited in the trough west of Tokyo and roughly coeval with the Upper Monobegava suite accumulated plant remains which have yielded *Acrostichopteris longipennis* and *Cupressinocladus japonicus* which are the young elements of the Early Cretaceous flora.

This overview makes it obvious that changes of flora within the Riosseki series proceeded very gradually. Such older elements as *Neocalamites*, *Klukia*, *Naktongia* and *Matonidium* that continued from the Jurassic, disappear from the upper sections particularly in the Upper Monobegava suite. At the same time this suite is deprived of remains of angiosperms generally emerging from the Lower or Middle Albian in many countries and from the Upper Albian elsewhere. This suggests that the upper boundary of the Upper Monobegava suite lies somewhere inside the Albian. Finds in the Lower Cretaceous deposits of the outer zone of Japan of fossil trunks should be mentioned (Nishida and Nishida, 1983). Thus, a fragment of a thin trunk of the bennettitalean *Bucklandia* and the arborescent fern *Cyathocaulis* were detected from the Chosi suite (Chiba prefecture) coeval with the Sebayasi suite dated as Aptian. Fragments of the trunks of *Cyathocaulis* were also found in other areas of Japan, situated on its outer side. These finds attest to the subtropical climate of Japan evidenced by flora composition determined from leaf impressions.

The Tethori group is split into two series (subgroups), i.e. lower Kuzuryu and upper Itosiro which stands out on the inner side of the Khonsyu island. The former of the two series is composed of marine terrigenous rocks enclosing ammonites of the Late Jurassic (*Kepplerites* (*Seymourites*) *japonicus*, *Reineckia yokoyamai*).

The Itosiro series with abundant plant remains was earlier ascribed to the upper Jurassic. However, the abundant floras containing typical Early Cretaceous species are found in the well-established Lower Cretaceous sediments of the southern Far East where they have underlying strata with a fauna of Berriasian and Valanginian age. The noticeable difference between the Itosiro and Kuzuryu series made us date the Itosiro series as Lower Cretaceous. The species found in the Itosiro series include the following species in common with those from the Riosseki series: *Ruffordia goeppertii*, *Coniopteris burejensis*, *Adiantopteris sewardii*, *A. yuasensis*, *Onychiopsis psilotoides*, *Weichselia reticulata*, *Acrostichopteris plurapartita*, *Cladophlebis exiliformis*, *Dictyozamites grossinervis*, *D. kawasakii*, etc.

The lower horizons of the Itosiro series (Neocomian) developed on the inner side of Japan, are characterized by the rich flora of Ogusi (Kimura and Hirata, 1975), which exhibited representatives of the following genera: *Equisetites*, *Todites*, *Gleichenites*, *Hausmannia*, *Klukia*, *Coniopteris*, *Eboracia*, *Birisia*, *Onychiopsis*, *Adiantites*, *Ruffordia*, *Sphenopteris*, *Sagenopteris*, *Cladophlebis*, *Ctenozamites*, *Otozamites*, *Dictyozamites*, *Pterophyllum*, *Neozamites*, *Ctenis*, *Nilssoniopteris*, *Ginkgoites*, *Ginkgodium*, *Phoenicopsis*, *Eretmophyllum*, *Czekanowskia*, *Podozamites*, *Elatocladus* and *Taeniopteris* (Fig. 3.22).

The peculiarities of the Ogusi flora are abundance of chiefly large-pinnuled *Cladophlebis* (13 species), a large number of *Coniopteris* (6) including *C. burejensis*, a variety of *Dictyozamites* (10), and a somewhat inferior diversity of *Ctenis* (4) and *Nilssonia* terminated by presence of Ginkgoales and Czekanowskiales.

The flora of Akaiva (Kimura and Sekido, 1976, 1978) collected from the middle part of the Itosiro series approximately coeval with the Lower Monobegava vegetation contains largely the same genera as the Ogusi flora but their specific composition is far less varied. A somewhat younger age of this flora is implied by the disappearance of *Klukia* and *Todites* and emergence of *Asplenium* cf. *dicksonianum*. The variety of *Dictyozamites* is diminished. This flora, just as that from Ogusi, contains *Ginkgoites digitata*, *G. sibirica*, *G. huttoni* and *Ginkgodium nathorstii*, as well as such Czekanowskiales as *Czekanowskia* sp., *Phoenicopsis* sp. and *Leptostrobus* sp.

The flora from Tamadani in the upper part of the Itosiro series roughly age-equivalent to the Upper Monobegava suite vegetation is much poorer in composition. It is totally devoid of Czekanowskiales retaining, however, certain Ginkgoales. The ferns are represented by *Osmundopsis* (?) sp., *Gleichenites porsildii*, *Adiantites* sp. and *Cladophlebis pseudolobifolia*, while new ferns include *Arctopteris*. *Dictyozamites* was not encountered but *Nilssonia* are present. Among the Ginkgoales, *Ginkgoites* and *Pseudotorellia* sp. were confirmed whereas conifers were represented by *Podozamites reinii* and *P.* cf. *eichwaldii*.

According to Kimura (1975a, b) the Riosseki series is characterized by widespread development of Matoniaceae (*Matonidium*, *Nathorstia*), *Weichselia* which is close to this family, and substantial distribution of *Onychiopsis psilotoides*. Segments of cycads and Bennettitales are usually represented by the genera *Otozamites*, *Zamites*, *Zamiophyllum*, *Pterophyllum*, *Nilssoniopteris*, *Ptilophyllum* (rarely) and *Nilssonia* having a narrow lamina of no great size. Ginkgoales are very rare, while Czekanowskiales are lacking.

The Itosiro (Tethori) series is dominated by bipinnate *Cladophlebis* with large pinnules and by Dicksoniaceae (*Coniopteris*, *Sphenopteris*), whereas Bennettitales include many and varied *Dictyozamites* and cycads such as *Ctenis* and *Nilssonia*. However, representatives of *Ptilophyllum* and *Zamiophyllum* are missing here although they are represented in the Riosseki series. One peculiarity of the Itosiro series flora is the occurrence of Ginkgoales (*Ginkgoites*, *Ginkgodium*) and rare Czekanowskiales (*Phoenicopsis*, *Czekanowskia*). The former are very rare while the latter are totally absent from the Riosseki vegetation. No remains of the squamifoliate conifers *Brachyphyllum* and *Pagiophyllum* were detected in this flora, either.

Palynologically the Early Cretaceous floras of Japan have been inadequately studied so far. The Aptian–Albian deposits (Miyaki suite) of north-eastern Japan have yielded spores and pollen which were dominated by *Classopollis* pollen (up to 70%). The spores featured *Cicatricosisporites*, *Appendicisporites* and *Gleicheniidites*. A single grain of *Tricolpopollenites* sp. was discovered corroborating the presence in the Miyaki suite of Albian deposits.

Kimura (1979, 1980) came to the correct conclusion on the affinity of the Riosseki flora, whose habitats are confined to the outer zone of Japan, with those of the Euro-Sinian (previously Indo-European) region, and on its major similarities to the Wealden vegetation from southern England and the flora of the Primorje area. However, it is hardly possible to assign the Itosiro (Tethori) flora to the Siberian–Canadian palaeofloristic region. Comparison of the typical Early Cretaceous floras of Siberia with the vegetation of Japan's inner zone reveals significant disparities. These include in the first place the absence in the former of varied *Dictyozamites*. The *Dictyozamites* described by Vakhrameev (1970b) from the Verchoyansk ridge, proved, after A. I. Kirichkova examined the cuticle, to belong to the genus *Ctenis*. In addition, the Itosiro series displays varied *Gleichenites*, sporadic *Otozamites*, *Cycadites*, numerous *Onychiopsis*, and *Ruffordia*, while Czekanowskiales are rare. Thus, the Itosiro flora appears to be a special plant community of the East Asia province.

At the same time the Siberian–Canadian and the Itosiro floras have something in common which is expressed in the presence of some common or similar species of *Coniopteris*, *Adiantopteris*, *Birisia*, *Neozamites*, *Nilssonia* and *Ctenis*. This renders the Itosiro flora typically ecotonal, transitional between the Siberian–Canadian and Euro-Sinian regions. But while the flora of the Amur province with a core of Siberian elements should be assigned to the Siberian–Canadian region, the flora of Itosiro, wherein distinctive southern elements prevail with the Siberian ones being secondary, should be referred to the East Asia province with its mixed composition.

We believe that the composition of floras of both the Southern Primorje area and the Itosiro (Tethori) and Riosseki area is related to the complicated inter-relation between land and sea. Additionally the occurrence of meridional or submeridional elevations and ridges probably served as areas of displacement in the intermediate sedimentation basins and, possibly, as barriers between plant associations. The most warmed sector of land was situated on the outer side of Japan washed by the warm waters of the Pacific. It was here that the Riosseki flora was growing, being virtually devoid of northern elements. The hot climate is also evidenced by the high proportions of *Classopollis* pollen (around 70%).

The floras of the inner zone separated from the ocean by the highlands of the central parts of Japan grew in cooler climates. It is noteworthy that, according to many geologists, the Japanese islands were nearer to the eastern edge of the Eurasian mainland being separated from it solely by a shallow-water offshore area, sometimes dried up. The meridional orientation of the highlands separating the sedimentation basins was conducive to heterogeneous plant associations sometimes located on the same latitude.

This is observable not only in comparing compositions of the Riosseki and Itosiro (Tethori) floras but also in doing the same with the floras from the Suchan and Suyfun basins in Southern Primorje, separated from each other by a meridionally elongated displacement area. The Suchan basin was invaded by the sea in the Valanginian, while the Suyfun basin was exclusively receiving continental deposition. This might have been the reason why the Suyfun basin recorded an abundance of

Bennettitales and cycads whereas the Suchan basin was dominated by conifers. Of course, it is hard to ascertain the real causes of distribution of plant communities or of individual plants in the majority of cases, but correlation between land and sea seems to be relevant, as does dissected relief, generally oriented meridionally. The latter factor predetermined possibilities of migration of plants from north to south and back. It is evidently not by chance that many Ginkgoales and even Czekanowskiales penetrated far south and are recorded in the Ogusi flora located near the inner edge of central Japan. Obviously it is not accidental that *Dictyozamites*, abundant in the internal zone floras of Japan opening inland (nine species in the Ogusi flora), continues to remain a noticeable element in the Early Cretaceous vegetation of the southern Far East (5 species) being, however, absent from the outer zone of Japan (the Riosseki floras). This suggests a relation between the Early Cretaceous floras of the southern Far East and those of the inner zone of Japan. At the same time *Dictyozamites* were not found within the typical floras of the Siberian–Canadian region which, according to Kimura, encompassed the Cretaceous vegetation of the inner zone of Japan. The species present both in the floras of the southern Far East and in the west and east of Japan is *Onychiopsis psilotoides*. This species is virtually unknown within the Siberian–Canadian region occurring only sporadically in the second half of the Early Cretaceous of Yakutia.

Such heterogeneity of coeval floras is inherent in the transitional zone ecotone especially in the instances when migration paths are determined by meridionally oriented sufficiently dissected relief and by rugged coastlines.

3.2.4.4 CHINA AND KOREAN PENINSULA

The localities of floras of the East Asia province of the Euro-Sinian region are to be found generally south of 40° (see Fig. 3.13, p. 123). The northernmost locality of this flora is situated on the Shandung peninsula (Anon., 1963). Fossil plants, insects, Estheriaceae and fishes were found here from the Layan suite deposited on the granite–gneisses of the Precambrian. The fossils were recovered from the middle part of the suite made up of variegated and sometimes paper shales intercalated with bituminous shales. *Thinnfeldia* sp., *Zamites* sp., *Baiera* sp., many shoots of the conifers *Araucarites* sp., *Brachyphyllum* sp., *Pagiophyllum* sp., *Sphenolepis* and *Palaeocyparis* sp. were recorded here. Complete absence of ferns is quite conspicuous while shoots of conifers with awl-shaped or closely clinging needles (*Brachyphyllum*) are relatively diverse. The latter supplemented by the presence of *Zamites* are indicative of the affinity of the complex with the East Asia province. Such a complex coupled with joint occurrence of fossils of plants, fishes, Estheriaceae and insects is peculiar to the sediments of large lakes. It will be noted that paper shales with such a selection of plant remains are quite widespread in the Lower Cretaceous (the Turga strata in the Transbaikalia, some formations in Mongolia).

The Gansyu province in the western part of the Khuan-Khe river bend represented, just as on Shandun peninsula, by paper shales intercalated with sandstones and marls, witnesses a very high content of gymnospermous pollen dominated by *Classopollis*.

A spore–pollen complex with roughly the same composition was encountered just north-west of Shanghai (Tsyansu province) in the Gegun suite wherein the high proportions of *Classopollis* pollen are concurrent with the shoots of *Brachyphyllum* and *Pagiophyllum*. The Lower Cretaceous age is established by the presence of the spores of *Cicatricosisporites*. In the south-eastern part of the Hupei province, some 900 km east of Shanghai, the spore–pollen complex isolated from the formation of sedimentary and volcanogenic rocks contains almost 50% of *Classopollis* pollen. Bisaccate pollen is rare. The presence of the spores of *Cicatricosisporites*, *Plicatella*, *Pilosisporites*, etc. implies the Early Cretaceous age of the deposits.

In a publication by Chow and Tsao (1977), eight new species were described from eastern China in the genera *Brachyphyllum*, *Manica* (*Pseudofrenelopsis*) and *Frenelopsis*. It is to be noted that the genera *Frenelopsis* and *Pseudofrenelopsis* are lacking in the Siberian–Canadian region appearing only towards its southern boundary. Unfortunately, no exact locality of these finds is indicated in the English translation of the Chinese abstract. The finds of the shoots of *Frenelopsis* and *Brachyphyllum* are known from the Lower Cretaceous of the northern part of the Korean peninsula on the Yalutszyan river along which the frontier between China and Northern Korea passes.

The southernmost locality of the Early Cretaceous plants is related to the Bantou suite (Futszyan province) lithologically close to the composition of the Layan suite on Shandung peninsula (Sze, 1945). The upper part of the Bantou suite is made up of intercalated paper shales and sandstones. The age of these sediments is estimated by Chinese geologists as the second half of the Early Cretaceous. Local finds include *Ruffordia* cf. *goeppertii*, *Cladophlebis*, *Sagenopteris*, *Otozamites*, *Ptilophyllum*, *Nilssonia*, *Baiera*, *Brachyphyllum*, *Pagiophyllum*, *Podozamites* sp. and *Sphenolepidium* sp.

In conclusion, mention should be made of a find of the fern *Weichselia reticulata*, a typical representative of the Euro-Sinian region and of the Equatorial flora (Cai, 1982). This find was reported from the province of Chzhedzyan (south of Shanghai) in the Moshishan suite. Another find of *Weichselia reticulata* is known from Tibet (Linbozong suite).

Comparing the boundary earlier drawn in East Asia between the Siberian–Canadian and Euro-Sinian regions by myself (Vakhrameev, 1964; Vakhrameev *et al.*, 1970) and by Kimura (1979) with its position presumed in this work we discover rather substantial differences. In the new interpretation this boundary passes south of its earlier position (see Fig. 3.13).

The boundary starts in the middle part of the Sikhote-Alin, then proceeds south-west extending roughly to the base of the Korean peninsula. It subsequently by-passes Beijing from the south, follows north-west across Mongolia and leaves its southern part within the East Asia province.

The basic difference from the original variant is that the area of the Amur province is expanded to include all of north-eastern and northern China as well as the Transbaikalia. Evidence in favour of such a boundary was cited above. I have changed my views because of the fresh data obtained on the Chinese floras after an almost 20-year break, and evidence on the Early Cretaceous flora of Mongolia previously almost unknown.

The boundary suggested by Kimura (1979) passes from the southern tip of Sakhalin, by-passing from the south and west the habitats of the Far East floras which are legitimately assigned to the floras of the Riosseki type (Euro-Sinian region). Then it makes a sharp turn to the north-west and, having reached Hokkaido, again abruptly bends southwards following the axial part of Japan. East of this boundary there is the Riosseki flora, west the Tethori. Further south this boundary passes Shanghai and proceeds westwards leaving to the south the localities of Early Cretaceous vegetation situated in the Tsyanen and Fudzyan provinces.

The discrepancies between the positions of the boundaries of the Siberian–Canadian and Euro-Sinian regions drawn by myself and Kimura are that I attribute the Itosirov and Riosseki floras to the Euro-Sinian region (the rationale was supplied above) which eliminates the abrupt bay-like bend in this boundary in the Southern Primorje area. I also drew this boundary within the Asiatic mainland farther north than does Kimura referring the floras of Korea, Shandung, Hupei and Gansu containing the shoots of *Frenelopsis* coupled with abundance of *Classopollis* pollen to the Euro-Sinian region. Still farther north this boundary crosses Mongolia. Here its position remains not quite clear since no Early Cretaceous vegetation has been reported between the habitats of floras of this country situated in its southern half and having similarities to the floras of the Euro-Sinian region and of Baikal lake.

3.3 Equatorial region

3.3.1 South America (see Fig. 2.23, p. 85)

The Equatorial region was located within the greater part of the South American continent (minus Patagonia and southern Chile), and northern, western and central Africa. In fact, we are familiar only with a minor part of the Early Cretaceous land surface of this gigantic belt whose major part extends through the Pacific and Indian Oceans. The Equatorial region also covers south-east Asia. However, the localities of the Early Cretaceous floras situated in this territory are very few and there is so far no ground for identifying a separate province here. South America and Africa, currently separated by the Atlantic Ocean, were adjacent to each other precisely in the Equatorial belt which contributed to the unity of the then floras, impeding recognition of segregated provinces. The Early Cretaceous floras of India are considered to be within the Austral (Notal) region since they are different in composition from the floras of the

Equatorial region within the limits outlined. In addition, according to palaeomagnetic maps, India was situated within the more southerly Austral region. First, we shall survey macrofossils and then move on to the palynologic characterization which is most striking for the Equatorial region.

In Colombia, 90 km south of the city of Medelin in the valley of the river Quebrakho-Campanas, there is a locality of plant remains related to marl deposits in which finds of *Holcostophanus* sp. and *Neohoploceras* indicate the base of the Late Valanginian. *Coniopteris martinezi, Piazopteris branneri, Sagenopteris* sp., *Nilssoniopteris major, Zamites* cf. *quiniae, Zamites* sp., *Podozamites* sp., *Cupressinocladus lepidophyllum, Cyparissidium* sp. and *Elatocladus* sp. were determined from here (Lemoigne, 1984).

Also in Colombia, the 'Valle Alta' suite, composed of shallow-water terrigenous deposits laid on eroded diorites, contains plant remains which are assigned by Lemoigne to the Upper Jurassic rather than to the Lower Cretaceous. This locality is situated 120 km south of Medelin in Central Cordillieras. From here Lemoigne described *Cladophlebis denticulata, C. exiliformis, C. (Klukia ?) koraiensis, Sphenopteris (Ruffordia)* cf. *goeppertii, Pachypteris* sp., *Sagenopteris* sp., *Nilssoniopteris major* (?), *Otozamites simonatoi, O.* cf. *peruvianus, Otozamites* sp., *Anomozamites minor, Zamites lucerensis, Ctenozamites* spp., *Ptilophyllum* cf. *cutchense, P.* cf. *distans* and *Desmiophyllum* sp.

Having reviewed this list we discover therein such species as *Sphenopteris (Ruffordia)* cf. *goeppertii* or *Cladophlebis exiliformis*, i.e. the species heretofore reported only from the Early Cretaceous. This does not agree with the conclusion of Lemoigne on the Late Jurassic age of this flora but makes us place it, at least tentatively, in the survey on Early Cretaceous vegetation.

Interesting data on Colombia are provided by Pons (1982a). The fullest samples were gathered by her from the eastern slope of the Quetam mountain near the town of Villavichenchio. The deposits of the Lower Cretaceous outcropping here permitted the recognition of three members (of rocks) containing plant remains. Here only the list of the macro-remains of plants is cited while the extended palynologic characterization will be given below for the entire Equatorial region (within Africa and South America). The lower member yielded *Cupressinocladus pompeckyi, C. leptocladoides* and *Podozamites* sp. Going by the absence of *Dicheiropollis* pollen, which is widely disseminated in the Neocomian of the Equatorial region, this member is post-Barremian or, most likely, Early Aptian. In the middle member the fern *Weichselia reticulata* was found together with the spores of *Dictyophyllidites* cf. *harrisii* (up to 75% of the total amount of spores and pollen detected in the samples) very similar to the spores identified by Alvin (1971) from the sporangia of *Weichselia. Weichselia* was found here in conjunction with *Phlebopteris* sp., and both seem to belong to the family Matoniaceae. It is noteworthy that *Weichselia* is very widespread in the Lower Cretaceous of the Cordilleras of South America. Pons indicates eight localities of this fern in Peru, seven in Colombia, and seven in Venezuela. It should be pointed out that all finds of *Weichselia* in South America are associated with Lower Cretaceous deposits whereas in northern Africa they appear in the Late Jurassic or, probably, Middle Jurassic persisting into the Cretaceous.

The same horizon witnesses emergence of angiosperms represented by the fragments of rather large leaves described as *Monocotylophyllum heterophylla*, *Montonia quetamensis*, *M. sinuata* and *Dicotylophyllum spatulatum* sp. Pons considered this complex to be Late Aptian but the presence of imprints of large although incomplete leaves of angiosperms suggests these deposits may be somewhat younger (Early Albian). The upper member supplied *Zamites gigas*, *Podozamites* sp., *Dicotylophyllum spatulatum* and *Dicotylophyllum* sp. Occurrence of the impressions of angiosperms along with miospores of *Elaterosporites* sp., *Elaterocolpites* and the pollen of the angiosperms *Tricolpites* and *Tetracolpites* points, according to Pons, to a Late Albian–Early Cenomanian age.

The second section of the Lower Cretaceous deposits (Pons, 1982b) is located in the northern part of Colombia along the river Lebrilla north of the town of that name. Here the deposits of the Giron series are composed of terrigenous often obliquely laminated sediments up to 5000 m thick; some thick members are reddish. Within the Giron series plant remains were located at various levels; the discoveries comprise *Piazopteris branneri* (this genus is very close to the genus *Phlebopteris*), *Sagenopteris* sp., *Brachyphyllum* sp. and woody *Protophyllocladoxylon* sp., as well as spores and pollen. The latter include such lower Cretaceous indicator forms as various *Cicatricosisporites* spp. (18.3%) and *Pilosisporites* spp. (7.4%). Apart from these, *Biretisporites potoniaei*, *Leptolepidites* sp., *Verrucososporites* spp., *Inaperturopollenites* sp. and *Araucariacites* sp. were encountered; there is relatively high *Classopollis* spp. pollen (20%). Presence of the varied species of *Cicatricosisporites* and *Pilosisporites* suggests that the Giron series wherein they were found belongs to the Lower Cretaceous. A gap at the base of this series that was deposited unconformably on the Lower Jurassic (Bokas suite) evidences that merely the lowest part of it may be assigned to the Jurassic (in all probability Upper).

3.3.2 Africa

Comparatively few plant remains of Early Cretaceous age were uncovered in northern Africa. In the Albian of southern Tunisia and Nigeria fossil remains of *Weichselia* were detected. The Wealden of Cameroon furnished *Frenelopsis* sp. In south-west Egypt on the Gill-Kebir plateau, Lower Cretaceous (possibly) continental deposits resting on the suite of *Lingula* shales with the fauna of Tithonian–Berriasian age (Nicol-Lejal, 1981) provided *Weichselia*, *Matonidium*, *Phlebopteris*, *Araucaria* and *Paleocyparis*. Still higher there are deposits of Cenomanian age with the imprints of angiospermous leaves, bones of dinosaurs and *Lingula* shells (the area of Suez witnesses exposed variegated sediments with *Phlebopteris* (?) and *Otozamites* (see Fig. 3.15).

3.3.3 Palynology

The Early Cretaceous flora is characterized far more extensively by palynologic data. The unique composition of spores and pollen of the Equatorial region is well shown in the works of Brenner (1976), Doyle *et al.* (1977, 1982), Kotova (1978), Herngreen (1974, 1975), Herngreen and Khlonova (1981) and many other researchers. One of the controversial sequences is one of the Cocoabeach series situated in north-eastern Gabon and made up of terrigenous rocks of continental origin without coals.

This series encompasses (Doyle *et al.*, 1982) part of the Neocomian (*sensu stricto*), Barremian and Early Aptian. The Barremian and the underlying deposits are characterized by the presence of *Dicheiropollis etruscus* pollen. The finds of this form in Italy where it was first described and in Switzerland are apparently related to the lithospheric plate drift northwards because this species is commonly not reported from the marine sediments north of Morocco. Somewhat higher in strata belonging to the Barremian or Aptian occur *Clavatipollenites* and *Retimonocolpites* pollen which are regarded as the primitive pollen of angiosperms. From the Aptian onwards the pollen of *Afropollis* and *Tricolpites* appears. A salt-bearing formation is deposited higher in the Gabon sequence.Some palynologists place it in the Upper Aptian on the strength of its position at the base of the Lower Albian, but I. Z. Kotova thinks it to be Albian by the presence of *Tricolpites* alone.

The superimposed supra-salt series Madiela, composed of shallow-water deposits, encloses a fauna of Albian and higher, of the Early Cenomanian. Here very typical miospores make their appearance being referred to the genera *Elaterocolpites* and *Elaterosporites* supplied with elater processes. Their distribution, like that of *Dicheiropollis*, is confined to the Equatorial region. *Afropollis* persists but disappears towards the end of the Early Cenomanian. Somewhat higher the pollen of *Hexaporotricolpites* and *Cretacaeiporites* appears. The same vertical dissemination of the above-mentioned characteristic forms is retained in other areas of western Africa (from Senegal and Angola), in adjoining parts of the Atlantic Ocean (the Morocco depression, Cape Verde depression), the northern part of the South Atlantic, the coastal areas of eastern Brazil and the north-western edge of South America.

There are only a few disagreements concerning estimation of host rock age rather than the sequence of their appearance. Doyle and collaborators (Doyle *et al.*, 1982), considering the age of the rocks outcropping in the sections of littoral areas of western Africa and eastern Brazil represented in their lower parts by continental formations, think that the typical pollen of angiosperms (*Tricolpites*) emerges for the first time in the Aptian; the salts are also dated by them as Aptian.

I. Z. Kotova, who investigated deep-sea borehole cores north of the Moroccan trough and the Cape Verde trough, came to a somewhat different conclusion. In these troughs the Lower Cretaceous is represented wholly by marine deposits characterized not only by spores and pollen but also by dinoflagellates and macrofauna. Salt deposits are missing here because the salt-bearing basin of the South Atlantic was separated from the north by a bridge existing at the junction of Brazil and western Africa. According to Kotova, *Tricolpites* pollen occurs only at the base of the Albian. Lower in the sequence only *Clavatipollenites* and *Retimonocolpites* have been reported from the Barremian or Aptian. She believes that the age of the salt appearing in the sequences of the Congo, Gabon and eastern Brazil should be dated as Albian. Of great interest is the range of *Classopollis* pollen abundant in the Lower Cretaceous of the Equatorial belt whose peculiarities will be dealt with a little later. *Classopollis* pollen is concurrent with a great frequency of *Ephedripites* pollen and its relative *Gnetaceapollenites* represented by various species.

Pons (1982a, b), who in the post-Barremian deposits of Colombia identified three complexes of spores and pollen of which the lowest age limit is determined by the absence of *Dicheiropollis*, indicates that the *Classopollis* content in the lower complex (upper Aptian) does not exceed 2.5%. *Araucariacites australis* pollen is plentiful (39%) as is *Inaperturopollenites* spp. (21.2%). The amount of pollen appears far greater than that of spores. The latter yielded representatives of the genera *Impardecispora*, *Cicatricosisporites* and *Chomotriletes*.

The middle complex is dominated by spores, with *Dictyophyllidites* being encountered most frequently (75%). They must have been produced by the fern *Weichselia*. New pollen comprises *Ephedripites* (= *Equisetosporites*) and *Afropollis* (*Reticulatosporites*) *jardinus*. Miospores with elaters and tricolpate pollen are lacking. *Classopollis* amounts to 2.2% and *Inaperturopollenites* to 4.5%. Absence of miospores and tricolpate pollen implies the pre-Albian rather than post-Aptian age.

In the upper complex the amount of the pollen *Inaperturopollenites* runs to 47%, *Araucariacites australis* to 8.2%, and spores of *Dictyophyllidites harrisii* to 11%. *Elatosporites* and *Elaterocolpites* emerge (Fig. 3.23) over and above tricolpate pollen of angiosperms indicating the Albian–Early Cenomanian age of this complex.

The similarities between the deposits of the Neocomian and their palynologic characterization on the opposite shores of the South Atlantic corroborates that at the time the South Atlantic had not yet separated Brazil from Africa. At the very end of the Jurassic or even at the outset of the Early Cretaceous major faults formed here producing a system of grabens outlining the division between Africa and South America. The system at that time was rather close to the contemporary one in eastern Africa. In grabens, lakes and river valleys began to be filled with terrigenous deposits assignable, according to palynologic data, to the Neocomian (including Barremian). In the Aptian further expansion ensued followed by depression of this weakened zone permitting the sea to penetrate from the north as far as the Congo, i.e. to the region of continental deposit development thus forming a narrow lagoon separated from the high seas by elevation of the Kitovy ridge. This lagoon accommodated a thick salt deposit, patches of which are observable both on the western coast of Africa (Gabon, the Congo, Angola) and in Brazil.

Subsequent expansion of the graben system caused the Albian transgression from the south leading to the formation of a marine basin of normal salinity separating Brazil

Table 3.1. *Percentage content of* Classopollis *pollen*

Age	South-west of Cape Town (borehole 361)	Falkland plateau	Argentina (Huatrin suite)	Angola basin (borehole 364)	Gabon, Congo	Moroccan depression (borehole 370)
Cenomanian	10–15	—	—	—	30–75	61
Albian (lower part)	80–90	17.5–55	62.5	dominant	50–80	65–82
Aptian	80–90	—	—	dominant	dominant	80
Neocomian (including Barremian)	—	42–81	—	—	frequent	70

from central Africa. However, complete unification of the southern and northern Atlantic was not accomplished in Albian time, as is indicated by discrepancies in composition of the ammonite fauna in these two basins (Vakhrameev, 1981a, b). It was not until the Cenomanian or even the Early Turonian that the South Atlantic merged with the northern part.

Earlier it was mentioned that significant proportions of *Classopollis* pollen (60 to 70% of the total content of spores and pollen) in some deposits of the Jurassic and Cretaceous are indicative of a semi-arid or arid climate prevailing at the time of their formation (Vakhrameev, 1980). *Classopollis* pollen was produced by the coniferous family Cheirolepidiaceae (Alvin, 1982) that became extinct at the end of the Cretaceous period. The majority of the Cretaceous Cheirolepidiaceae carried scaly (*Frenelopsis, Pseudofrenelopsis, Brachyphyllum*) or hook-like leaves (*Pagiophyllum*) having an xeromorphic nature. Members of this family are established to have been thermophilic trees or bushes withstanding arid climates.

Table 3.1 presents the percentages of *Classopollis* pollen in the deposits of the Lower Cretaceous and Cenomanian from north to south along the South Atlantic from borehole 361 situated 300 km south-west of Cape Town, and the Falkland plateau up to the Moroccan depression (Kotova, 1978, 1983; McLachlan and Pieterse, 1978; Morgan, 1978).

The boreholes were drilled by the *Glomar Challenger* as part of the Deep Sea Drilling Project. In Argentina (Volkheimer and Salas, 1975) and Gabon (Boltenhagen and Salard-Cheboldaeff, 1980) exposures and ground borehole cores were studied, although for some areas palynologists have pointed out prevalence of *Classopollis* pollen without providing its percentage content.

It is seen from the table that the content of *Classopollis* pollen in marine and lagoon deposits of the Lower Cretaceous, notably the Aptian and Albian, starting with the southern margin of Africa (borehole 361) up to the Moroccan depression (borehole 370) attains very high values ranging from 40 to 80 through 90%. It is not until the

Cenomanian that the percentages begin decreasing abruptly, fluctuating in some samples from 10 to 75%. In the Turonian, *Classopollis* pollen virtually vanishes except for the presence of sporadic grains.

Besides this, the palynospectra of the Lower Cretaceous contain very few spores of ferns requiring moist conditions for growth. A noticeable quantity of spores is observable only in the south (Falkland plateau) and north (Moroccan depression), i.e. practically outside the Equatorial belt. These data contradict the supposition about the existence in the Early Cretaceous and Cenomanian of a dry climate prevailing in the area of formation of the South Atlantic and adjacent parts of Africa and South America, and of widespread humid tropical forests (Ronov and Balukhovsky, 1981). Dominance of the dry climate is indicated, too, by the formation in the Aptian of the above-mentioned thick salt deposit stretching from the Kotovy ridge as far as Cameroon. Its remains have been detected both in Gabon and the Congo and in south-eastern Brazil. Taking into account the dryness of climate it may be assumed that the forests composed of Cheirolepidiaceae producing *Classopollis* pollen were not sufficiently developed since shortage of moisture must have inhibited emergence of a dense canopy. Moisture shortage is likely to have interfered with trees reaching full height.

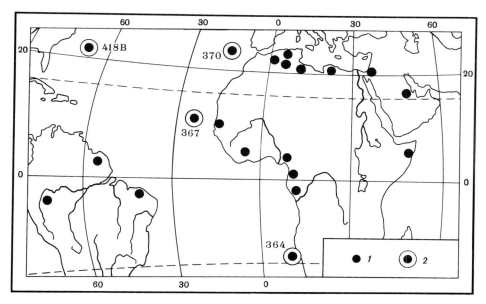

Fig. 3.23 Distribution of elater-bearing pollen (*Elaterosporites, Elaterocolpites, Elateroplicites, Sofrepites*) in Albian–Cenomanian rocks of Africa, South America and in Deep-Sea Drilling Project boreholes in the Atlantic Ocean. 1, Localities on continents according to G. Herngreen (Herngreen and Khlonova, 1981), 2, Deep-Sea Drilling Project boreholes in Atlantic palynologically studied by Kotova (1978, 1983).

Existence of the dry climate in Brazil, Argentina, western, central and, possibly, southern Africa in the Early Cretaceous is attributable to the fact that the above regions were chiefly situated in the central part of the vast mainland which had not yet disintegrated into Africa and South America. Remoteness of these areas from both the Pacific and Indian Oceans apparently prevented moist monsoon winds from penetrating here. Precipitation was also inhibited by the absence of any more or less major elevations.

Another picture is observed in the vicinity of the western coast of South America. Thus, carbonaceous deposits are there represented in the Neocomian of Peru. In Colombia near Bogota (Pons, 1982a, b) the Aptian and Albian deposits do not contain more than 2.2 to 2.5% of *Classopollis* pollen. On the other hand, they witness a drastic increase in the proportions of the pollen of *Araucarites* and *Inaperturopollenites* (in the Aptian, 39 and 21%, respectively). A rise is also noted in the number of fern spores. The upper Albian yielded an imprint of a large leaf of *Platanus*. This testifies to the existence in the Early Cretaceous of a moist climate in the vicinity of the Pacific Ocean coast of South America. However, eastwards it quickly became dry.

In concluding this section let us consider the composition of the spore–pollen complexes of the Early Cretaceous in the north-eastern part of Africa, i.e. in Egypt. The most ancient sediments of the Lower Cretaceous referred to the Neocomian were exposed by a borehole near Abu-Subeikh situated near the Mediterranean shore somewhere midway between Cairo and Libya (Aboul Ela, 1979). The deposits of the Neocomian are represented by shaly clays and sandstones intercalated with dolomite and, in the lower strata, with limestone. The content of *Classopollis* pollen fluctuates between 70% in the lower part and 40% in the upper. Among spores *Gleicheniidites*, *Cyathidites* and *Trilobosporites* dominate. Small quantities of *Concavissimisporites* and *Matonisporites* were also encountered.

In the central part of Egypt the Maukhoub West-2 borehole was drilled into the Abu Billas suite. Therein two compositionally similar palynocomplexes were identified: one from a depth of 628 to 634 m the other from 592 to 602 m. The most frequently occurring pollen in these palynocomplexes includes *Retimonocolpites* and *Ephedripites* spp. (total amount of 17 to 24%). Other species comprise *Afropollis* sp., *Clavatipollenites hughesii*, *Asteropollis* cf. *asteroides*, *Eucomiidites* cf. *troedssonii* and *Classopollis* spp. The typical spores yielded *Deltoidospora*, *Cyathidites* sp., *Cicatricosisporites* sp., etc. (Shrank, 1982). The age of the deposits is most probably Early Albian since there is a tricolpate pollen of angiosperms (*Tricolpites*). By the composition of pollen this complex is close to those from Algeria and Tunisia.

The Tahrir borehole located a little west of the Nile delta penetrated the deposits likely to belong to the Albian (Sultan, 1978) which is indicated by the presence of *Tricolpites* pollen emerging, as a rule, from the Albian. Apart from this the spores of *Cyathidites*, *Cicatricosisporites*, *Appendicisporites*, etc. were found supplemented by the pollen represented by *Afropollis*, *Classopollis*, *Araucariacites*, *Inaperturopollenites*, *Eucommiidites*, *Ephedripites* and *Gnetaceaepollenites*. Rare grains of *Elaterosporites* were encountered; a find of the pollen with elaters (*Elaterocolpites*) from the upper Albian exposed by the borehole situated near the eastern shore of Saudi Arabia (Srivastava, 1984) shows its distribution belt to range from South America to the Persian Gulf.

Comparison of the spore–pollen complexes from different boreholes reveals a curious pattern. In the Abu-Subeikh borehole located near the Mediterranean coast which is a relict of the Tethys Ocean the Neocomian deposits are represented by littoral formations (see above) containing major amounts of *Classopollis* pollen (40 to 70%). The inter-relation between these deposits and the sea is implied by the presence in them of dinocysts. Cheirolepidiaceae producing *Classopollis* must have lived close to the coast. Major proportions of this pollen point to a dry and hot climate.

The Aptian and Albian deposits enclose far less *Classopollis* pollen with the role of spores being increased, though. The comparison with the coeval complexes (Aptian–Albian) of western Africa and Brazil shows *Classopollis* content in the latter to be far superior to that in Algeria, Tunisia and Sudan. This is likely to be associated with the fact that the northern part of South America and the countries of northern Africa accommodated a relatively moist vegetation belt corresponding to the Equator (Doyle *et al.*, 1982). According to the palaeomagnetic maps, Africa was turned counterclockwise relative to South America (Herngreen and Khlonova, 1981). With such a position of the two mainlands the Equator crossed Africa roughly from the Ivory Coast through the southern Sahara passing beyond its confines somewhat south of the Sinai peninsula. As far as South America is concerned it remained as good as unchanged. On the whole *Classopollis* pollen content is inconstant along the section of the Equator passing through northern Africa. It attains peak values in the Neocomian of Egypt (see above) while in the Aptian and Albian deposits its amount does not exceed 20%.

Thus, in the Mali-Nigerian syneclise situated on the Equator during Cretaceous time *Classopollis* content is drastically changed across the sequence (Trofimov *et al.*, 1969). In the Neocomian (Tazoli-Tigedi strata) spores of ferns (*Cyathidites* sp. up to 50%, *Cicatricosisporites* and *Appendicisporites*, etc.) are obviously predominant but higher in the Aptian (Elraz strata) *Classopollis* pollen reaches 64% in some samples. In the upper part (Eshkar-Farak strata), dominated by the pollen of gymnosperms, *Classopollis* content does not surpass 8% whereas the key role is passed over to the pollen of *Gnetaceapollenites* and *Ephedripites*. The age of this complex is estimated as Albian based on the appearance of *Tricolporopollenites* pollen.

Summing up the review of the Equatorial region floras within South America and Africa in the Early Cretaceous epoch we are entitled to state the following. *Piazopteris* and *Weichselia* (Matoniaceae) (see Fig. 3.15) are most frequently encountered among fern macro-remains, although the ferns are very diverse according to spores. Ancient Pinaceae producing bisaccate pollen are virtually missing. Macro-remains include a wide range of Bennettitales. Regrettably enough, their pollen is not distinctive among gymnosperms. According to macrofossils Czekanowskiaceae and even Ginkgoaceae are considered to be absent. The latter, as is known, are widespread in the subtropical part of the Northern Hemisphere. *Nilssonia* is a rare find. The Equatorial region is characterized by spores and pollen belonging to the genera of uncertain systematic affinity, i.e. *Dicheiropollis* (Neocomian), *Afropollis* (Aptian–Albian), *Elaterosporites*, *Elateropollenites* and *Galeacornea* (Albian). Only *Afropollis* somewhat transgresses the confines of the region in question. *Classopollis* pollen is very broadly disseminated, often in great amounts. To a lesser extent this applies to *Ephedripites* and its relative *Gnetaceaepollenites*, with the latter displaying a great variety of morphological features. *Inaperturopollenites* pollen is common.

3.3.4 Asia

South-eastern Asia includes several moderately rich habitats of the Early Cretaceous whose age cannot at this stage be assessed more accurately. The fullest summary of the results of the relevant research bty Kon'no (1967, 1968, 1975) and Smiley (1970) belongs to Asama *et al.* (1981).

In Thailand, near the border with Malaya (Trang locality), *Gleichenoides giganensis*, *G. panitiens*, *Otozamites giganensis*, *Podozamites pahangensis* and *Frenelopsis* sp. were found.

In Malaya several habitats (Gagay, Maran, Ulu-Endau, Panta) contain remains of the Early Cretaceous plants yielding *Equisetites burchardti*, *Gleichenoides giganensis*, *G. malanensis*, *G. pantiensis*, *G. serratus*. *G. stenopinnula*, *Otozamites giganensis*, *O. malayana*, *O. kondoi*, *Ptilophyllum agobanum*, *Zamites* cf. *buchianus*, *Weltrichia* cf. *whitbiensis*, *Nilssonia* sp., *Podozamites pahangensis*, *Frenelopsis malaiana*, *Sphenolepis kurriana*, *Cupressinocladus acuminifolia* and *Conites spinulosus*. It is probable that the number of species of Gleicheniaceae is less in actual fact because the fragmentariness of remains which served as a basis for the description of fossil species permits no distinct differentiation among the species.

The overall generic composition, despite its poverty, certainly is suggestive of a subtropical or, more likely, a tropical nature of the flora which is indicated by the presence of Bennettitales (*Otozamites, Ptilophyllum, Zamites, Weltrichia*). Occurrence of Gleicheniaceae, and particularly *Frenelopsis*, suggests that these floras refer to the Early Cretaceous vegetation of the Northern Hemisphere because within Australia and New Zealand this genus has so far not been recorded. It is possible that the Early Cretaceous floras of south-east Asia will allow identification of a new province of the Equatorial region after further research is accomplished. A difference between these floras and the age-equivalent floras of northern Africa is the absence of the fern *Weichselia*. It should be pointed out that northern Africa, like Arabia, was separated from south-east Asia by the marine expanses of the Tethys Ocean.

Abundance of red rocks within the composition of the Lower Cretaceous deposits, very small sizes of the leaves of Gleicheniaceae and fairly small leaves of *Podozamites* and Bennettitales, coupled with the presence of *Frenelopsis* and *Cupressinocladus* with their scaly leaves closely clinging to the stem, all indicate a definite dryness of the climate prevalent in the region. Smiley (1970) reminds us that at present Gleicheniaceae in the territory of Malaya commonly grow on sun-exposed surfaces, often on the fringes of tropical forests. He thinks that the Early Cretaceous flora of Thailand and Malaya lived under climates with prolonged dry seasons. In physiognomy the landscape of these regions suggests a present-day savannah with scattered trees (probably *Frenelopsis* and *Cupressinocladus*) interspersed with ferns and low bennettitaleans. Asama (Asama *et al.*, 1981) emphasizes that the leaves of plants assigned to the same genera but growing in the Early Cretaceous of Japan proved far larger which, in our opinion, is accounted for by a moister marine climate of that country.

3.4 Austral (Notal) region

Based on its position relative to the Equator this region corresponds to the Euro-Sinian region in the Northern Hemisphere. However, there is a major difference between the two. While the Euro-Sinian region embraces a broad strip of land in Eurasia and northern America, most of the subtropics of the Southern Hemisphere, i.e. the Austral region, is occupied by the ocean. Therefore, within its confines we may single out only three provinces, i.e. Patagonian, Australian and Indian. Their Early Cretaceous floras have been studied chiefly during the last two decades.

The Early Cretaceous flora of South Africa is so far studied very inadequately although it has obvious features of subtropical climate, which permits its description within the Austral region. At the same time dearth of material does not allow us to identify a separate province there.

Just as for the Triassic and Jurassic periods, we fail to find in the Early Cretaceous any analogues to the vegetation of the moderately warm climate, i.e. analogues of the Siberian–Canadian region of the Northern Hemisphere. In the Antarctic we are familiar with the exposures of the lower Cretaceous with plant remains on Alexander island adjacent to the Antarctic peninsula which is somewhat advanced northwards. These remains belong to the plants common for the subtropical area of the Southern Hemisphere. The greater part of the Antarctic is hidden beneath the present glacial shield. The common features of the Austral region are the abundance of *Podocarpidium* remains (notably, trisaccate pollen of *Microcachryidites* and *Trisaccites*), Araucariaceae and not infrequently of Cheirolepidiaceae, the paucity of Ginkgoales and *Nilssonia*, and the absence of Pinaceae and Czekanowskiaceae. Many ferns, particularly certain Schizeaceae producing the spores of *Cicatricosisporites*, *Appendicisporites* and *Pilosisporites* as well as Gleicheniaceae, are in common with those from the Northern Hemisphere. The generic composition of cycads, especially Bennettitales, was largely identical for both the Northern and Southern Hemispheres, which is indicative of their unimpeded migration across the Equatorial zone.

3.4.1 Patagonian province

Habitats with plant remains belonging to this province are located *par excellence* in Patagonia. The largest of these are situated in the Santa Cruz province of Argentina with only a few to the west of it in the territory of Chile. Less enriched habitats are also reported from the northern part of Argentina (Chubut province). During the last 20 years or so they have been investigated mainly by Archangelsky (1963, 1965, 1966, 1967, 1976; Archangelsky and Baldoni, 1972) and Menendez (1966) who were mostly preoccupied with outward morphology and also the structure of the epidermis. The composition of the spore and pollen complexes was also established.

The province of Santa Cruz has the broad range of the Baquero suite divided into two subsuites. The lower one is composed of dark, primarily grey terrigenous rocks containing mummified remains of plants. The colour of the upper subsuite is pale from white to yellow; no cuticle was preserved on the plant remains found in the upper subsuite.

Ferns in the Early Cretaceous flora of the Patagonian province are represented relatively poorly. Local finds include occasional imprints of *Hausmannia* and *Gleichenites* as well as diverse *Sphenopteris* and *Cladophlebis*. Among *Sphenopteris* we should note *Sphenopteris psilotoides* which is similar to *Onychiopsis psilotoides* in the structure of sterile pinnules. Another *Sphenopteris* may be identified with the sterile pinnules of *Ruffordia* cf. *goeppertii*. Both these species are typical for the Neocomian of the Northern Hemisphere.

Gymnosperms are rather varied being of uncertain systematic affinity and approximated by Archangelsky to pteridosperms or cycads. He described new genera and species, i.e. *Mesodescolea plicata*, *Mesosingeria coriacea*, *M. herbstii*, *M. mucronata*, *M. striata*, *Rufflorinia sierra*, *Ticoa harrisii*, *T. lamellata*, *T. magallanica* and *T. magnipinnulata*. The dissected leaves of these gymnosperms resemble the pinnules of *Cladophlebis* and *Sphenopteris* but the epidermal structure testifies to the affinity of these genera with the gymnosperms. Outside Patagonia representatives of these genera have not yet been found. We should also note a find of an earlier undescribed fructification of *Ktalenia circularis* which is closer to *Caytonia* than to anything else and is commonly uncovered from the strata with *Ticoa*. Along with these endemic genera a representative of the widespread genus *Pachypteris* was encountered (local species of *P. elegans*).

Bennettitales are rather widely distributed in Patagonia. They are represented by *Dictyozamites areolatus*, *D. crassinervis*, *D. latifolius*, *D. minusculus*, *Otozamites archangelsky*, *O. grandis*, *O. parviauriculata*, *O. sanctaecrucis*, *O. waltonii*, *Zamites decurrens*, *Ptilophyllum angustus*, *P. antarcticum*, *P. hislopi*, *P. longipinnatum*, *Pterophyllum trichomatosum*, *Cycadolepis coriacea*, *C. involuta*, *C. lanceolata*, *C. oblongata*, *Williamsonia bulbiformis* and *W. umbonata*.

A striking feature is the specific variety of *Dictyophyllum*, *Otozamites* and, in part, *Ptilophyllum*, along with the rare occurrence of *Pterophyllum*. In comparison with Bennettitales, cycads are far more impoverished. They include only two species of

Pseudoctenis, i.e. *P. crassa* and *P. dentata*. Archangelsky points out that the genera *Ticoa* and *Mesodescolea* identified by him possess stomata which are very similar to those of contemporary Cycadales. A typical feature of the Santa Cruz flora inherent also in other floras of the Austral region is absence (or almost complete absence) of the genus *Nilssonia*, which is widespread in the Early Cretaceous of the Northern Hemisphere.

Ginkgoales are represented exclusively by the genera *Ginkgoites* (*G. tigrensis*) and *Karkenia* with the species *K. incurva*. It should be noted here that this name belongs to the described fructification having a central axis with attached short processes and oval closely adhering ovules with micropyles directed towards the axis. Structurally they are quite distinct from the fructifications of the present-day Ginkgoaceae carrying orthotropic ovules furnished with a so-called collar. Along with *Karkenia*, dissected leaves were found which are indistinguishable from the leaves of Ginkgoales. These fructifications are presumed to have belonged to some initial form possessing *Ginkgo*-like leaves whence the contemporary genus *Ginkgo* descended. However, recognizing this we ought to decline to refer the Mesozoic *Ginkgo*-like leaves directly to the genus *Ginkgo* retaining them within the form-genus *Ginkgoites* because it is not known to which of the two above-reviewed genera (*Karkenia* or *Ginkgoites*) they may belong.

Conifers encountered in the Baquero suite are quite diverse, comprising *Araucarites baqueroensis*, *A. minimus* (cones), *Brachyphyllum brettii*, *B. irregularis*, *B. mirandai*, *B. mucronatum* (shoots) and *Podocarpus dubius* as well as *Apterocladus lanceolatus* and *Trisacoladus tigrensis* (shoots and cones) approximated to Podocarpaceae. *Tomaxiella biformis* and *T. degiustoi* are also present.

The shoots of *Tomaxiella biformis* carried female and male cones (Archangelsky, 1968) whence *Classopollis* pollen was extracted. The scales of the female cones are very similar to those of *Cheirolepidium*. All this suggests assignment of the genus *Tomaxiella* to the family Cheirolepidiaceae. The latter apparently embraces the shoots of the form-genus *Brachyphyllum*. The rock with *Brachyphyllum mirandai* shoots also yielded male cones whence·pollen was extracted close to *Classopollis*. It is not excluded that certain shoots described as *Brachyphyllum* may actually belong to *Araucaria* whose cones, as has already been mentioned, were found in the Baquero suite in Patagonia.

In Chile's Springhill suite extending also into Patagonia, the same plants were uncovered as from the Baquero suite, but their preservation proved far poorer (Archangelsky, 1976). The presence of three species of *Ptilophyllum* is characteristic. A very scarce flora containing some species familiar from the Baquero suite was found north of Patagonia in the Chubut province (Argentina).

The South Shetland islands are located at the tip of South America within the confines of the Antarctic. One of these (Livingstone) has deposits referred to the Lower Cretaceous where the following are recorded: *Cladophlebis* cf. *patagonica*, *Gleichenites sanmartinii*, *Taeniopteris* sp., *Ptilophyllum* sp, *Mesodescolea* sp., *Ticoa* (?) sp., *Araucarites* cf. *baqueroensis*, *Ruffordia* (?) and *Williamsonia* sp. Specifically unidentifiable fragments of conifers apparently belong to Podocarpaceae. In older marine deposits the fauna of the Tithonian–Barremian was found (Hernandez, Pedro and Azcarate, 1971).

Alexander Island, situated in the immediate vicinity of the west of the Antarctic peninsula (Jefferson, 1982), exhibited tufogenic deposits of Lower Cretaceous hosting remains of silicified trunks located at various stratigraphic levels. Some of them were buried when alive. Yearly increments of annual rings of the woods are substantial and different from those of the woods of present-day high-altitude forests. They have a closer similarity to the annual rings of woods from warm-climate regions with a prolonged vegetation period. The same deposits yielded leaf imprints which, judging from the oral communication of N. F. Hughes, belong to *Phlebopteris*, *Cladophlebis*, *Pterophyllum*, *Otozamites*, *Dictyozamites*, *Ptilophyllum*, *Elatocladus conferta*, *Araucarites cutchensis*, *Brachyphyllum*, *Pagiophyllum*, etc. Looking at this listing of tentative definitions we find major similarity between the floras of Patagonia (Baquero suite) and Alexander Island.

The palynologic characterization of the Lower Cretaceous of the southern part of Argentina and Chile is given in a miscellany of works by S. Archangelsky, A. Baldoni, V. Volkheimer, J. Seilor, D. Hammer and A. Salas. Our survey will be started in the north with Neuken province (Argentina) situated on the borderline between the Equatorial and Austral regions. The sequence of the Lower Cretaceous rocks is represented by the suites Mulichinko (Berriasian–Middle Valanginian), Agrio (Hauterivian–Barremian) and Khutrin (Aptian–Albian). Ages of the suites chiefly made up of marine sediments are established according to ammonites.

The palynofloras of all suites are dominated by *Classopollis* pollen with content in some samples running up to 90% (Archangelsky, 1980). Among the spores peculiar to the Lower Cretaceous we should note *Taurocusporites segmentatus*, *Interulobites triangularis*, *Densoisporites velatus*, *Trilobosporites apiverrucatus* and *Balmeiopsis limbatus*. The pollen of conifers includes *Callialasporites*, *Microcachryidites* belonging to the Podocarpaceae, *Ephedripites* and others in addition to *Classopollis*. From the Barremian onwards rare pollen of the angiosperms is encountered (*Clavatipollenites* and *Liliacidites*) while the Khutrin suite sees the appearance of tricolpate pollen of *Stephanocolpites*.

More to the south, in the province of Chubut of the northern part of the Santa Cruz province (Archangelsky and Seiler, 1978; Archangelsky and Baldoni, 1981) the Lower Cretaceous was encountered in boreholes. The examination of spores and pollen using palynologic data permitted recognition of three formations (A, B, C). The A formation (Berriasian–Valanginian) is dominated by *Classopollis* and *Callialasporites* pollen occurring in approximately equal proportions with the amount of each fluctuating between 30 and 40%. They are accompanied by saccate pollen and spores. *Cicatricosisporites australiensis* was reported as part of the latter.

In the lower part of the middle formation (B) encompassing the whole of the Hauterivian, Barremian and Lower Aptian, *Classopollis* content declines to 25–30% whereas the amount of the pollen of *Cyclusphaera* and *Balmeiopsis* increases. *Balmeiopsis* pollen is likely to belong to the Araucariaceae. In the middle and upper parts of formation B *Classopollis* pollen content again rises to 60–75%. In formation C (Aptian) *Classopollis* is abundant (up to 80%). This level witnesses emergence of the monocolpate pollen of the first angiosperms (*Clavatipollenites* and *Liliacidites*).

As was mentioned earlier, a rich leaf flora was uncovered in the Baquero suite in the Santa Cruz province slightly south of the boreholes described above. This suite yielded spores and pollen (Archangelsky and Gamerro, 1966; Archangelsky, 1980) in which quantitatively *Classopollis* pollen retained domination. At the same time the systematic diversity of spores increased; they are represented by the genera *Densoisporites*, *Cyatheacidites*, *Aequitriradites*, *Polypodiaceoisporites*, *Trilobosporites*, *Foraminisporites*, *Taurocusporites*, *Cicatricosisporites* (6 species), *Cyathidites*, *Osmundacites*, *Contignisporites*, *Concavissimisporites*, etc.

Gymnospermous pollen includes *Podocarpidites*, *Araucariacites*, *Trisaccites*, *Ginkgo-cycadophytus* and *Caytonipollenites* in addition to the prevalent *Classopollis*. Mono-colpate pollen of angiosperms (*Clavatipollenites hughesii*) was encountered. Its presence as well as the existence of such species as *Aequitriradites baculatus* and *Taurocusporites segmentatus* implies a Barremian or Barremian–Aptian age for the Baquero suite. Its upper boundary does not appear to include the Albian due to the lack of tricolpate pollen and of macro-remains of angiosperms. It is typical that the composition of spore–pollen complexes includes the pollen of Podocarpaceae (*Callialasporites*, *Podocarpidites*, *Trisaccites*), Araucariaceae (*Araucariacites*) and Cheirolepidiaceae (*Classopollis*). The presence of the same families was established according to plant macro-remains. The pollen of Bennettitales appears to be represented by the genus *Ginkgocycadophytus*.

The southernmost area wherein the Lower Cretaceous deposits are widespread is Tierra del Fuego and the adjoining northern coast of the Magellan strait. The palyno-complex studied from the Springhill suite (Archangelsky, 1976; Baldoni and Archangelsky, 1983) correlates with those from formation A in the boreholes situated in the southern Chubut and in the northern part of the Santa Cruz province. Its age is estimated as Berriasian–Valanginian.

The palynocomplex from the Springhill suite is characterized by prevalent pollen of Podocarpaceae, i.e. *Callialasporites* (30–50%), spores (30–50%) and *Classopollis* pollen (15–17%). The spores featuring major systematic diversity are *Cyathidites*, *Lygodiumsporites*, *Cicatricosisporites* (*C. australiensis* (Cookson) Pot.), *Converruco-sisporites*, *Osmundacites*, *Klukisporites*, *Trilobosporites*, *Contignisporites*, *Densoisporites*, *Aequitriradites*, etc. The earlier mentioned pollen of *Callialasporites* and *Classopollis* is concomitant with *Microcachryidites*, *Vitreisporites*, *Araucariacites*, etc.

Kotova (1983) examined the cores from borehole 511 from the Deep Sea Drilling Project located on the underwater plateau west of the Falkland Islands (Malvinian). In palynocomplex II characterizing the interval from 518 to 555 m, *Classopollis* pollen is prevalent (48–51%) with *Vitreisporites pallidus* pollen being plentiful (3–26%) and *Inaperturopollenites limbatus* and *Cyclusphaera* pollens present. Spores are not numerous just as in the stratigraphically lower complex (interval 555 to 632 m) associated with the Tithonian. Rare *Cicatricosisporites*, *Polypodiaceoisporites* and *Taurocusporites* have been observed.

Upward in the sequence within the interval 495 to 518 m (palynocomplex III) *Clavatipollenites* pollen appears coupled with rare *Tricolpites* and the rising variety of the spores of the genus *Cicatricosisporites*. At the same time *Classopollis* pollen content is declining, fluctuating within a rather wide range (17.5–55%). This palynocomplex is fairly similar to the one identified from the Baquero suite but the presence in it of a number of Albian forms, notably *Crybelosporites* aff. *striatus* and *Laevigatosporites ovatus* known from the Lower Albian of Australia makes I. Z. Kotova assign it to the Albian.

Comparison of the Early Cretaceous palynofloras from the Neuken province in the north to Tierra del Fuego in the south reveals that *Classopollis* content is declining in this direction. While in Neuken in the Muchilinko suite (Berriasian–Middle Valanginian) the *Classopollis* pollen content normally exceeds 50%, in Chubut and the northern part of the Santa Cruz province it is somewhat reduced (30–40%), and in Tierra del Fuego it comes down to 15–17%. This change is certainly related to the cooling of climate as one moves over to the more southern regions, i.e. climatic zonality. The same pattern has been ascertained for the Northern Hemisphere (Asia), the only difference being that in the Southern Hemisphere no warm-temperate climatic belt is evolving whence *Classopollis* disappears from the palynocomplexes of the Early Cretaceous. Investigation showed the Lower Cretaceous of the northern part of Siberia to be virtually devoid of *Classopollis*. Distribution of the pollen *Classopollis* throughout the sequence of the Early Cretaceous deposits in South America reveals that it keeps on changing although no regularity for all sequences has so far been discovered.

Comparison of the systematic composition of the Early Cretaceous floras of the Australian and Patagonian provinces even at the present state of the art elucidates tangible differences which may be said to be of critical importance. The Patagonian flora relative to its macrofossils is slightly poorer in ferns. Notably it is deprived of the easily identifiable genus *Coniopteris* and the endemic genera *Aculea*, *Alamatus*, *Amanda* so far not known outside Australia. But then the flora of Patagonia is richer in gymnosperms of uncertain systematic affinity (genera *Mesodescolea*, *Mesosingeria*, *Ruflorinia*, *Ticoa*).

The appreciable difference of the Patagonian flora is its abundance of varied Bennettitales (*Dictyozamites*, *Otozamites*, *Ptilophyllum*) while the Australian flora contains largely *Ptilophyllum* represented by small-sized leaf species and rare *Otozamites*. Ginkgoaceae, both in the Australian and Patagonian provinces, include the genus *Ginkgoites*. *Baiera* is reported from the former province.

The composition of conifers is more or less identical. It includes Cheirolepidiaceae, Podocarpaceae and Araucariaceae, as is indicated both by macrofossils and palynologic data. However, an important difference is the *Classopollis* pollen content. In the Lower Cretaceous of Patagonia it is far greater (in certain samples as high as 50%) whereas in Australia it rarely attains 5%; this must be due to the drier climate of Patagonia.

3.4.2 Australian province

The best-studied rich Early Cretaceous floras of Australia are related to the Victoria province. Two basins, Otway and Gippsland, made up of coal-bearing deposits are situated within its confines, south-west and south-east of Melbourne, respectively. Originally these two were considered to be Jurassic but later palynologic research (Dettmann, 1963) proved their Early Cretaceous age. Douglas, who described plant macro-remains in two monographs and a number of articles (1969, 1970, 1973; Douglas and Williams, 1982), confirmed this estimate.

Relying on these works below we shall give a general characterization of the Early Cretaceous flora of Victoria age-wise embracing the whole of the Early Cretaceous. As we shall see later its composition changes but little upwards in the sequence pointing to stable conditions and absence of any substantial gaps in sedimentation of the continental formation containing plant remains. The sediments reach a thickness of 4000 m and are composed of tufogenic sandstones and argillites with members of quartz sandstones and conglomerates, coal beds and horizons of fossil soils. Their deposition represents a littoral lowland; in all likelihood we are dealing with the deposits of the upper part of a delta.

Spore plants are represented by remains of liverworts (*Hepaticites*), lycopods (*Lycopodites victoriae*), Equisetales (*Equisetites wonthaggiensis*, *Phyllotheca* sp.) and numerous ferns comprising *Coniopteris* aff. *hymenophylloides*, *C. fructiformis*, *C. nanopinnata*, *Adiantites lindsayoides*, *A. dispersus* (according to the classification effective in the USSR these should be referred to the genus *Adiantopteris*), *Hausmannia bulbo-formis*, *Stachypteris* (?) sp., *Cladophlebis australis*, *Sphenopteris warragullensis*, *S. mccoyi* and *S. lacerata*. Of special interest are the representatives of the three new genera identified by Douglas so far devoid of analogues in the floras of other continents. These are *Aculea bifida*, *Alamatus biforius* and *Amanda floribunda*. The fronds of the ferns belonging to these genera are deeply dissected into thin ribbon-like many-fold dichotomizing segments. In *Aculea bifida* the sori are situated on both sides of the middle vein of the fertile segments. Sterile and fertile features in *Amanda floribunda* are different from each other. The sterile segments resemble the pinnae of '*Asplenium*' distributed in the Cretaceous of the Northern Hemisphere while the fertile segments are reduced, possessing irregular pecopteridal pinnules with sporangia on the abaxial surface. Many sphenopteroid leaves remained undefined.

Palynologic research (Dettmann, 1963) shows that among the ferns there were Osmundaceae, Schizeaceae, Cyathaceae and Gleicheniaceae. The imprints of leaves belonging to the first three families must have been described by Douglas under the form-generic names of *Cladophlebis* and *Sphenopteris*. Osmundaceae are also represented by silicified trunks reaching the length of several metres uncovered from many localities. This testifies to the presence of arborescent fern forms within Early Cretaceous vegetation of Australia. Pteridosperms yielded *Xylopteris difformis*, *Stenopteris* (?) *williamsonis*, *Phyllopteroides dentata* and *P. lanceolata* as well as *Thinnfeldia hastata, T.* cf. *chunakalensis* and *Pachypteris austropapillosa*. Among these notice should be taken of the representatives of the genera *Xylopteris* and *Phyllopteroides*. These are peculiar to the Triassic deposits of the Southern Hemisphere and may be regarded as relicts. Within pteridosperms Douglas incorporates taeniate leaves with bifurcate lateral veins attributed to *Taeniopteris daintreei*.

Above, when describing the Jurassic floras of Australia and the entire Southern Hemisphere, it was noted that after the works of Douglas the lower strata of the Cretaceous in the Victoria province furnished (Drinnan and Chambers, 1985) female strobili *Carnoconites cranwellii* and microstrobili *Sahnia laxiphora*, assigned to the order Pentoxylales whose representatives have been reported only from the Southern Hemisphere. The researchers describing these finds conjectured that the leaves of *Taeniopteris daintreei* belonged to these plants having the above-mentioned fructifications typical for Pentoxylales.

Douglas indicates numerous bennettitaleans such as *Ptilophyllum* (*P. elongatum, P. castertonensis, P. spinosum, P. cantherifera, P. boolensis, P. fasciatum, P. irregulare, P. gladiatum*). The number of new species is likely to have been exaggerated but their frequent occurrence and diversity are doubtless typical for the flora of Victoria. Small size of leaves is another conspicuous character. Some of the *Ptilophyllum* species described, judging from the illustrations supplied in the work of Douglas (1969), should be transferred to the genus *Otozamites* (e.g. *Ptilophyllum irregulare* having a well-developed auricle). The Early Cretaceous flora of Victoria has so far failed to furnish undoubted cycads of which, as is known, the widest-occurring Mesozoic form was *Nilssonia*.

Ginkgoaceae are represented by *Ginkgoites australis* and *Baiera delicata*. The earlier reports of the presence of the genera *Phoenicopsis* and *Czekanowskia* were not confirmed by Douglas's investigations.

The description of conifers whose remains are prevalent in many localities has not yet been published by Douglas but from preliminary data (Douglas and Williams, 1982) they are represented chiefly by Podocarpaceae and Araucariaceae. Pinaceae are lacking which is characteristic of both this and the floras of the entire Southern Hemisphere. Frequent finds of fossil tree-trunks imply that conifers were made up of woody forms. Presence of annual rings is indicative of a seasonal temperate-warm climate. According to Dettmann (1963) conifers are represented by Araucariaceae (*Araucarites*) and monosaccate (*Tsugaepollenites*) and bisaccate (*Podocarpidites*) forms as well as by a certain amount of *Classopollis* pollen; the content of the latter fluctuates in per cent from fractions up to 4%, surpassing this figure very rarely.

Remains of angiosperms appearing only in the upper part of the coal-bearing formation (zone D) are determined as Angiosperm sp.A., *Hydrocotylophyllum lusitanicum, Hemitrapa* (?) sp. and *Lappacarpus austata*. The two latter forms are reproductive organs, probably belonging to aquatic plants. Angiosperm sp.A. is represented by a fragment of an entire lancet-shaped leaf with pinnate venation. *Hydrocotylophyllum lusitanicum* resembles a small fern leaf of *Hausmannia* but its venation (digital–pinnate) is different, indicating its affinity to angiosperms.

Douglas identified four zones in the continental formation going by the data on the distribution of plant macro-remains (Douglas, 1973, p. 171). The lower zone A contains a moderately rich complex represented mainly by horse-tails, ferns (*Cladophlebis australis, Coniopteris* aff. *hymenophylloides*) and pteridosperms, and several species of *Ptilophyllum* are also present in the upper parts.

The zone B is typified by *Taeniopteris daintreei* and various *Ptilophyllum* coupled with *Ctenis coronata*. In zone C, just as in zone B, there are many *Taeniopteris daintreei* but their leaves become smaller. Specific variety and frequency of occurrence of *Ptilophyllum* is obviously less. *Ginkgoites australis* becomes a typical plant of this zone. Such ferns as *Aculea bifida* and *Adiantites lindsayoides* are found in both zones (B and C).

The upper zone D sees the emergence of the remains of angiosperms with *Pterophyllum* becoming rare. The zone is the only one to include the ferns *Alamatus bifarius, Amanda floribunda* and *Adiantites dispersus*. Such frequently encountered ferns as *Cladophlebis australis* and *Coniopteris* aff. *hymenophylloides* persist throughout all the zones. Remains of conifers are disseminated through the entire continental formation but their number shows a slight increase toward its upper part.

Douglas dated the zone A as intermediate between the Jurassic and Early Cretaceous, zone B as the greater part of the Neocomian and, finally, zone C wherein angiosperms appear as Albian.

In Queensland carbonaceous deposits of the Lower Cretaceous (Burrum suite) and the underlying Marybarrow suite yielded *Gleichenites gleichenoides, Sphenopteris erecta, Pagiophyllum jemmetti* and *Podozamites* cf. *lanceolatus*. A greater diversity was displayed by spores of the genera *Dictyophillidites, Osmundacites, Ceratosporites, Klukisporites, Lycopodiacidites, Lycopodiumsporites, Cicatricosisporites, Gleicheniidites, Appendicisporites, Pilosisporites, Foraminisporis, Contignisporites, Aeqitriradites, Trilobosporites* and others. Pollen proved rather rare and less varied, and was manifested mainly in *Classopollis classoides* (Hill, Playford and Woods, 1968). The last area of Australia where plant remains were recorded in Cretaceous sediments is the Canning basin located north of Perth. *Ruffordia, Hausmannia, Otozamites, Ptilophyllum* and *Brachyphyllum* were found here.

Summing up we see that the composition of the Early Cretaceous flora of Australia proves rather homogeneous. Abundance and variety of ferns is conspicuous evidence for a moist climate.

Other most distinctive features of the Australian province as against other provinces of the Austral (Notal) region include the presence of the so far endemic ferns described by Douglas as *Aculea, Alamatus* and *Amanda*, the abundance and diversity of *Ptilophyllum* represented by original small-sized leaves, and the almost total absence of other Bennettitales. Low proportions of *Classopollis* pollen appear to be also characteristic indicating restricted distribution of Cheirolepidiaceae and a moist climate.

Megastrobili of *Araucarites* and *Podocarpus* (Mildenhall and Johnston, 1971; Mildenhall, 1976) have been discovered from the Lower Cretaceous of New Zealand.

3.4.2.1 SOUTH AFRICA

Within South Africa plant remains are recorded from the Uitenhage suite situated at the base of a Lower Cretaceous sequence of the extreme south of this continent as well as from marine shallow-water deposits of Pondoland and Zululand somewhat to the north (Du Toit, 1953). The richer flora from the Uitenhage suite is represented by the genera *Onychiopsis*, *Dictyozamites*, *Cycadolepis*, *Nilssonia*, *Pseudoctenis*, *Araucarites*, *Brachyphyllum* and *Taxites*. The same, but more impoverished composition is encountered in the floras of Pondoland and Zululand.

3.4.3. Indian province

In describing the Jurassic floras of India their similarity to the Early Cretaceous vegetation was mentioned and an impossibility to distinguish between them was indicated in a number of cases, at least with the present state of knowledge. It may be maintained that transition between the floras of this age was gradual and, therefore, no major climatic changes at the Jurassic–Cretaceous boundary occurred.

One of the most interesting sequences of Upper Jurassic–Lower Cretaceous rocks is situated on Kutch peninsula in the north-eastern corner of India. A thick Umia formation is well developed here (up to 1000 m) containing late Tithonian ammonites in the lower part. Somewhat higher strata with *Trigonia* are related to the Valanginian (Venkatachala, 1969) and higher strata contain Aptian ammonites (*Austarliceras*, *Tropaeum*, etc.). The upper third of the Umia series (Bhuj strata) hosts plant remains (Sitholey, 1954). They are represented by *Pachypteris indica*, various species of *Ptilophyllum*, *Williamsonia blanfordii*, *Taeniopteris vittata*, *Araucarites* spp., *Brachyphyllum expansum* and *Elatocladus tennerrima*. Presence of *Pachypteris* implies proximity of a coastline, corroborated by presence in the Umia series of marine deposits.

Another area accommodating Lower Cretaceous deposits containing plant remains is the Bansa and Sehora regions (Madhya Pradesh state) situated in the north of India. The Jabbalpur series is well developed here composed of terrigenous continental formations lying on the Precambrian base. The ferns *Gleichenia rewahensis*, *Cladophlebis* sp., *Onychiopsis paradoxus*, *O. psilotoides*, *Matonidium goeppertii* and *Weichselia reticulata* were found here.

Gymnosperms exhibited several species as follows: *Cycadopteris* and *Ptilophyllum*, *Williamsonia indica* and many conifers such as *Araucarites*, *Brachyphyllum*, *Elatocladus* and *Pagiophyllum*. Occasionally *Ginkgo* and *Podozamites* are encountered (Sitholey, 1954; Bose and Sukh-Dev, 1959, 1961; Bose, 1960). The basic background is constituted by a variety of *Ptilophyllum* as well as by conifers with scaly or awl-shaped leaves.

It is important to note the presence of such species as *Onychiopsis psilotoides* (*O. paradoxa* is, most probably, its synonym since it reveals no marked differences) and *Weichselia reticulata*. They are normally Early Cretaceous and only in northern Africa (Equatorial region) do they appear in the Middle or Late Jurassic. However, total absence of these species just as, incidentally, of other representatives of these genera in the Jurassic of the Raj Mahal highlands, is most likely indicative of the Early Cretaceous age of these plants.

The flora known from the basin of the Godovari river (Andhra Pradesh state) located in the central part of India is also of Early Cretaceous age. Previously the Lower Cretaceous deposits now singled out as the Gangapur suite were not segregated from the Lower Jurassic sediments with remains of fishes and reptiles, referred to as the Kota suite. In the Gangapur suite such forms typical of the Early Cretaceous as *Cicatricosisporites* and *Aequitriradites* were found, with other forms (Rao, Ramanujam and Varma, 1983) in common with those from the Bhuj strata, resting on the upper part of the Umia series on Kutch peninsula (Venkatachala, 1969) and being dated as Early Cretaceous.

Both palynologic complexes exhibit major similarities to those from the countries of the Southern Hemisphere. Notably they are dominated by such genera as *Microcachryidites*, *Callialasporites*, *Podocarpidites*, etc. well known from the Early Cretaceous palynocomplexes of Australia, Argentina and South Africa.

✿ 4 ✿

Late Cretaceous floras

In the Late Cretaceous, phytogeographic zonation is changed only in details in comparison with the Early Cretaceous period. However, the names of many of the provinces are changed. Just as in the Early Cretaceous, the Siberian–Canadian, Euro-Sinian, Equatorial and Austral (Notal) regions are identified. In an earlier work (Vakhrameev *et al.*, 1970) the European–Turanian region was recognized for the Late Cretaceous because of a total absence of data on China, although even then it was possible to assume that China together with Europe, Kazakhstan and Middle Asia constituted, as was the case before, the same region. Emerging palaeobotanical data on southern China (Hsü, 1983) confirm its affinity to the subtropical belt climate, and this also appears to hold for the central regions of the USSR. Therefore, I now use the same name both for the Early and Late Cretaceous, i.e. Euro-Sinian region.

Let us consider provincial division of individual regions. The Siberian–Canadian region is divided into the South Urals (existing only in the Late Albian and Cenomanian), East Siberian, Okhotsk–Chukotsk and Canadian–Alaskan provinces. The change of name of the latter province (from Canadian to Canadian–Alaskan) was caused by the existence of a number of Late Cretaceous floral localities in Alaska, whereas Early Cretaceous vegetation had not been recorded therefrom.

The Euro-Sinian region is split up into the European, Potomac, Middle Asia, Japanese and South-China provinces. The Euro-Sinian region also includes the Greenland province because its Late Cretaceous flora is more subtropical than warm-temperate. This is due to the more southerly position of Greenland in the Late Cretaceous, when it was situated in the subtropical belt of the Northern Hemisphere, at its northern boundary. The Equatorial region is typified merely by spores and pollen and does not yet lend itself to division.

In the Austral (Notal) region it is possible to identify the Patagonian, Australian and Antarctic provinces. The latter emerges only from the Senonian indicating the appearance of a belt of moderate-warm climate in the Southern Hemisphere. Later, in the Palaeogene, it becomes an independent region. Before the Senonian the Antarctic province cannot be outlined according to the composition of the vegetation

180

although it stands out as far as foraminifera are concerned. There is no basic contradiction here since ecologic requirements of various groups of organisms differ. The ecologic situation is also different with relation to marine protozoans, on the one hand, and land plants, on the other. It is not excluded that the central parts of the Antarctic might have witnessed the existence of the Jurassic and Early Cretaceous plants that lived under conditions of warm-temperate climate but that now their remains are buried beneath the thick glacial shield.

The Indian province may be identified only with certain reservations because the Early Cretaceous is represented here by thick volcanic (trappean) formations. Only one point in the east of India allowed a study of a Senonian spore–pollen complex from littoral deposits.

The wide distribution of angiosperms at the very end of the Early Cretaceous epoch effected a marked transformation of the physiognomy of the floras throughout the world. The main criterion for phytogeographic zonation shifts to the composition of angiosperms while in the Early Cretaceous emphasis was laid on the composition of the ferns and gymnosperms. Due to the emergence of numerous angiosperms already in the floras of the Late Albian we shall include in our review the flora of this period, too, although according to the generally accepted scale of geological time the Late Albian refers to the Early Cretaceous.

While in the Euro-Sinian region situated south of the Siberian–Canadian region the transition from Early to Late Cretaceous is concomitant with a drastic reduction in conifers, and forest vegetation composition comes to be dominated by the dicotyledons, the confines of Siberia and Canada at this boundary emerge with evolving coniferous–deciduous forests wherein conifers might have been prevalent numerically rather than in systematic diversity. This gave Krassilov (1972a) reason to believe that at the outset of the Late Cretaceous the angiosperms had not yet become a dominant group of plants; this is contrary to the accepted views. This can hardly be subscribed to (Vakhrameev, 1981b) because the conclusions of V. A. Krassilov are founded only on the vegetation of the warm-temperate belt while on the whole the systematic diversity of angiosperms in the Cenomanian and anyhow in the Turonian far exceeds that of conifers. Even today the belt of temperate and cold-temperate climate occupied by taiga continues to be dominated by woody conifers numerically belonging largely to the Pinaceae. However, hardly anybody will rely on this to debate the fact that angiosperms are currently, just as in the second half of the Late Cretaceous, the most varied group both systematically and from the point of view of adaptability to changing ecologic environments.

4.1 Siberian–Canadian region

This region, situated in the belt of warm-temperate climate, is typified by mixed coniferous–deciduous forests with prevalence among the angiosperms of leaf-shedding species suggestive of climatic seasonality. Gigantic plataniferous leaves are widely distributed within its limits (*Platanus, Pseudoprotophyllum, Paraprotophyllum*) as well as morphologically miscellaneous *Trochodendroides* accompanied by the fructifications *Trochodendrocarpus*. *Macclintockia* is common in the second half of the Late Cretaceous. The prevailing leaves of dicotyledons possessed a toothed margin which is typical of a region of temperate or warm-temperate climate.

The generic composition of angiosperms in the Late Cretaceous of the Siberian–Canadian region has been subjected to revision of certain definitions effected by the palaeobotanists who had originally described these floras. These are mainly concerned with the Platanaceae described within this province as varied *Protophyllum* and *Aspidiophyllum* and now transferred into the genus *Pseudoprotophyllum*. The above-mentioned genera (Vakhrameev, 1976a) established for the Upper Cretaceous of the USA (Potomac province) are substantially different from the overwhelming majority of leaf imprints from the Upper Cretaceous of northern Asia equated with these genera. This revision revealed notably that the genuine *Pseudoprotophyllum* first described by Hollick (1930) from the upper Cretaceous of Alaska does not exist outside the Siberian–Canadian region and serves as one of its indicators. Herman (1984a), having studied the leaves of *Protophyllum*, singled out therefrom a new genus *Paraprotophyllum* differing in the absence of the shield-like base, a more elongated leaf blade and numerous suprabasal veins of the second order. The majority of angiosperms were represented by deciduous woody or bushy plants.

Pinaceae are numerous among conifers (*Pityostrobus, Pityospermum*) with Taxodiaceae being especially plentiful. The latter are dominated by *Sequoia* and *'Cephalotaxopsis'* (= *Taxites*). From the second half of the Late Cretaceous onwards, *Glyptostrobus* and *Metasequoia* emerge in great numbers. *Libocedrus* (*Andrivettia*) and *Cupressinocladus* (*Thujites*) assigned to the family Cupressaceae are also widely disseminated, although to a lesser degree. *Protophyllocladus* occurs from the Turonian and rarely from the Cenomanian. Among the conifers *Metasequoia* seasonally shed its shoots. *'Cephalotaxopsis'* might very well also have been a deciduous conifer, since the bedding planes are not infrequently densely strewn with shoot impressions belonging to this genus. The sequoias must have belonged, just as their contemporary relatives, to the evergreen plants.

While at the beginning of the Late Cretaceous the ferns were still numerous and diverse (*Anemia, Birisia, Onychiopsis* etc.) as is evidenced by the presence of imprints in almost every significant locality, the end of this period witnessed their considerable reduction. The most frequently occurring fern is a very typical species tentatively referred by O. Heer (*A. dicksonianum* Heer) to the genus *Asplenium* as far back as the last century, although only its sterile shoots were found. Recently, V. A. Krassilov uncovered fertile pinnules along with the sterile remains of this fern from the upper Cretaceous of Sakhalin. These pinnules resemble the fertile pinnae of *Anemia*. Krassilov suggested that subsequently the remains of this fern (both fertile and sterile) be assigned to the above genus. Krassilov's supposition seems to be well-based, but until fertile leaves of better preservation are found we shall stick to the generic name 'Asplenium' placing it in inverted commas. Another thing is that in a good number of works it is described as Asplenium dicksonianum Heer. Finds of the aquatic fern Azimia and Salvinia were reported; the former is described from the fossil remains of sporocarps from the Sym suite on the river Taz.

Onoclea (Aspidiaceae) appears at the end of the Cretaceous. Palynologic analysis reveals abundance of the fabiform spores of the Polypodiaceae. The areas adjacent to the Pacific Ocean retain many relicts represented by genera widespread in the Early Cretaceous but vanishing from the remainder of the Siberian—Canadian region with the onset of the Late Cretaceous epoch (Vakhrameev, 1981b). These comprise Ginkgoales (Sphenobaiera), Czekanowskiaceae (Phoenicopsis), cycads and Bennettitales (Ctenis, Nilssonia, Pseudocycas, Pterophyllum etc.) and Caytoniales (Sagenopteris).

The Siberian—Canadian region is divisible into several provinces, viz., South Urals (for the Late Albian—Cenomanian), East Siberian, Okhotsk—Chukotska and Canadian—Alaska. This province-wise segregation is distinctly different from the one I suggested some time ago (Vakhrameev et al., 1970). Since that time a great deal of material has been accumulated and new generalizations made (Krassilov, 1975a; Budantsev, 1983). Boundaries among the provinces identified may only be delineated tentatively due to the rarity or total absence of floral localities in many large areas. Within these provinces there are also certain dissimilarities, but if we embark on a more detailed division then it will be the floras of separate vast basins, e.g. Chulym—Yenisei or Vilyui, that will be dealt with while many localities with scarce species will remain unassociated with the floras of these smaller phytochoria. Therefore I prefer to consider the peculiarities of the rich floras of both the basins within the framework of the larger East Siberian province, thus underlining their differences.

4.1.1 South Urals province

Within the Siberian—Canadian region this province is recognized only for the Late Albian and Cenomanian. It includes the localities of floras of this age situated primarily north and east of the Aral Sea. Previously it was termed the Urals province. However, this name should be updated because all known phytochorial localities adjoin the southern margin of the Urals (Mugodjaram and Chushkakul highlands). The northern and middle Urals have no localities of floras of this age. The main localities of the Late Albian floras is Kuldenentemir while the Cenomanian is represented by Terektysai and Tasaran. The composition of the Late Albian and Cenomanian floras is obviously dominated by Platanus featuring particularly numerous Platanus cuneifolia (= pseudoguillelmae) (Vakhrameev, 1952; Nikitin and Vassilyev, 1977; Shilin, 1983). In the majority of the localities of this age Platanus leaf impressions amount to no less than 80–90% of the total number of imprints. *Platanus* also includes the leaves earlier ascribed to the genus *Credneria*.

It appears typical that such representatives of the Platanaceae as *Pseudoprotophyllum* so broadly propagated in more northerly provinces of the Siberian–Canadian region, e.g. in the Chulym–Yenisei, are encountered here very rarely. This genus may possibly be said to comprise the three-lobed leaf described as *Pseudoaspidiophyllum kazakhstanicum* from Kuldenentemir. Later this genus was placed within the synonymic range of *Pseudoprotophyllum* (Vakhrameev, 1976a).

In a number of localities we find intercalations consisting of closely superimposed imprints. It is doubtless that we deal here with the 'buried' leaf-fall, i.e. with the leaves shed by *Platanus* at the end of the vegetation period. Platanaceae formed clear, apparently open forests covering broad alluvial valleys. This conclusion is borne out by burial of well-preserved, sometimes intact leaves in obliquely laminated sandstones or in subordinate clayey lenses. *Platanus* are concurrent with the finds of the genera *Araliaephyllum*, *Anacardites*, *Cissites*, *Dalbergites*, *Daphnophyllum*, *Diospyros*, *Lindera* (*Sassafras*), *Magnolia*, *Myrica*, *Sterculia* and *Zizyphys*. The imprints previously assigned to *Sassafras* were studied in detail by Imkhanitskaya (1968) and referred to her by the genus *Lindera* (= *L. jarmolenkoi*). Like *Anacardites* they were not found in the Kuldenentemir locality dated as Upper Albian; *Trochodendroides* is a very rare find.

Remains of conifers in the upper Albian and Cenomanian deposits of the South Urals province normally merged into the Altykuduk suite, occur rarely. They are represented by the twigs of *Sequoia* (commonly *S. reichenbachii*), more rarely of *Cyparassisium* and *Glyptostrobus*. '*Cephalotaxopsis*' and the seeds of Pinaceae (*Pityospermum*) are seldom detected. Isolated imprints of *Ginkgo* ex gr. *adiantoides* were recorded.

The strata projecting on the river Kuldenentemir yielded *Nilssonia* and Bennettitales (*Otozamites jarmolenkoi*); the latter featuring small size. Presence of such relics and their correlation with marine sediments located to the west (Vakhrameev, 1952) makes us assign the lower part of the Altykuduk suite to the upper Albian. Ferns are scarce, being dominated by individual species of '*Asplenium*', *Cladophlebis*, rarer *Gleichenia* and *Onychiopsis*.

Paucity of the remains of ferns and conifers in the localities fails to conform to the composition of the spore–pollen complexes extracted from these deposits (Bolkhovitina, 1953). The content of coniferous pollen in them fluctuates between 25 and 50%. This may have been caused by the great productivity of conifers whose pollen is disseminated by wind and should, therefore, be more abundant, but also by the existence on the declivities of water divides of coniferous-to-broad-leaved deciduous forests. Their canopy might have housed the herbaceous ferns whose existence is indicated by the appreciable proportion of spores in palynospectra.

The Late Albian and Cenomanian floras of the South Urals province are compositionally similar to the coeval vegetation of the Chulym–Yenisei basin. Both these floras are dominated by the Platanaceae, but while the former are represented by miscellaneous species of the genus *Platanus*, the Chulym–Yenisei basin is dominated by *Pseudoprotophyllum*. The latter includes notably less entire leaves and no *Lindera*, *Daphnophyllum* or *Diospyros*, as well as somewhat more diverse conifers with *Podozamites* in particular.

Humid conditions and, possibly, some cooling as compared with the Aptian was conducive to expansion of broadleaved, notably deciduous, forests not only within the warm-temperate belt but also in the adjoining part of the subtropics of the Albian whose climate in the pre-Albian featured dryness and apparently higher temperatures inhibiting the local development of moisture-loving vegetation as is obvious in Kazakhstan. This brought about the disappearance of the arid belt at least within the Asiatic part of the USSR. Coal-bearing formations emerge even in Mongolia in the upper part of the Lower Cretaceous sequence (Khukhtyk horizon). The climatic change traced by replacement of various lithologic formations is also corroborated by the dramatic decrease in *Classopollis* pollen content during the transition from the Aptian to the Albian expressed in the sequences of Middle Asia. Albian deposits contain minimal amounts of this pollen while in the second half of the Cenomanian its proportion again starts increasing (Vakhrameev, 1980) attaining appreciable values in the Turonian.

Climatic change has led to a tangible shifting of the southern boundary of the Siberian–Canadian region southwards as far as the Aral Sea, encroaching upon the areas of the South Urals and western Kazakhstan. Its southern projection formed a province which we named the South Urals. With the gradual warming and certain drying of the climate that set in as far back as in the Cenomanian, the floras of the Turonian and especially Senonian physiognomy of this region acquired a subtropical aspect becoming part of the Euro-Sinian region whose boundary shifted northwards at this time.

Not all researchers agree with assigning the floras of the Late Albian and Cenomanian of the South Urals province to the Siberian–Canadian region. Thus, Shilin (1986) is inclined to incorporate them within the Euro-Sinian (European–Turanian) region on the strength of certain similar features with the age-equivalent floras of Central Europe (Czechoslovakia, East Germany) clearly belonging to the subtropical belts of the Northern Hemisphere. The common feature is the presence in both floras of *Platanus*. However, while they are clearly prevalent in the floras of the South Urals province, in Central Europe they constitute merely one of the elements. For instance, the Peruč flora in Czechoslovakia (Cenomanian) displays a great number of narrow-leaved, entire forms of angiosperms as well as the remains of arborescent fern stems which are missing from the floras of Albian and Late Cenomanian age in the South Urals province and which probably imply a subtropical climate. A more detailed coverage of the composition of Cenomanian floras of western Europe will be given when characterizing the European province.

4.1.2 East Siberian province

The floras of the East Siberian province include the flora from the Chulym–Yenisei basin, the basin of the river Khatanga, the Vilyui trough, the Novosibirsk islands and the Zejya-Bureya depression. The floras of the Kolyma basin described here in conjunction with those of the Okhotsk–Chukotka province appear to be intermediate between the East Siberian and Okhotsk–Chukotka provinces.

4.1.2.1 CHULYM–YENISEI BASIN

Among the floras of the East Siberian province the oldest are those from the Kiy suite of the Chulym–Yenisei basin and from the Agrafenovsk suite of the Lena basin. Both these suites are of continental origin. The composition of the Kiy suite changes slightly in different localities owing either to the varying stratigraphic position of such localities within a sequence or to differing ecologic situations. In the locality situated on the river Kiya near the mouth of the river Serta the impressions of *Platanus* were found along with *Sphenobaiera* and numerous conifers (*Elatocladus*, *Pinus*, *Abies*, *Sequoia*, *Thuites*) which suggests association with the upper strata of the Early Cretaceous (Late Albian). Elsewhere (Kubayevo village) imprints of angiosperms are prevalent comprising *Nelumbites*, *Araliaephyllum* (*Aralia*), *Platanus*, *Pseudoprotophyllum*, *Trochodendroides*, *Dalbergites* and 'Rulac'. These occurred with the shoots of *Sequoia* and rarer *Glyptostrobus* cones and seed scales of both pine and cedar, *Asplenium dicksonianum* and *Danaeites kiensis*. The age of the entire Kiya complex should be estimated as Late Albian–Early Cenomanian.

In the Vilyui basin it corresponds approximately to the complex of plants from the Agrafenovo suite (Agrafenovsk stratoflora according to Kirichkova and Samylina (1978)). It is also dominated by angiosperms of the genera *Platanus*, *Pseudoprotophyllum*, *Araliopsis*, *Cissites*, *Celastrophyllum*, *Macclintockia* and *Dalbergites*. The large, well-formed leaves of angiosperms were found along with small-sized forms, i.e. *Cissites microphylla*, *Crataegites* cf. *borealis* and *Celastrophyllum ovale*. At the same time, relatively numerous ferns include '*Asplenium*', *Birisia* and *Coniopteris*. The conifers are represented by *Araucarites*, *Brachyphyllum*, '*Cephalotaxopsis*', *Cryptomerites* and *Sequoia*. Other occurrences comprised very rare representatives of the Early Cretaceous flora (*Sphenobaiera* and *Podozamites*).

As was noted by Budantsev (1979) and Samylina (1974), the floras transitional from Early to Late Cretaceous may be represented in various localities by fossil plant associations (oryctocenoses) which are quite different from each other in systematic composition. Thus, in the Vilyui basin the fossils associated with clayey rocks are dominated by Early Cretaceous elements, i.e. ferns, Ginkgoales and rare small-leaved angiosperms. Other deposits usually confined to sands and sandstones have prevalent large-leaved angiosperms represented normally by platanoid forms. Plant assemblages must have been more diverse at the time of floral restructuring accompanying the transition from Early to Late Cretaceous, because certain localities continued to be inhabited and even dominated by the Early Cretaceous elements that disappeared later.

The younger deposits of the Simonov suite from the Chulym–Yenisei basin are represented by two floristic complexes which are similar to each other. The older is confined to the lower part of the Simonov suite, assigned to the upper Cenomanian, and is often named the Chulym complex. The upper complex is linked with the second half of the Simonov suite (Turonian), and is also said to comprise the flora of the Kass suite developed on the river Kass which is a left tributary of the Yenisei river. The second complex is termed Kass (Lebedev, 1955, 1962).

The Chulym flora is compositionally close to the Cenomanian vegetation of western Kazakhstan, notably to the flora from the Terektysai locality. It is made up of *Platanus*, including *Platanus cuneifolia*, *P. cuneiformis*, *P. embicola* and others, as well as *Credneria*-related forms. Common species also include *Anacardites neuburgae* and *Lindera jarmolenkoi* (= *Sassafras polevoii*). It incorporates representatives of the genera *Araliaephyllum*, *Menispermites*, *Magnolia*, *Dalbergites* and 'Rulac' known also in the Cenomanian flora of western Kazakhstan. Perhaps, the basic difference is the appearance of a number of new species of *Pseudoprotophyllum* and *Macclintockia* missing from the flora of western Kazakhstan. These genera typify the floras of the East Siberian and Okhotsk–Chukotka provinces being common for the warm-temperate climatic belt. The composition of conifers and rare ferns is similar to that indicated for the Kiya suite.

The flora of the upper half of the Simonov suite differs from that of the lower part by emerging *Trochodendroides*, *Zizyphoides* and *Juglans*. The composition of conifers is somewhat impoverished featuring only '*Cephalotaxopsis*' and *Sequoia*, while ferns are represented by the genus '*Asplenium*'. The complex from the Kass river exhibits *Glyptostrobus*, *Taxodium* and *Protophyllocladus* among the conifers, and these may be regarded as younger elements. *Viburnum* is another newcomer. Appearance of entire leaves in the Kass suite is to be pointed out, with *Laurophyllum* sp. (defined as *Laurus plutonia*) and *Diospyros* sp. in the first place, possibly signalling climatic warming. The core of the Kass flora is still made up of platanoid leaves.

While the Chulym complex should be referred to the Late Cenomanian, the Kass complex, comprising both the Kass and upper Simonov floras, is of Turonian age. This is supported by the emergence in the Kass complex of the above-mentioned young elements, notably *Protophyllocladus* which appears only in the Turonian in western Kazakhstan.

4.1.2.2 THE VILYUI BASIN

The Vilyui basin, above the Agrafenovo complex confined to the deposits of the suite of that name, includes the Nizhny Chyrimin floristic complex whose lower age boundary is drawn at the base of the Turonian and the upper roughly through the upper strata of the Senonian (Coniacian–Santonian). According to the new concept of Budantsev (1979) the Nizhny Chyrimin complex comprises the flora of the Upper Agrafenovo suite as understood traditionally and the vegetation of the lower part of the Chyrimin suite (Budantsev, 1968). It conforms to the middle complex of Vakhrameev (1958); thus, the independent status of the Upper Agrafanovo complex is eliminated.

In this complex ferns are sporadic (*Asplenium dicksonianum*, *Anemia arctopteroides*), Ginkgoales are represented exclusively by *Ginkgo* ex gr. *adiantoides*. The numerous conifers described include *Araucarites* (sporadic), '*Cephalotaxopsis*' (Figs. 4.1, 4.2), *Taxus*, *Florinia*, *Sequoia*, *Metasequoia*, *Glyptostrobus*, *Cryptomeria*, *Taiwania*, *Cuninghamia*, *Cupressinocladus* (*Thuja*), *Libocedrus*, etc. (Sveshnikova, 1967). The predominant angiosperms are *Trochodendroides*, *Zizyphoides* and *Viburnum*, with *Platanus*, *Pseudoprotophyllum*, *Macclintockia*, *Menispermites*, *Dalbergites* and '*Rulac*' also being widespread. The number of leaves generally represented by large sizes is reduced as compared with the Agrafenovo complex. The peculiarity of this complex is prevalence of narrow and small-leaved forms belonging to genera manifested in the Chulym–Yenisei basin by bigger leaves. This is particularly applicable to *Trochodendroides*, *Macclintockia*, some *Menispermites* and *Zizyphoides* which have been described by Budantsev (1968) with new generic names.

These differences give Budantsev a reason to assign the Chulym-Yenisei and Vilyui floras to separate phytochoria (1979). This can be subscribed to, but I prefer to retain

A

B

Fig. 4.1 Plant fossils of Late Cretaceous age from the Vilyui river basin; natural size. A, *Cephalotaxopsis heterophylla* Hollick; B, *Viburniphyllum whymperi* (Heer) Herman.

them within a single province because the generic composition remains virtually the same. Within this province it is possible to recognize smaller phytochorial sub-provinces or districts. The occurrence of small and narrow leaves in the majority of representatives of various genera in the Nizhny Chyrimin complex is most probably due to the drought conditions reigning in the Vilyui basin in comparison with those in the Chulym–Yenisei basin or in the Khatanga basin and Novosibirsk islands whose

Fig. 4.2 Plant fossil of Late Cretaceous age from the Vilyui river basin; natural size; *Protophyllum sternbergii* Lesquereux.

characterization will follow. All this is likely to be associated with the more continental climate featuring a dry summer which was prevalent in the region of the Vilyui basin located far from the sea whereas the rest of the regions enumerated were situated either near or not far from marine basins and their climates were uniformly moist. It will be noted that Transbaikalia was situated south of the Vilyui basin, and further south was Mongolia whose climate was arid in the Late Cretaceous.

According to the supposition of Budantsev (1979) the age of the Nizhny Chyrimin complex may encompass a wider temporal interval (Turonian–Lower Senonian) as against the Kass flora (Turonian). This is supported by the presence in the former of numerous *Trochodendroides* and *Macclintockia* which are rare in the Kass flora. If this is correct then there is a gap in the sequence of the Chulym–Yenisei basin and the left tributaries of the Yenisei river.

The youngest Sym suite developed on the river Sym (a left tributary of the Yenisei river) and referred to the upper Senonian (Campanian–Maastrichtian) or even directly to the Maastrichtian, contains rare ferns, primarily *'Asplenium'*. Among the conifers *Taxodium*, *Sequoia* and *Pinaceae* (*Pinus*, *Cedrus*, *Pseudolarix*) are dominant with more rarely occurring *'Cephalotaxopsis'*, *Glyptostrobus*, *Libocedrus*, *Protophyllocladus* and *Cupressinocladus* (*Thuja*). Angiosperms are represented mainly by *Trochodendroides*, with *Nordenskioldia*, *Zizyphoides*, *Viburnum* and *'Acer' arcticum* also being present. Only *Platanus* remains among the platanus-like leaves; Amentiflorae (*Betula* ?) appear. The flora of the Sym suite is close to one reported in the Antibes suite of the Chulym–Yenisei basin dominated by varied *Trochodendroides*. The accompanying elements are *Platanus*, *Zizyphoides* and *Viburnum*. Conifers are manifested chiefly in *Sequoia*. Budantsev (1979) combines the Sym and Antibes floras into the Antibes complex.

In the Vilyui trough the Sym complex is approximately equated with the Vekhny Chyrimin complex. The latter for the first time features *Onoclea sensibilis* L.f. *fossilis* (Vakhrameev, 1958) in addition to *Anemia* and *'Asplenium'*. The variety of conifers is diminished; in particular *'Cephalotaxopsis'* disappears. Angiosperms are dominated by *Trochodendroides* accompanied by *Trochodendrocarpus*. *Zizyphoides heterophylla* and *Macclintockia borealis* become quite widely distributed. *Populus*, *Juglans* and *Rhamnites* (i.e. the representatives of the genera later widespread in the Palaeogene), occur. *Platanus*-like forms become rare. The age of the Verkhny Chyrimin complex is most probably Late Senonian (Campanian–Maastrichtian) and this is corroborated by palynologic data.

The two regions considered here which are richest in fossils allow identification of three floristic complexes, i.e. Late Albian–Cenomanian, Turonian–Early Senonian, and Late Senonian (Campanian–Maastrichtian). The age boundaries are generalized because it is not possible to collect leaf flora stratum by stratum. Furthermore, in the regions indicated the Upper Cretaceous sequence is composed exclusively of continental deposits primarily of alluvial origin with their correlation with marine remains being elusive due to the absence of the latter in the Vilyui basin.

4.1.2.3 RIVER KHATANGA BASIN

Let us review the Late Cretaceous floras of other areas, notably of the river Khatanga basin. Four complexes are identified here (Samoilovich, 1980; Abramov, 1983). The lowest of these, characterizing the lower and middle parts of the suite, hosts *Anemia*, *Arctopteris*, *Anomozamites* sp., *Ginkgo* ex gr. *adiantoides*, *Sphenobaiera* sp., *'Cephalotaxopsis' intermedia*, *Menispermites* sp., *Dalbergites sewardiana* and *Cissites comparabilis*. Although in agreement with other geologists L. N. Abramov assigns it to the Cenomanian–Turonian, the presence of such Early Cretaceous forms as *Arctopteris*, *Anomozamites* and *Sphenobaiera* does not suggest an age above Cenomanian. The Kiya and Agrafenovo complexes are its closest relatives.

The second complex records complete disappearance of relicts with the conifers being represented by *'Cephalotaxopsis'*, *Sequoia* and *Taxodium* while angiosperms are dominated by *Pseudoprotophyllum* and its close associates. Also present are *Viburnum*, *Zizyphoides* and *Trochodendroides*. This complex should be referred to the Turonian–Coniacian.

The third complex associated with the Kheta suite is very impoverished in composition and devoid of platanoid leaves. Its composition must have been determined by local conditions more than anything else. Going by the deposition of host rocks below the Mutino suite with upper Santonian–Campanian inocerami, its age is Early Santonian; it may be united with the second complex. The Mutino suite proper yielded *Viburnum* (?) and impressions of leaves of the aquatic plants *Pistia* (?) *marginata* and *Quereuxia angulata* in addition to numerous *Pseudoprotophyllum*. The age of this compositionally poor complex is determined by the above-mentioned fauna. Remains of the aquatic plants indicate that this fern grew in littoral lowlands covered by lakes and situated near sea coast.

4.1.2.4 NOVOSIBIRSK ISLANDS

Terrigenous continental deposits with secondary coal beds are exposed on the southern coast of the Novosibirsk islands in the Derevyanniye mountains; they reveal a rich flora. The presence therein of *Hausmannia*, *Sphenobaiera*, *Podozamites* and numerous conifers common for the lower part of the Upper Cretaceous along with *Pseudoprotophyllum*, *Platanus*, *Trochodendroides*, *Zizyphoides*, *Cissites* and *Macclintockia* imply, according to the authors who described it (Sveshnikova and Budantsev, 1969), its affinity to the Turonian.

4.1.2.5 ZEYA–BUREYA AREA

The south-westernmost area of the distribution of the floras of the East Siberian province is the Zeya-Bureya depression. The lower half of the upper Cretaceous referred to as the Zavitinsk suite was revealed only by boreholes; therefore, plant remains extracted from cores are not numerous. They feature '*Asplenium*', *Onychiopsis*, *Nilssonia* (*N. alaskana*), '*Cephalotaxopsis*', *Sequoia*, *Platanus*, *Trochodendroides*, *Viburnum* and *Quereuxia* (Gorbachev and Timofeev, 1965). The age is estimated as Cenomanian–Turonian.

The rich flora from the sediments of the upper part of the Upper Cretaceous exposed on the surface and termed the Tsagayan suite was studied by many investigators. The most important works are by Poyarkova (1939), Krishtofovich and Baykovskaya (1966) and Krassilov (1976). The lists of plants of the Tsagayan flora supplied on the one hand by Krishtofovich and Baykovskaya and, on the other, by Krassilov, are dramatically different on the level of species because the latter researcher strove toward significantly narrower contents of some genera, considering many discrepancies as resulting from intraspecific variability. For instance, he extraordinarily extended the scope of the species *Platanus raynoldsi* which is quite unacceptable. The generic composition quoted by this palaeobotanist also differs from that cited by Krishtofovich and Baykovskaya. It should be pointed out that Krassilov excluded from the list of the Tsagayan flora the following: *Glyptostrobus*, '*Cephalotaxopsis*', *Ficus*, *Zelkova*, *Grewiopsis*, *Pterospermites* and *Tetracentron*.

After deletions of the above genera made by Krassilov, the generic composition of the Tsagayan flora embraces *Ginkgo*, many conifers such as *Podocarpus*, *Araucarites*, *Pinus*, *Pseudolarix*, *Sequoia* (rare), *Libocedrus* (*Androvettia*), *Cupressinocladus* (*Thuja*) and especially numerous *Metasequoia* and *Taxodium*. The following aquatic plants are present: *Nelumbo*, *Potamogeton*, *Arundo*, *Phragmites*, *Limnobiophyllum* and *Quereuxia*. Angiosperms are dominated by *Trochodendroides* and the co-occurring *Trochodendrocarpus*, *Platanus* and *Tilliaephyllum* (*Tilia*). The rarer occurrences include *Myrica*, *Macclintockia*, *Menispermites*, *Nyssa*, *Viburnum* (*Viburniphyllum*) and *Nordenskioldia*. *Celtis* and *Alnus* appear, these being peculiar to the Palaeogene flora.

The age of the Tsagayan flora and its host deposits is determined in different ways. Initially it was wholly referred to the Danian stage (Krishtofovich and Baykovskaya, 1966). Then Bratseva (1969) substantiated its Maastrichtian age after examining the composition of pollen from the Tsagayan suite. Later Krassilov (1976) noticed certain change in systematic composition during transition from the middle to the upper subsuite. This change consisted of the increase of *Taxodium*, *Platanus raynoldsi* and *Trochodendroides* in the middle subsuite and *Tiliaephyllum tsagaianicum* in the upper. The correlation effected by this author chiefly with the sequences from Canada and the USA brought him to the conclusion that the middle and upper subsuites of the Tsagayan suite (Krassilov, 1976, Table 2) are of the Danian age while the lower, whence he virtually failed to collect any flora, was assigned to the Maastrichtian.

At the same time G. M. Bratseva palynologically characterized the lower and middle subsuites related to the Maastrichtian. It proved impossible to extract spores and pollen from the upper subsuite. Thus, the dispute boils down to the determination of the middle subsuite. It is not excluded that the boundary between the middle and upper subsuites is drawn by the researchers in question on the basis of different premises; Bratseva chiefly made use of core material, while Krassilov used samples from outcrops.

The Tsagayan suite is composed largely of alluvial deposits devoid of distinctive marker members needed for a well-based segregation and correlation between separate sections. To adopt one member of conglomerates as a stratigraphic marker, as was done by Krassilov, is hardly reliable because conglomerates reflect merely intra-formational wash-outs inherent in the series of alluvial origin rendering them very inconsistent along the strike.

I am inclined to believe that the upper subsuite of the Tsagayan suite is represented by a single sedimentary cycle whose lower part is made up of clastic rocks and upper part of a coal-bearing member. The latter furnished a rich flora different from the Tsagayan proper by emergence of Palaeogene elements (*Woodwardia, Alnus, Ulmus*). The richest floral complex was discovered by Naryshkina (1973) in the clays in the roof of the 'Verkhny' (Upper) bed attributed to the Kivda suite. The local finds comprise *Woodwardia, Osmunda sachalinensis, Cladophlebis* cf. *arctica, C. oerstedtii, Ginkgo adiantoides, Taxodium dubium, Metasequoia* sp., *Glyptostrobus europaeus, Ginkgo cretaceae, Alnus* sp., *Ulmus pseudobraunii, U.* cf. *longifolia, Trochodendroides* ex gr. *arctica, Protophyllum* sp., *Platanus* sp., *Viburnum* sp., etc. We date the flora of the Kivda suite as Danian, referring the Danian to the Palaeogene.

4.1.2.6 NORTH-EAST CHINA

Localities of the Tsagayan flora are available in China, too (Hsü, 1983), on the right bank of the Amur river (Kheilunizyan province). Shoots of conifers are numerous here being represented by *Metasequoia* and *Sequoia; Thuja cretacea* is encountered more rarely. The angiosperms present include *Protophyllum* cf. *microphyllum, Pseudoproto-phyllum* cf. *dentatum, Betula prisca, Alnus* sp., *Populus carneosa, Ziziphus phoshporia, Mahonia* cf. *furnaria, Menispermites borealis, M. obtusiloba, M. kuliensis, Ampelopsis aceri-folia, Debeya tikhonovichii, Trochodendroides arctica, Tetracentron* sp., *Sorbaria* sp., *Tiliae-phyllum* cf. *tsagajanicum, Viburnum cupanioides, V. antiquum, V. asperum, Bauhinia* sp., *Cissus marginata, Rhus* cf. *turcomanica* and *Cyclocarya* sp.

On the whole these species represent leaf-shedding trees and bushes. However, there are apparently evergreen elements, too (*Mahonia*). A surprising feature is a combination of the forms typical of the second half or the end of the Cretaceous (*Pseudoprotophyllum, Debeya*) with those emerging as late as the Danian—Palaeocene (*Betula, Alnus*). It is not excluded that leaf impressions from different strata might have become included in the processed collection.

The presence of such thermophilic elements as *Debeya, Bauhinia* and *Protophyllum* attests to the affinity of the Late Cretaceous flora of north-eastern China to the eco-tonal zone situated between the floras of moderate-warm and subtropical climatic belts.

4.1.3 Okhotsk–Chukotka province

The boundaries of the Okhotsk–Chukotka province are the basin of the Kolyma river in the west and the Arctic and Pacific Oceans in the north and east, respectively. From the south this boundary shifted meridionally several times owing to climatic change. While at the beginning of the Cretaceous it passed through the southern tip of Sakhalin, in the Santonian and Early Campanian the boundary advanced northwards leaving Sakhalin and the southern Far East coastal area in the south. The most comprehensive sections of the province in question are to be found in the vicinity of the Okhotsk Sea, Penzhina bay, and in the basin of the river Anadyr.

According to the geologic structure and nature of the relief two large areas are identifiable here, i.e. the Okhotsk–Chukotka volcanogenic belt (henceforward – belt) which in the first half of the Late Cretaceous was a montane volcanic system and, east of it, a littoral plain bordering on the marginal seas of the Pacific Ocean. The transgressions occurring repeatedly during the Late Cretaceous inundated this plain depositing marine sediments with invertebrates (chiefly *Inoceramus*). During regressions this plain became a venue of continental formation, primarily deltas with plant remains. Alternating sequential marine formations containing marine fauna and continental deposits permit assessment of the age of the floristic complexes.

The basic works studying the Late Cretaceous floras of the territories framing the north-western part of the Pacific Ocean are those of Krishtofovich (1958a, b), Yefimova and Terekhova (1966), Vassilevskaya and Abramova (1974), Filippova (1975, 1978, 1979, 1982), Vakhrameev (1976b), Terekhova and Filippova (1983) and Herman (1984a, b; 1985) who monographically described the flora of Penzhina bay and Ugolnaya harbour. The most important research on the stratigraphy of the upper Cretaceous deposits was performed by Pergament (1961, 1978, etc.). The Late Cretaceous floras of the Okhotsk–Chukotka volcanogenic belt proper were investigated mainly by Samylina (1963, 1984) and Lebedev (1979, 1982, 1983). A monograph on the floras of the Far East coastal areas was written by Ablave (1974) with the latest data on these floras having been obtained by Nevolina (1984). The floras of Sakhalin have been described by Krishtofovich (1937), Krishtofovich and Baykovskaya (1960), and Krassilov (1979). Reviews of the Cretaceous floras of the Far East were made by Vakhrameev (1966, 1976b), Krassilov (1975a), and Krassilov, Nevolina and Filippova (1981), as well as by Samylina (1974).

Since the Late Albian and the most ancient Late Cretaceous flora up to the Cenomanian have been reported only from the Okhotsk–Chukotka belt, we start our description with these. A difficulty in age estimation of these floras is their restriction to the volcanogenic formations totally devoid of rocks of marine origin.

Richer localities of later floras are distributed along the coastal zone of the north-eastern part of the Pacific Ocean. Therefore, from the Late Cenomanian to Early Turonian characterizations of the floras under discussion are founded on the localities of littoral zones whereas more impoverished coeval floras of 'the belt' will be only briefly referred to. The oldest of those considered here is the Arinda stage comprising the floristic complex confined to the lower part of the Ulya suite evolved in the southern part of the Okhotsk–Chukotka belt. At this stage the flora is mixed because the Early Cretaceous elements (*Gleichenia, Birisia, Taeniopteris*) are concurrent with the newly appearing conifers *Elatocladus smittiana*, '*Cephalotaxopsis*' *heterophylla* and *Cupressinocladus* (*Thuja*) *cretacea* which are inherent in the Late Cretaceous. But the main newcomers are the large leaves of angiosperms of platanoid type and *Menispermites* sp. appearing in the sequence for the first time. This correlation of forms and particularly the presence of the large-leaved angiosperms is indicative, most likely, of Late Albian age.

The flora of the Arman suite may well also belong to this age. The above suite is developed in the Okhotsk sector of the belt. It is also typified by mixed Early Cretaceous elements represented by ferns (*Birisia, Arctopteris, Acrostichopteris,* Czekanowskiaceae (*Phoenicopsis, Czekanowskia*) and Podozamitaceae, along with conifers which were widespread in the Late Cretaceous (*Cupressinocladus*, miscellaneous *Sequoia*) and rather diverse angiosperms (*Menispermites, Dalbergites, 'Zizyphus',* *Cissites, Araliaephyllum*, platanoid forms, etc.).

A peculiarity of the Arman suite underlined by Samylina (1974) amounts to a unique combination of Early and Late Cretaceous elements. The narrow leaf angiosperms so peculiar to the Buorkemyus flora vanishes in Arman time. The deposits of the Arman suite normally represent both the localities enclosing remains of the Early Cretaceous plants, primarily ferns, Czekanowskiaceae and Podozamitaceae, and the localities with remains of conifers and fairly large leaves of angiosperms of Late Cretaceous physiognomy. V. A. Samylina believes that the process of flora restructuring during transition from the Mesophytic to the Cenophytic might have been concomitant not only with the penetration of younger elements into phytocenoses composed of the Early Cretaceous plants but also with the emergence of phytocenoses of a new and younger type including new conifers and angiosperms. These phytocenoses might have coexisted, taking up different ecologic niches. Depending on the changing physical and geographic situation they might have been replacing each other in time and in sequences until younger phytocenoses ousted the older ones.

As far as the Arman flora is concerned there is no generally accepted opinion; Krassilov (1975a) followed by G. G. Filippova thinks it to be Cenomanian. It is noteworthy that the presence of large leaves of angiosperms in the Upper Albian is a regular rather than accidental feature since the deposits of this age in western Kazakhstan and the USA (lower strata of Dakota suite) exhibit many impressions of *Platanus, Protophyllum, Menispermites*, etc.

The Amka stage embraces the flora of the middle part of the Ulya series of the belt located above the Arinda strata. Ferns become less varied here. *Metasequoia, Araucarites* and *Libocedrus* (rarely) occur among the conifers. *Sphenobaiera* and *Arctobaiera* have been reported in addition to *Ginkgo* in the Ginkgoaceae. Czekanowskiaceae are represented by *Phoenicopsis*. Angiosperms see the emergence of *Trochodendroides* and

Quereuxia angulata. Lebedev (1982) recognized three complexes replacing one another in the Amka flora of the Ulya trough. Of these the middle one is richest in conifers whose remains are obviously prevalent in some localities. This suggests the period of time corresponding to this complex as a colder one. According to Y. L. Lebedev the Arkagala flora whose localities are situated somewhat west of the belt may correspond only to the upper complex of the Amka stage.

The Arakagala flora was meticulously studied by V. A. Samylina who thinks that coeval floras are known not only from the Ulya depression but also from the Pervomaysk coal deposit and eastern Chukotka (basins of the rivers Amguema and Leurvaam). It is different from the preceding Arman flora in the predominance of conifers represented mainly by Taxodiaceae (*Sequoia, Metasequoia, 'Cephalotaxopsis'*) and rare Cupressaceae (*Thuja*); shoots of the latter are often defined as *Cupressinocladus*. Angiosperms feature *Cinnamomoides, Cocculus, Trochodendroides, Zizyphus, Diospyros* and *Celastrophyllum*. Remains of a characteristic aquatic plant, i.e. *Quereuxia angulata*, are constantly found. Platanoid leaves have not been recorded. ·

Based on the number of species, angiosperms take up slightly more than 20%. However, occurrence of the leaves of angiosperms proves very inconstant and they are often lacking in several localities. Certain species are represented by single impressions. One of the frequently encountered Early Cretaceous relicts is *Phoenicopsis angustifolia* whose remains are often associated with swamp facies where their number is appreciably increased.

The representatives of the genera *Osmunda, Coniopteris, 'Asplenium', Hausmannia, Cladophlebis, Lobifolia* and *Sphenopteris* were detected among ferns. Of these *'Asplenium'* with cladophleboid pinnules and *Cladophlebis septentrionalis* are encountered rather often. Cycadophytes are very rare and represented by the form-genus *Taeniopteris*. Ginkgoaceae have yielded *Ginkgo, Sphenobaiera* and *Pseudotorellia*.

The age of the flora of the Amka stage comprising the Arkagala complex is estimated as Early Cenomanian. Possibly, it also embraces the uppermost strata of the Albian. It should be noted that judging by the results of determination of absolute age obtained recently, the duration of the Albian age attains 15 million years whereas that of the Cenomanian is 6.5 million years, the Turonian 2.5 million and the Coniacian merely 1 million years (Harland *et al.*, 1982). Krassilov (1979) thinks that the Arkagala suite should be dated as Turonian estimating it as younger than the Grebenka suite, but this is hardly acceptable going by the far greater variety of angiosperms in the Grebenka flora.

The study of the floras of the littoral zone has permitted recognition of a number of phases in its development whose analogues are also discovered in adjacent belt regions. The oldest Grebenka phase is represented by the flora collected from the Krivorechensk suite on the right bank of the river Anadyr. *Inoceramus* recorded both in the suite itself and in the overlying deposits and typical of the *Inoceramus nipponicus* zone imply a Late Cenomanian–Early Turonian age. The complex (Figs. 4.3, 4.4, 4.5) is already dominated by angiosperms manifested in three frequently occurring species of

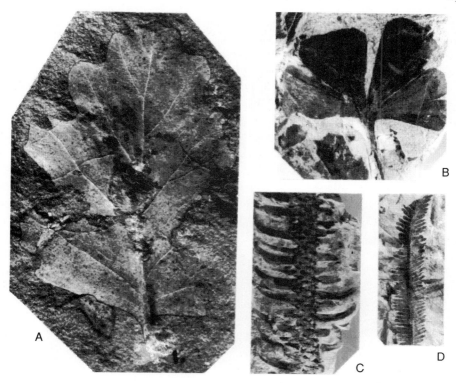

Fig. 4.3 Plant fossils of Cenomanian age from the Anadyr river basin; natural size. A, *Dalembia vakhrameevii* E. Lebedev and Herman, isolated lateral leaflet; B, *Ginkgo* ex gr. *adiantoides* (Unger) Heer; C, *Araucarites anadyrensis* Krishtofovich; D, *Cycadites hyperborea* (Krisht.) E. Lebedev.

Fig. 4.4 Plant fossils of Cenomanian age from the Anadyr river basin.
A, *Gleichenia zippei* (Corda) Heer, fertile pinnule, × 2; B, *Coniopteris grebenkaensis* Filippova, sterile and fertile pinnules, × 2; C, *Cephalotaxopsis intermedia* Hollick, natural size; D, *Grebenkia anadyrensis* (Krysht.) E. Lebedev, natural size.

Fig. 4.5 Plant fossil of Cenomanian age from the Anadyr river basin; natural size; *Platanus* aff. *newberryana* Heer.

Menispermites, diverse platanoids (*Platanus*, *Credneria*), representatives of the new genus *Grebenkia*, and multi-leaved *Dalembia*, as well as representatives of the genera *Dalbergites* and *Leguminosites*. The rarer occurrences are *Magnoliaephyllum*, *Celastrophyllum*, *Araliaephyllum* (?), *Sorbites*, *Lindera* (?), *Myrtophyllum*, *Cissites*, *Sapindopsis* and '*Zizyphus*'.

The ferns *Coniopteris*, *Cladophlebis*, *Birisia* and *Gleichenia* are numerous while '*Asplenium*', *Hausmannia*, *Arctopteris*, etc. are rarer. Cycadophytes are present with *Cycadites* and three species of *Nilssonia* (*N. serotina*, *N. yukonensis*, *N. alaskana*) being frequently encountered. *Taeniopteris* is found rarely. Absence of cuticle does not allow more precise establishment of the generic affinity of the latter. Three species of Ginkgoaceae have been reported featuring *Ginkgo* ex gr. *adiantoides*, *Sphenobaiera biloba* and *Desmiophyllum* sp. Conifers are represented by the genera *Elatocladus*, *Araucarites*, '*Cephalotaxopsis*', *Sequoia*, *Pityocladus*, *Athrotaxopsis*, *Cupressinocladus*, *Podozamites*, etc. Of these the commonest are *Araucarites anadyrensis* forming assemblages and *Elatocladus smittiana* coupled with *E. gracillimus*. Very similar compositionally but more impoverished floras were uncovered in the basin of the river Ubienka (left tributary of the river Anadyr) as well as on the river Levaya Beryozovaya (left tributary of the river Penzhina).

The flora detected in the middle part of the Ginterovo suite of Ugolnaya harbour corresponding to the middle part of the Cenomanian (Pergament, 1978) turned out to be roughly age-equivalent to the Grebenka complex. This suite also furnished *Baiera* cf. *gracilis* and *Nilssonia yukonensis*. Within the belt the Grebenka phase conforms to the Dukchanda complex known from the central part of the Ulya trough (upper part of the Ulya series) containing numerous common forms.

The Penzhina phase (Herman, 1984a) is most fully represented by a flora from the lower portion of the Valizhgen suite from the vicinity of Konglomeratovy Cape and Yelistratov peninsula (Penzhina bay). This age level embraces the Volchinsk flora found in the basin of the Ubiyenka river. Leaf impressions proved to be concurrent here with Turonian fauna (Devyatilova, Nevretdinov and Filippova, 1980). The similarities between the floras of the Volchinsk formation and the lower part of the Valizhgen suite of Konglomeratovy Cape support the view of Pergament (1961, 1978) of the Turonian age of the lower strata of the Valizhgen suite.

The flora of the Penzhina phase features an abundance of large-leaved platanoids (*Platanus*, '*Credneria*', *Zissania*, *Pseudoprotophyllum* and *Paraprotophyllum*); the latter was recently established by Herman (1984a). *Menispermites* persists but the diversity and frequency of occurrence of the representatives of this genus are reduced. Appearances at this level of *Paraprotophyllum ignatianum*, *Viburniphyllum whymperi* and '*Zizyphus*' *smilacifolia*, as well as of the frequently occurring *Trochodendroides*, are typical. Angiosperms continue to be found being assigned to the recently established genera *Penzhinia* and *Dalembia*; these angiosperms are reported from all the floras of the Grebenka phase.

Protophyllocladus generally not emerging earlier than Turonian should be particularly noted among the conifers. *Sequoia* and *'Cephalotaxopsis'* continue to be numerous, with new but so far rare species of *Glyptostrobus* and *Metasequoia*. Ferns are abundant and represented by the genera *Arctopteris*, *Gleichenites*, *Onychiopsis*, *'Asplenium'*, *Cladophlebis*, etc. supplemented with persisting *Ginkgo* ex gr. *adiantoides*. Cycadophytes have not been recorded. Within the Okhotsk—Chukotka belt (Okhotsk sector) the Penzhina phase seems to correspond to the floras from the rivers Tal and Kananyga. Angiosperms here are dominated by the large-leaved *Platanus* and *Pseudoprotophyllum*, as well as *Trochodendroides* and *'Zizyphus'*. *Protophyllocladus* has been reported among the conifers.

The Kayvayamsk phase first identified by A. B. Herman is represented by the flora collected from the second, third and fourth cycles of the Valizhgen suite (Herman, 1984a) in the sequence of Konglomeratovy Cape and from the upper part of the flora-bearing member of Yelistratov peninsula. A similar flora has been recorded in the Poperechnaya suite of the Pekulney ridge. The age of this suite is dated as Coniacian (Terekhova and Filippova, 1983) due to its deposition between marine sediments with *Inoceramus multiformis* (Upper Turonian) and *I. yokoyama* (Upper Coniacian—Lower Santonian). A flora from the lower part of the Arkovo suite of western Sakhalin named by Krassilov (1979) as the Ayan suite may be referred to the same phase. The above flora is situated in an ecotonal zone between the Japan and Okhotsk—Chukotka provinces.

Just as the Penzhina phase, the Kayvayamu phase is characterized by domination of large-leaved platanoids notably the genus *Paraprotophyllum* featuring appearance of the species *P. pseudopeltatum*. *Trochodendroides sachalinensis*, *Magnoliaephyllum magnificum*, *Viburniphyllum whymperi* and related species prove to be common. Representatives of the genera *Araliaephyllum*, *'Zizyphus'*, *Cissites* and *Dalembia* are also encountered. As compared with the floras of the Penzhina phase, *Menispermites*, *Platanus* and *Celastrophyllum* become less numerous. The conifers are manifested largely in *'Cephalotaxopsis'* and *Sequoia*. Less seldom encountered are *Metasequoia*, *Glyptostrobus*, *Elatocladus* and *Cupressinocladus*. The relics include *Ctenis* sp.

The systematic composition of the Kayvayama phase is in fact very similar to that of the vegetation of the preceding Penzhina one, differing only in emergence of certain species and quantitative correlation between representatives of some genera. Such similarity is easily accounted for by the short duration of the Coniacian age (1 million years). It is not excluded that in the sequence of Penzhina bay the strata hosting this flora may encroach upon the Santonian or part of it.

The Barykovsk phase (Vakhrameev, 1976b; Herman, 1984b) is most comprehensively represented by the floras from the upper Bystrinsk and, perhaps, the upper strata of the Valizhgen suite of north-western Kamchatka (Figs. 4.6, 4.7, 4.8, 4.9, 4.10) and by the flora from the Barykovo suite of Ugolnaya harbour. The age of the Barykovo flora and the entire phase is well expressed in the sequence of Ugolnay harbour wherein plant remains are confined to the lower part of the Barykovo suite. Its upper marine part displays *Inoceramus schmidtii* while *I. patootensis* has been reported from a lower part. Thus, the age of the flora and the whole of the Barykovo phase is

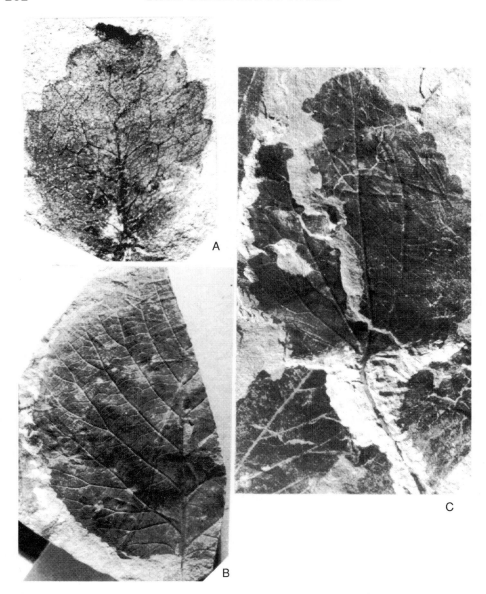

Fig. 4.6 Plant fossils of Turonian age from north-west Kamchatka. A, *Celastro-phyllum rectinerve* Herman, × 2; B, *Trochodendroides* ex gr. *arctica* (Heer) Berry, natural size; C, *Paraprotophyllum ignatianum* (Krisht. and Baik.) Herman, natural size.

Fig. 4.7 Plant fossil of Coniacian age from north-west Kamchatka; natural size; *Dalembia pergamentii* Herman and E. Lebedev.

Fig. 4.8 Plant fossil of Coniacian age from north–west Kamchatka; natural size; *Ternstroemites harwoodensis* (Dawson) Bell.

Fig. 4.9 Plant fossil of Coniacian age from north–west Kamchatka; *Paraproto-phyllum pseudopeltatum* Herman, × 0.5.

Fig. 4.10 Plant fossils of Late Cretaceous age from north-west Kamchatka; natural size. A, *Nilssonia* aff. *alaskana* Hollick, Campanian age; B, *Sagenopteris variabilis* (Velen.) Velenovsky; C, *Macclintockia ochotica* Vakhrameev and Herman, Campanian age.

estimated as Late Santonian (?)–Early Campanian. It is typified by domination of *Macclintockia* (especially *M. ochotica*) and *Quercus tschucotica* in the angiosperms, with insignificant involvement of platanoids most often featuring *Paraprotophyllum ignatianum*. Presence of the representative of *Grewiopsis*, *Rhamnites* and *Vitis* is characteristic. *Phoenicopsis* sp., *Nilssonia* aff. *serotina*, *N. yukonensis*, *Pterophyllum validum* and *Cycadites hyperbores* are present. The composition of conifers (preliminary definitions) remains unchanged in comparison with the Kayvayama phase. Fern remains are not many but in the Barykovo flora they feature the presence of *Onoclea sensibilis* L.f. *fossilis* normally appearing from the Campanian onwards.

This time level in the Okhotsk–Chukotka belt appears to incorporate the Ust–Emuneret, Kavralyan, Delokachan and Murgal floras. However, they are much less representative in composition than those mentioned earlier. It is noteworthy that the floras enclosed within the belt feature *Phoenicopsis* missing from the floras of the littoral zone.

In the Campanian or at the end of the Santonian the boundary between the warm-temperate and subtropical climate shifted to the north leaving Sakhalin in the sub-tropical belt. Accordingly, the Santonian–Campanian flora of Sakhalin will be considered when describing the Japan province.

The youngest phase (Vassilevskaya and Abramova, 1966) is marked by a flora found in the vicinity of the Amaam lagoon located south of Ugolnaya harbour. Local finds in the deposits resting higher than Maastrichtian with *Inoceramus* ex gr. *balticus* comprise *Onoclea* sp., *Woodwardia* sp., *Ginkgo* ex gr. *adiantoides*, '*Cephalotaxopsis*' sp., *Metasequoia* sp., *Glyptostrobus* sp., *Cupressinocladus* (*Thuja*) ex gr. *cretacea*, *Trochodendroides* ex gr. *arctica*, *Trochodendrocarpus* sp., *Paraprotophyllum* aff. *ignatianum*, *Corylus* spp., *Vitis rarytkinensis*, *Cissites* sp., *Celastrus* spp. and *Platanus* spp.

The deposition of these strata with such a flora on the marine Maastrichtian permits dating them in the Danian stage. Comparison with the floras of the upper strata of the Barykovo suite reveals that the vegetation of the Amaam lagoon witnesses disappearance of *Nilssonia*, *Macclintockia* and '*Rulac*' *quercifolium* abundant in the former as well as *Pseudoprotophyllum*. At the same time various species of *Corylus* and *Woodwardia* make their appearance.

The Maastrichtian may also be said to comprise a rather impoverished flora situated far south in the lower reaches of the Amur river (abandoned settlement Pad, upwards of Susanino settlement). *Equisetum* sp., *Onoclea sensibilis* L.f. *fossilis*, *Paradoxodium* sp., *Trochodendroides* ex gr. *arctica* and *Magnolia* sp. were recorded here. A flora from the Malomikhaylovka suite which is situated still more downstream in the Amur river and near the settlement of Malomikhaylovka (Akhmetyev, Bravseva and Vakhrameev, 1976) appears to be younger still. The overall list of the plants found therein includes the following: *Equisetum arcticus*, *Onoclea sensibilis* L.f. *fossilis*, *Woodwardia* sp., *Dennstaedtia* sp., *Metasequoia* sp. (abundant), *Glyptostrobus europaeus*, *Cupressinocladus cretacea*, *Libocedrus* sp., *Pityostrobus* sp., *Trochodendroides* ex gr. *arctica* (abundant) and *Corylus* sp. (in places abundant). The presence of *Corylus* sp. against the background of *Metasequoia* domination coupled with the finds of *Woodwardia* points to a Maastrichtian and most probably Danian age of the deposits.

This description suggests the following basic changes in the composition of the floras of the Okhotsk–Chukotka province throughout the Late Cretaceous epoch. The oldest Amka (Arkagala?) flora corresponds to the end of the Albian–Early Cenomanian. Its distribution restricted by the Okhotsk–Chukotka montane ridge (Amka flora proper) rendered it somewhat cooler which is expressed in an abundance of conifers (notably in the middle part of the Amka series) and total absence of cycado-phytes despite occurrence of numerous other relict forms of ferns, Czekanowskiaceae and Ginkgoaceae. Angiosperms are not numerous. No analogues of this flora in the littoral zone were found.

The floras of the Late Cenomanian–Early Turonian (Grebenka phase) of the littoral belt contain various *Menispermites*, platanoid leaves, *Dalbergites* and new species of *Dalembia* and *Grebenkia*. Of cycadophytes, *Nilssonia*, *Cycadites* and rarely *Taeniopteris* are there. The coeval floras of 'the belt' failed to exhibit cycadophytes. The floras of the Penzhina and Kayvayama phases (Upper Turonian, Coniacian and perhaps Lower Santonian) feature platanoids in the first place coupled with frequently encountered *Trochodendroides* and occasional *Protophyllocladus*. Cycadophytes were not featured. The Barykovo phase (Early Campanian) is typified by the broad range of distribution of *Macclintockia* and *Quercus tchucotica*, abundance of cycadophytes (*Nilssonia*, *Cycadites*, rarer *Pterophyllum*); *Pterophyllum* was also discovered in the belt (Delokachan suite).

Characteristically, no cycadophytes were encountered within the belt (with the exception of the Delokachan suite). But then here, incidentally just as in the deposits of the littoral zone, *Phoenicopsis* was found. It is logical to relate the emergence of cycadophytes to climatic warming spells. Their almost complete absence in the belt is easily explained by its more moderate climate conditioned by climatic zonality (Lebedev, 1982). Climatic warming must have coincided with the end of the Cenomanian–Early Turonian (Grebenka phase) and particularly with the Early Campanian (Barykovsk phase) when cycadophytes appear even within the belt. The intermediate cooling encompasses the Coniacian and Santonian (perhaps in part). The last phase of evolution of the Late Cretaceous floras having no name of its own coincides with the Danian era characterized by complete vanishing of cycadophytes, emergence of *Corylus* and *Woodwardia*, abundance of *Metasequoia*, *Trochodendroides*, and by rare occurrence of platanoid leaves. No coeval flora has so far been reported from the belt. In comparison with the Campanian, the Danian witnesses a new cooling.

Presence of a great number of Early Cretaceous flora relicts proves to be a peculiarity of the entire Okhotsk–Chukotka province just as of the Japanese province situated in the subtropical belt. These relicts include Ginkgoales *Baiera* and *Spheno-baiera* and *Ginkgo* with the lamella dissected into several lobes, frequently encountered *Phoenicopsis* and, possibly, some *Czekanowskia*. There are numerous *Nilssonia* expanding here into the lower boundary of the Palaeogene. *Cycadites* are also known coupled with the rarer *Pterophyllum* (Vakhrameev, 1981b). Not a single representative of the above genera can be found west of the Kolyma river (Fig. 4.11).

4.1.4 Canadian–Alaskan province

The oldest Cenomanian floras of this province have been reported from the basin of the river Yukon (Alaska) and the south of western Canada (Dunvegan). The Yukon basin accommodates well-developed Melozi and Clatag suites of continental origin rich in plant remains. They are Cenomanian or, possibly, Early Turonian. The boundary between the two and the underlying Nulato suite with the Albian and at one point Cenomanian (*Turrilites acutus*) fauna seems to be diachronic (Patton, 1973). It is not excluded that the lower age boundary of the deposits hosting the flora of the lower reaches of the Yukon river, if considered at this point as a single entity, may well pass somewhat lower elsewhere at the Late–Middle Albian boundary. Patton admitted that the deposits earlier identified as the Melozi suite are actually the continental facies of the Nulato suite which is basically Albian.

The rich flora (Hollick, 1930) is characterized by prevalence of large platanoid

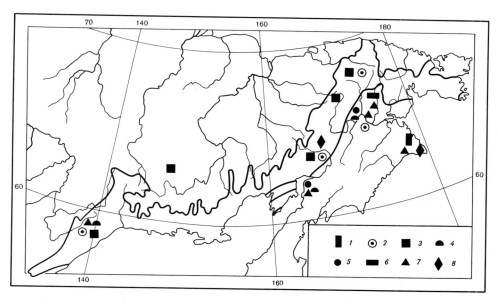

Fig. 4.11 Distribution of certain gymnosperms in the Late Cretaceous of the north-east of the USSR. 1, *Baiera*; 2, *Sphenobaiera*; 3, *Phoenicopsis*; 4, *Taeniopteris*; 5, *Cycadites*; 6, *Ctenis*; 7, *Nilssonia*; 8, *Pterophyllum*. The black line delineates the limits of the Okhotsk–Chukotka volcanogenic belt.

leaves (*Platanus, Pseudoprotophyllum, Paracredneria, Credneria*) found along with *Menispermites, Castallites, Araliaephyllum, Cissites*, etc. Angiosperms and numerous conifers represented largely by *Sequoia*, '*Cephalotaxopsis*', *Glyptostrobus, Protophyllo-cladus* and *Nageiopsis* are concomitant with a broad range of such relict forms as *Nilsso-nia, Sagenopteris* and *Podozamites* as well as rarer *Pterophyllum*. There are many impressions of *Ginkgo* ex gr. *adiantoides*; ferns are few.

The Cenomanian flora of the northern shore of Alaska defined tentatively by Smiley (1969) and collected from is III–IV zones is similar to the age-equivalent floras from the basin of the lower reaches of the Yukon river. Smiley refers the Zone III to the Late Albian–Early Cenomanian and Zone IV to the Early Cenomanian. It features few ferns and also contains the same genera of conifers which include varied Podozamitaceae. From the angiosperms Smiley singles out the forms with platanoid leaves as well as *Cissites, Menispermites*, '*Zizyphus*', etc. No illustrations and descrip-tion of plant remains have so far been published. The presented lists were compiled on the basis of preliminary definitions.

The composition of the Cenomanian flora of Dunvegan (Bell, 1963) whose localities are situated in the southern part of western Canada is distinctly different from the coeval flora from the Yukon basin. Platanoid forms (Fig. 4.12) were encountered here as well (*Platanus*, '*Credneria*', *Pseudoprotophyllum*). The leaves identified by Bell as *Aspidiophyllum* were transferred by this author to *Pseudoprotophyllum*. These leaves coexist with the noticeable numbers of emerging smooth-margined leaves referred to *Magnolia, Liriodendron, Laurophyllum*, '*Cinnamomum*', *Diospyros*, etc. which are unknown within the age-equivalent Yukon flora. *Menispermites, Castallites, Araliae-phyllum* and *Dalbergites* were also detected.

Among the conifers the dominants are *Sequoia*, '*Cephalotaxopsis*' and *Elatocladus*, with *Brachyphyllum* and *Protophyllocladus*. Notable relicts comprise *Pseudocycas* and *Pseudoctenis* and *Baiera* sp. The ferns are represented mainly by the form-genera *Cladophlebis* and *Sphenopteris*.

The great number of entire leaves coupled with reduced numbers of platanoids as well as the appearance of *Brachyphyllum* is indicative of a more thermophilic aspect of the Dunvegan flora as compared with the Yukon vegetation. This is easily accounted for by the far more northerly position of the latter (almost 20° latitude).

The younger floras of Alaska are allied to the deposits of Chignik suites on the Alaskan peninsula assigned to the Cenomanian. Plant remains were collected from the middle and upper parts of this suite (Hollick, 1930). The lower part composed of marine sediments has yielded ammonites *Pachydiscus* sp. and *Inoceramus undulato-plicatus*. In the Chignik suite one fails to find platanoid leaves which are so numerous in the Caltag and Melozi suites. Ferns are very few within the Chignik suite. The relicts established include *Nilssonia* and *Sagenopteris* while Podozamitaceae, so abundant in the Cenomanian, have not been reported. The composition of conifers on the generic level remains almost unchanged yet *Protophyllocladus* was not found.

Ulmus, Quercus, Viburnites and *Hollickia* ('Rulac') *quercifolia* are there. *Trochoden-*

droides have been described occasionally under the name of *Populus*, e.g. *P. elliptica*. Many entire leaves were defined with the generic names of recent plants (*Magnolia, Cornus, Diospyros, Ficus*, etc.). *Menispermites* and *Cissites* disappear. Krassilov (1979), comparing the floras of the middle and upper parts of Chignik suite which are slightly different, with those from Sakhalin (Guilyak and Zhonkyer), assigns the middle part flora to the Santonian and the upper to the Campanian. But in all probability both the floras should belong to the Campanian since the lower part of the Chignik suite accommodates the ammonites *Pachydiscus*, implying Campanian age.

In western Canada Campanian floras are confined to the Nanaimo series disintegrating into a number of suites exposed on the north-western shore of Vancouver Island (Bell, 1957; Muller and Jeletsky, 1970). Plant remains are not infrequently associated with carbonaceous deposits. Compositionally the forms found here are subtropical which is indicated by a find of a palm (*Geonomites*) which makes us refer this flora to the Potomac province situated to the south and to be characterized below. Growth of palms and other subtropical plants was due to the Campanian warming advancing the boundary of subtropical areas 15 to 20° northwards.

The youngest flora of western Canada (Alberta province), exhibiting a warmtemperate aspect is the flora from the Edmonton series of Maastrichtian age. Stratigraphically the Pascapu suite referred to the Palaeocene (including Danian) rests

Fig. 4.12 Plant fossil of Cenomanian age from western Canada; natural size; *Pseudoprotophyllum boreale* (Dawson) Hollick (from Bell 1963, Plate 35, Fig. 4).

above. From the Edmonton series Bell (1949) determined *Ginkgoites*, *Nilssonia*, 'Cephalotaxopsis', *Metasequoia*, *Cupressinocladus* (*Thuja*), *Trochodendroides*, *Quereuxia*, *Vitis* and *Platanus*. It is readily seen that the discrepancies between the floras of the Okhotsk–Chukotka and Canadian–Alaskan provinces are rather slight. This appears to be due to the existence in the Late Cretaceous of the Bering land bridge contributing to large-scale migration of plants.

The above-mentioned differences are mainly concerned with specific compositions whereas almost all genera are common for both provinces. Exceptions are certain relict genera (*Phoenicopsis*, *Czekanowskia*, *Baiera* and *Sphenobaiera*) which have either become extinct at the Lower–Upper Cretaceous boundary or are not yet found in North America. Podozamitaceae are far more often encountered in the lower half of the Upper Cretaceous of Alaska than in the north-east of Asia.

Earlier, in reviewing the phytochoria of Asia alone, (Vakhrameev *et al.*, 1970) I identified the north-east and Far East of the USSR without the Pacific Ocean province leaving aside North America. When phytochoria are recognized on the surface of the entire globe this name proves unacceptable. Its distribution on the territory of Alaska and western Canada is hardly reasonable because the phytochoria situated in the coastal Pacific region of the USSR and the phytochoria embracing Alaska and Canada highlight certain differences which are found to become more profound as they are looked into.

By the example of the Okhotsk–Chukotka and Canadian–Alaskan provinces which have a substantial meridional elongation it is obvious how the physiognomy of coeval floras changes from warm-temperate to subtropical southwards. The number of species with entire leaves rises considerably from north to south.

The first half of the Late Cretaceous was characterized by the large magnitude of development of broad-leaved forests wherein angiosperms were dominated by deciduous trees with platanoid leaves. The Campanian saw emergence of an appreciable number of evergreen trees and even palms in the south of Canada. This was facilitated by the warming in the Campanian conducive to a shift of the subtropical climate boundary northwards. At the very end of the Cretaceous a certain cooling proved responsible for the flora acquiring a more warm-temperate aspect with *Trochodendroides* especially widespread in the Danian coming to be of noticeable importance.

4.2 Euro-Sinian region

This region encompasses the subtropical belt of the Northern Hemisphere of Late Cretaceous time. It extends through western Europe, the south of the USSR, the Caucasus, Middle Asia, south and central China, Japan, the USA and the southern half of Greenland. During the Campanian warming this region embraced Sakhalin and a south-western area of Canada (Vancouver Island). The characteristic features of this region include the palms occurring from the Turonian and the prevalence of smooth-margined, sometimes narrow-leaved forms among the leaves of angiosperms.

4.2.1 European province

Embarking on a description of the floras of the European province stretching from France and the Iberian peninsula to the Caucasus it is noteworthy that using the works of the last and the beginning of the present century we occasionally effect some corrections in generic names as compared with those cited by researchers in their lists.

Many generic names made use of in classifying contemporary plants have acquired the endings 'ites' or 'phyllum', suggestive of the tentative nature of their affinity to the present-day taxa. Thus, *Aralia* was replaced with *Araliaephyllum*, *Myrtus* with *Myrto-phyllum*, *Quercus* with *Quercophyllum*, *Banksia* with *Banksites*, *Cunninghamia* with *Cunninghamites*, etc. This is notably applicable to the genera making part of the family Proteaceae growing at present in Australia. The works of contemporary palaeo-botanists reveal that assignment of leaf impressions found in the Late Cretaceous deposits of Europe to the genera currently disseminated in Australia (e.g. *Banksia*, *Dryandra*, *'Eucalyptus'*) is borne out by more detailed research.

The genus *Dewalquea* was everywhere changed for *Debeya*, according to the priority rules in effect (Knobloch, 1973). Certain generic names were taken in inverted commas, for instance *'Asplenium'*, since the affinity of this very widespread form to the present-day genus appeared dubious.

The Late Cretaceous floras of Portugal, Spain, Italy, France, West Germany (Aachen), East Germany (Harz, Niederschena), Czechoslovakia, Romania, Bulgaria, Poland, Ukraine and the Caucasus (Daralages) are the representatives of the European province. It is natural that flora could not remain homogeneous over this expanse especially if we take into account that throughout the Late Cretaceous this territory was a combination of small and large islands and peninsulas whose outlines underwent constant change owing to transgressions and regressions.

The most ancient and at the same time the best studied are the Cenomanian floras of Bohemia (Peruč strata), Moravia and the north adjacent part of East Germany (Niederschena). According to new data (Knobloch, 1971) these floras comprised miscellaneous ferns represented by the genera *Anemia*, *'Asplenium'*, *Drynaria*, *Gleichenites* (several species), *Phlebopteris*, *Osmundophyllum*, *Onychiopsis*, etc. Special attention should be paid to the finds of dendrolith stems of ferns of the genera *Dicksonia*, *Oncopteris* and *Tempskya*. These ferns display a close similarity to those inhabiting Europe in the second half of the Early Cretaceous.

The gymnosperms, over and above the prevailing conifers, feature rare entire taeniate leaves of *Nilssonia bohemica*, leaf fragments of *Sagenopteris* and lancet-shaped *Nehvizdya* previously described as *Podozamites obtusus*. The revision studies under-taken by Hluštik (1974a) showed that these leaves are closest to the genus *Eretmo-*

phyllum and are likely to belong to Ginkgoaceae. Conifers are represented by the cones and shoots of various *Sequoia*, *Pinus*, *Frenelopsis alata*, *Cunninghamites oxycedrus* (*'Krannera mirabilis'*) and *Brachyphyllum squammosum*. *Ceratostrobus echinatus* should be noted among the conifers of uncertain systematic affinity. *Dammarites albens* (= *Krannera mirabilis*) presents considerable interest; its remains were recently subjected to a revision study (Hluštik, 1974b, 1977), indicating greatly expanded parts of shoots carrying ribbon-shaped leaves which are a continuation of the thickened scales. After the leaves are dropped the scales form a sort of cone. Hluštik considers that this formation is a genuine cone enclosing the stamens with anther sinuses of lower scales, and ovules in the upper. Up to now such formations have been recorded only in the Peruč series in Bohemia.

Angiosperms feature *Araliaephyllum* (*Aralia*), *Banksites*, *Cocculophyllum*, *Debeya* (*Dewalquea*), *Comptonia* (*'Dryandra'*), *Hederophyllum*, *Magnoliaephyllum*, *Myrica*, *Myrtophyllum*, *Platanus* (*P. cuneiformis*, *P. rhomboides*), *Proteophyllum*, *Sterculiphyllum*, etc. The angiosperms of the Cenomanian from Czechoslovakia are typified by a combination of narrow-leaved forms (*Myrtophyllum*, *Proteophyllum*, *Comptonia*, *Myrica*) with relatively large-leaved forms (*Platanus*, *Debeya*). Abundance of ferns including the dendritic forms implies a hot but moist climate.

Some localities present asymmetrical leaves of angiosperms with imperfect venation reminiscent of the leaves described from the Albian of the east coast of the USA (Patapsco suite) and western Kazakhstan (Kyzylshen suite). A case in point is *Proteophyllum araliopsis* whose leaves change shape from whole through two-lobed to three- and five-lobed. The lobes of these leaves are usually asymmetrical. Occurrence of such forms makes one presume that the lower part of the Peruč strata resting with unconformity on the folded base and forming cavities on its surface may be as old as Albian in age.

By and large the Cenomanian flora of Czechoslovakia and the related but more impoverished flora from the southern part of East Germany (Niederschena) feature a number of peculiarities. In particular, *Platanus* which, as is known, are also widespread in the belt of warm-temperate climate (Siberian–Canadian region) co-occur with the narrow-leaved, often entire forms of *Myrtophyllum*, *Comptonia* (*'Dryandra'*) and *Myrica*, suggesting a subtropical climate. The Peruč strata of Bohemia yield *Dammarites albens* and *Nehvizdya*[1] unfamiliar from other habitats of the European province.

Among other Cenomanian floras of Europe we should point out the Anjou locality (France) wherein plant remains are associated with lagoon deposits. This locality is dominated by remains of *Frenelopsis alata* and *Eretmophyllum*.[1] The structure of the epidermis (sunken stomata, cutinized enclosing cells) indicates a dry climate. The composition of the Late Cenomanian from Cuenca (Spain) is devoid of ferns but miscellaneous conifers appear (*Frenelopsis*, *Sphenolepidium*, etc.). *Debeya* and a palm (*Palaeophoenix*) were detected as well.

---

[1] D. Pons (1979) notes that *Eretmophyllum andegavense* actually may belong to the genus *Nehvizdya*. M.A.A. – Ed.

In the Crimea, localities of plant remains were uncovered (Krassilov, 1984) from the littoral deposits of Upper Albian and Lower and Middle Cenomanian age as characterized by fauna. In comparison with the Late Albian, the Lower Cenomanian witnesses a reduction in the number of fern remains accompanied by vanishing *Ruffordia*, an increasing number of *'Asplenium'*, and emerging *Osmunda*. The remains of conifers are few (*Geinitzia*, *Sequoia*). The diversity of angiosperms is somewhat enhanced (*Graminophyllum* sp., *Myricaephyllum* sp., Nymphaeaceae).

In the Late Cenomanian the composition of ferns is enriched by *Nathorstia* (*Phlebopteris*) *pectinata*. A new genus of cycadophytes appears, i.e. *Taurophyllum*, coupled with *Pterophyllum* sp. There are many shoots of conifers encountered in the Lower Cenomanian joined by *Sciadopitys*. The number of remains of the above-mentioned angiosperms appears increased, featuring the appearance of *Macclintockia cretacea* and *Celtoidophyllum* sp. In the works of V. A. Krassilov the remains of Cretaceous ferns earlier referred to the genus *'Asplenium'* have been transferred to the genus *Anemia*, the Early Cretaceous *Phlebopteris* to the genus *Nathorstia*, while the ferns previously assigned to the present-day genus *Gleichenia* are described within the framework of the form-genus *Gleichenites*.

The Crimean flora of the Upper Albian–Cenomanian should be regarded as another stage of development of the Aptian flora of the southern half of the European part of the USSR, i.e. the floras of the Voronyezh region (localities near Devitsa, Latnoye and Kriushi villages). Just as in the Aptian floras, the Upper Albian and Cenomanian of the Crimea furnish numerous *Ruffordia*, *Gleichenites* and *Nathorstia* (*Phlebopteris*) *pectinata*. However, the Aptian is still devoid of angiosperms appearing in the Upper Albian and gaining wider distribution in Kazakhstan. To a certain extent the flora of the Crimea is similar to the vegetation of the Albian in Georgia, very impoverished in composition. The latter is noted for the presence of *Sagenopteris*, *Zamites*, *Nilssonia*, *Sequoia* and *Sphenolepidium*, while remains of angiosperms were not found here.

An interesting Cenomanian flora has been reported from the littoral deposits of the Transcaucasian area (Daralagez) (Vakhrameev, 1952). It features rare ferns (*Gleichenia*), numerous conifers (*Sequoia*, *Araucaria*, *Brachyphyllum*) and angiosperms represented by small, often narrow leaves (*Comptonia*, *'Eucalyptus'*, *Platanus*, *Lindera*, *Cocculus*, *Araliaephyllum*, *Myrica*, *Proteophyllum*). The small sizes of leaves including *Platanus*, coupled with the narrowness of leaf of a number of forms are indicative of a dry climate of the coast where these plants grew. I tentatively incorporate this locality within the European province but it may equally belong to the Middle Asia province featuring a drier climate.

Turonian floras are known from Bulgaria, Romania and France. The upper Turonian of Bulgaria (Tenčov and Cernjavska, 1965) furnished for description ferns (*'Asplenium'*, *Gleichenia*), a bennettitalean (*Zamites* sp.), conifers (*Elatocladus*, *Widdringtonites*) and angiosperms (*Magnolia*, *Araliaephyllum*, etc.). Plant remains have been collected from inter-coal beds implying a moist climate in the area of growth. In

Romania (North Dobruja, basin of the river Babadag) fern remains are collected in the shallow-water carbonate marine deposits (*Cladophlebis* and *Weichselia*). The same deposits yielded the conifers *Geinitzia*, *Brachyphyllum*, *Cyparissidium*, *Sphenolepidium* and angiosperms *Ficophyllum*, *Magnoliaephyllum*, *Comptonia* ('*Dryandra*'), *Debeya*, *Myrtophyllum*, etc.

Plant remains were collected from southern France (Sabran quarry, the river Seize, right tributary of the Rhône river), namely from the member of obliquely laminated sands enclosing two lignite beds resting inside terrigenous marine deposits of the Upper Turonian (Ducreux, Gaillard and Samuel, 1982). Fern remains are rare and fragmentary. The conifers are represented by shoots and cones of *Sequoia*. Angiosperms are prevalent including *Debeya* (= *Dewalquea*), *Araliaephyllum*, *Celastrophyllum*, *Dryophyllum*, *Sassafras* (= *Lindera?*), *Ficus* and *Sterculites*. The presence of lignite beds and absence of such conifers as *Frenelopsis*, *Pseudofrenelopsis* and *Brachyphyllum* as well as absence of *Classopollis* pollen are suggestive of a moist climate.

Localities of the Cenomanian floras have been reported from East Germany (Quedlinburg), southern Bohemia, West Germany (Aachen) and Portugal. The floras of the Carpathian Ukraine, Poland and Romania are the youngest, of Maastrichtian or even Danian age. From as early as the beginning of the nineteenth century the vicinity of Quedlinburg has been known for a locality containing leaf impressions largely belonging to angiosperms (Mägdefrau, 1956). These feature prevalent *Credneria* with *Myrica* and *Debeya* as well as representatives of Moraceae and Vitaceae. Gymnosperms are represented by *Geinitzia formosa*. Spores of Schizeaceae (*Cicatricosisporites*, *Appendicisporites*) and *Gleichenia* have been recorded. *Cunninghamites*, *Sequoia* and numerous *Credneria* and *Quercophyllum* were described from littoral deposits of the Campanian in Westphalia (West Germany). The Campanian of Upper Pfalz (West Germany) exhibited *Debeya*. It is relevant to record that many imprints of *Platanus* were incorrectly assigned to the genus *Credneria*. The differences between the two are brought out in the work of Vakhrameev (1952).

Rich Senonian floras were recently described from southern Bohemia (Němejc, 1961; Knobloch, 1964; Němejc and Kvaček, 1975). Gymnosperms are represented by the shoots of *Brachyphyllum*, *Geinitzia*, needles of *Pityophyllum* and cone scales of *Dammarites*. The most numerous group is that of angiosperms comprising *Araliaephyllum*, *Cinnamomophyllum*, *Cocculophyllum*, *Credneria*, *Debeya* (= *Dewalquea*), *Grevilleophyllum*, *Laurophyllum*, *Myricophyllum*, *Proteophyllum* and *Quercophyllum*. Many small seeds were found. Ferns and horse-tails are represented by small fragments not allowing evaluation of their systematic affinity.

The south-westernmost locality of Senonian floras is situated in Portugal (Esgueira). Plant remains represented by rare ferns (*Onychiopsis*), gymnosperms (*Zamites* sp., *Frenelopsis*, *Protopodocarpoxylon*) and fragments of leaves of angiosperms were detected in shallow-water littoral formations. It will be noted that a *Frenelopsis oligostomata* found here is the youngest representative of this genus becoming extinct somewhere at the Cretaceous Palaeogene boundary. A stem of the tree-like fern *Neopsaronius* was found in a Senonian flysch deposit in Austria.

Of great interest are the floras from the uppermost strata of the Upper Cretaceous (Maastrichtian–Danian). In the Carpathian Ukraine in the vicinity of Potylitsa village the following impressions of small-leaved angiosperms were found at the turn of the century: *Dryophyllum*, *Debeya* and *Myrica*. Representatives of the genera *Cunninghamia*, *Geinitzia*, *Pinus*, *Sequoia*, *Debeya*, *Dryophyllum*, *Ficus*, *Laurus*, *Magnolia*, *Myrica*, *Myricophyllum* and *Platanus* were described (Karczmarz and Popiel, 1971) from the marine Maastrichtian of the Lublin highland in Poland.

The marine deposits of the Maastrichtian of Romania (Ruska Montana, Carpathians) have yielded many plant remains. The ferns are represented by diverse Gleicheniaceae and '*Asplenium*' (Petrescu and Dusa, 1980). The angiosperms feature palms, various *Pandanus*, and *Cinnamomophyllum*, *Credneria*, *Debeya*, *Ficus*, *Myrtophyllum* and *Platanus*. Conifers are represented only by pollen. Lack of shoots of these plants indicates that conifers grew on elevations more remote from the sea where their pollen appeared. Remains of palms (*Palmophyllum*), *Protophyllum*, *Myrica* and the fern '*Asplenium*' have been collected from another locality situated in Transylvania and also confined to the upper Cretaceous (Maastrichtian–Danian).

Composition change of the Late Cretaceous floras of Europe from locality to locality or from area to area is related not so much to age but rather to different conditions of growth and sedimentation of host deposits. The latter were mostly of littoral origin which is indicated by the finds of marine invertebrates in conjunction with plant remains. Thus, in the majority of instances, we are concerned with vegetation that was growing in the vicinity of coastlines of numerous small and large islands situated in the place of the present-day European continent. Transportation distances for plant remains hardly exceeded tens or perhaps hundreds of metres. Of special interest are the numerous leaf remains of *Pandanus* in the Maastrichtian of Romania (Rusca, Montana) currently growing, as a rule, along sea coasts in subtropical or tropical areas.

Burial of plant remains in certain areas (northern and southern Bohemia) happened on coastal plains in alluvial–lacustrine or deltaic sediments. As a rule, we always find in them numerous ferns whose tender leaves fail to withstand long transportation. It is only in the upper Turonian of Bulgaria that we come across genuine carbonaceous deposits that evolved apparently on a marshy coastal lowland.

The subtropical nature of the Late Cretaceous climate of Europe is indicated by the presence of dendroid fern stems and palm remains finds coupled with abundant narrow often entire leaves. A typical feature is the distribution of the leaves of *Debeya* (= *Dewalquea*) missing from the floras of the warm-temperate belt of Eurasia (Siberian–Canadian region). Only the extreme north-east of Europe could constitute part of the region of a warm-temperate climate.

In considering the composition of the European Late Cretaceous floras we are at this point unable to discern distinct change, indicating a stable climatic fluctuation.

This is apparently associated with the fact that temperature changes occurring throughout the Late Cretaceous made but a slight impact on the largely insular coastal vegetation of this time growing under conditions of mild subtropical climate. As we know, in Asia of which the major part was a vast continent at the time, the above changes were more clearly manifested in alteration of vegetation composition.

Tracing the vertical distribution of the Late Cretaceous plants based on the analysis of their generic composition reveals that the Senonian sees the appearance of the typical *Credneria*, *Dryophyllum* and *Pandanus*, with *Frenelopsis* becoming very rare and the diversity of ferns being drastically reduced (*Tempskya*, Matoniaceae, etc. vanish). Analysis of changes in specific composition at the present state of systematics of Late Cretaceous angiosperms described from various countries by different researchers for over a hundred years will hardly bring about well-based conclusions. This analysis should be carried out only after a revision of the systematics of some groups founded on comparative studies of material emanating from a number of European countries.

4.2.2 Potomac province

This province occupies chiefly the territory of the USA encroaching upon the southern part of western Canada during the Campanian warming. Its southern boundary passes south of Texas, perhaps into Mexico. To ascertain its limits precisely is impossible since this country has not yet exhibited any localities of Early Cretaceous floras. The Jurassic floras situated in the southern part of this country have been referred already to the Equatorial belt.

The monographs devoted to the description of the Late Cretaceous leaf floras were published mainly in the last and at the beginning of the present century. Only the floras of the upper Late Cretaceous (Lance, Medicine Bow) with localities situated in the Rocky Mountains were described in the 1940s. The tendency among palaeobotanists of the time to recognize as many species as possible led to the compilation of extremely long lists of specific taxa. Another peculiarity was the urge to incorporate the species identified within recent genera. This was true in the first place of angiosperms being the youngest and most numerous elements of Late Cretaceous floras of which reproductive organs are very seldom preserved.

The second half of the twentieth century ushered in a new phase in investigations of Cretaceous plants of the USA, now gaining great momentum. On the whole studies deal with separate most curious plant remains, i.e. reproductive organs, woods, leaf cuticle, etc., rather than entire floras. These features render it difficult to give a neat characterization of entire floras limiting us to some very general statements. The Raritan suite is situated above the Potomac series (Early Cretaceous) of the Atlantic coast of the USA. This suite is of continental origin being assigned to the Cenomanian and is rich in plant remains. According to the data of Berry (1916) the Raritan suite is poor in ferns (*Asplenium dicksonianum*, *Gleichenites*). Gymnosperms are more varied, comprising *Andrivettia* (*Libocedrus*), *Brachyphyllum*, *Cupressinocladus* (*Thuja*), *Frenelopsis*, *Pinus*, *Protophyllocladus*, *Sequoia* and '*Widdringtonites*'. *Podozamites* and *Williamsonia* were encountered as relicts.

Angiosperms feature the greatest diversity, being attributed by Berry (1916) to the genera *Andromeda*, *Araliopsis*, *Bauhinia*, *Celastrophyllum*, *Cissites*, *Dalbergites*, *Debeya* (*Dewalquea*), *Diospyros*, '*Eucalyptus*', *Ficus*, *Laurophyllum*, *Leguminosites*, *Liriodendron*, *Magnolia*, *Myrica*, *Platanus*, *Protophyllum*, *Salix*, *Sassafras*, *Viburnum* etc. The abundance of entire leaves including narrow forms is quite conspicuous. Some of them were assigned to the genera *Salix* and '*Eucalyptus*' which we cannot subscribe to at present. *Debeya* and *Liriodendron* are also there being characteristic of the subtropics of the Late Cretaceous epoch. *Platanus* nowadays grow both in warm-temperate and subtropical climates.

The Magothy suite is deposited higher in the sequence, and is dated as Turonian. The composition of the plant remains found in it is little different from that of the Raritan suite. The ferns feature *Osmunda* and *Onoclea*, the gymnosperms *Araucarites*. More diverse at this time are the angiosperms with *Cinnamomum*, *Ilex*, *Juglans*, *Rhamnites*, *Sapindus* (not encountered lower in the sequence) and, what is more important, the leaves of a fan-like palm *Sabalites*.

Continental deposits of the Cenomanian–Turonian age stretch along the coasts of New Jersey and Maryland states encroaching upon the states of North Carolina, Georgia and Alabama. The systematic composition of the plant remains hosted therein (on the generic level) remains virtually unchanged. The younger Senonian deposits are composed of marine sediments. In the south of the USA in Texas (Berry, 1916) there is an outcropping suite, the Woodbine, referred to the Cenomanian. The composition of the plant remains enclosed in it proved close to that from Raritan and Magothy suites.

The Dakota suite, made up of sandstones of continental origin, is developed in the central part of the USA, in the states of Nebraska, Kansas and South and North Dakota, as well as southwards, in Colorado. The lower part of this suite in Kansas is still of Late Albian age while in South and North Dakota its base is located at the Lower–Upper Cretaceous boundary.

One peculiar feature of the Dakota flora is paucity of ferns represented by fragments of '*Asplenium*' *dicksonianum*, *Pteris* sp. and *Gleichenites* sp. The conifers are not well represented either featuring *Araucarites*, *Brachyphyllum*, *Protophyllocladus* and *Sequoia*. The relicts include *Encephalartos*, *Podozamites* (?) and *Williamsonia*. *Nilssonia* was not encountered either here or on the Atlantic coast of the USA. Angiosperms display a great diversity even on the generic level. A great number of entire, occasionally narrow, leaves referred to the genera *Magnolia*, *Ficus* (?), *Diospyros*, *Laurus*, *Sassafras*, *Liriodendron*, *Liriophyllum* and '*Salix*' are concurrent here with a broad range of large-leaved forms with a toothed edge. The latter comprise *Platanus*, *Protophyllum*, *Aspidiophyllum*, *Cissites* and *Aralia* (*Araliaephyllum*).

Protophyllum, widespread in the Dakota suite, illustrates quite a different venation pattern as compared with *Pseudoprotophyllum* from the age-equivalent deposits of Alaska and Siberia. The former is confined in distribution mainly to the subtropical belt whereas *Pseudoprotophyllum* is common for the belt of warm-temperate climate (Vakhrameev, 1976a). However, several researchers dealing with the Late Cretaceous flora of Sakhalin assigned the leaves of *Pseudoprotophyllum* to *Protophyllum*, thus missing a vital diagnostic feature used by myself to separate a flora of a moderate-warm belt from a subtropical one.

Menispermites, *Viburnum*, *Quercus* (?), *Rhamnites* and *Populites* were also encountered in the Dakota suite. There are illustrations of five-lobed entire leaves with long, sharp-pointed lobes referred to *Sterculia*. Such leaves were absent from the coeval floras of the Atlantic coast of the USA and of Texas.

Present-day research by American palaeobotanists (Crane and Dilcher, 1984; Dilcher and Crane, 1984) who have studied flower remains from the Dakota and Woodbine suites reveals their close similarity to Magnoliaceae. The fruit of *Archaeanthus* from the Dakota suite (Kansas) and the leaves of *Liriophyllum kansense* turned out to belong to one plant. Morphologically close leaves were described from the same deposits as *L. populoides*. Another flower, also of Magnoliaceae type, was assigned to the established genus *Lesqueria*. Presence of Magnoliaceae is confirmed by a find of a stem belonging to the Magnoliaceae in the upper Cretaceous of central California which was subjected to a detailed anatomical examination (Page, 1984).

The differences between the Cenomanian floras of the Dakota suite and the age-equivalent floras of the Atlantic coast of the USA amount to the rarer occurrence and lesser diversity of ferns and conifers and broader range of distribution of the large-leaved *Protophyllum*, *Aspidiophyllum* and *Sterculia*. Broadleaved, fairly green forests mingled with conifers growing on the bottom and declivities of wide valleys must have been dominant in the first half of the Late Cretaceous in Kansas, Dakota and the adjoining parts of other states. The presence of conifers is corroborated by evolved sandstones of alluvial origin constituting the Dakota suite.

Younger floras in the central part of the USA are reported further south, in the basin of San Juan (Tidwell, Ash and Parker, 1981) situated mostly in the state of New Mexico. Here the Mesoverde series lies above the sandstones of the Dakota suite. This series is divided into three suites. The middle of these, Menefee, is composed of coal-bearing deposits containing plant remains. Ferns are represented by *Asplenium dicksonianum* and *Anemia hesperia*. The representatives of the genera *Araucaria*, *Sequoia*, *Metasequoia* and *Protophyllocladus* were found among the conifers. The angiosperms found include *Ficus*, *Trochodendroides* (?), *Platanus*, *Dryophyllum*, *Cinnamomum*, *Lauro-phyllum*, *Rhamnites*, *Cissus* and *Vitis* coupled with the fan-like (*Sabalites*) and pinnae-form (*Phoenocites*) palms. The same suite furnished remains of large dinosaurs. The age of the Menefee suite is apparently Turonian.

Two suites are located stratigraphically higher, i.e. the Fruitland and overlying Kirtland. They host floras of similar composition enabling us to give their common characterization. Ferns are not common but include '*Asplenium*', *Osmunda*, *Anemia* and *Cyathea*; *Onoclea neomexicana* appears for the first time and is very close to the contemporary *O. sensibilis*. Conifers exhibit representatives of the genera *Araucaria*, *Brachyphyllum*, *Sequoia*, *Metasequoia* and *Cupressinoxylon*. Angiosperms feature various species of *Pistia*, '*Salix*', *Carya*, *Quercus*, *Ficus*, *Nelumbo*, *Trochodendroides* (?), *Menispermites*, *Magnolia*, *Laurus*, *Platanus*, *Leguminosites*, *Rhamnus-Vitis*, *Cissus*, *Myrtophyllum*, *Dillenites*, *Cornus*, *Viburnum* and *Dombeyopsis*, as well as the leaves of the palms *Sabalites*, *Phoenocites* (?) and the well-preserved silicified trunk sections (*Palmoxylon* spp.) including their bases with diverging roots.

Occurrence of palms testifies to the affinity of the Late Cretaceous flora of the San Juan basin to the subtropical belt. Numerous coal-beds and aquatic plant remains suggest the existence of marshy littoral lowlands perhaps separated by drained areas which accommodated planes, ancient oaks, magnolias, laurels and other plants. The age of the deposits containing the Fruitland and Kirtland floras is Senonian.

Accurately dated Early Campanian floras are known from the very south of western Canada near the border with the USA on the western fringes of Vancouver Island. The subtropical aspect of these floras suggests their incorporation into the Potomac rather than the Canadian province. The plant remains are confined to the Nanaimo series composed of both marine and continental deposits. The formation with plant remains divided into a number of suites is underlain by marine formations of the Santonian and overlain by Campanian (Muller and Jeletzky, 1970). This flora, studied by Bell (1957), comprises relatively numerous ferns, i.e. *Anemia*, *Saccoloma*, *Davallites*, *Dryopteris*, '*Asplenium*', *Sphenopteris* and several species of the form-genus *Cladophlebis*. Some of the illustrated forms may belong to *Gleichenites*. *Onoclea hebridica* was also found. Relict forms may include *Pseudoctenis* and several species of *Nilssonia*. *Ginkgo* ex gr. *adiantoides* was found. Conifers are represented by *Metasequoia*, *Glyptostrobus*, *Cupressocladus* (*Thuja*), *Amentotaxus* and *Protophyllocladus*.

Angiosperms proved to be the most numerous featuring *Dryophyllum*, *Arctocarpus*, *Trochodendroides*, *Menispermites*, *Liriodendron*, *Platanus*, *Cinnamomum*, *Laurophyllum*, *Bauhinia*, *Dalbergites*, *Leguminosites*, *Celastrophyllum*, *Ternstroemites*, '*Zizyphus*', *Debeya* (*Dewalquea*), *Sapindus*, *Rhamnites*, etc. Presence of palms (*Geonomites*) appears to be an important factor for evaluation of climatic conditions.

The flora in question combines such subtropical elements as the palms, *Debeya*, *Dryophyllum* and *Liriodendron* which are numerically prevalent, with *Trochodendroides* in a warm-temperate flora. This confirms the peripheral position of the Nanaimo floras with relation to the subtropical belt. The presence of abundant conifers (notably *Metasequoia*) and rather frequently occurring *Nilssonia* reflects confinement of these floras to the Pacific Ocean coast indicating a marine moist climate.

The floras of Nanaimo are appreciably different from the floras of the central parts of the USA (Dakota) which are almost deprived of ferns and feature fewer conifers. The climate of the central part of the USA might have been less humid (possibly with a drier summer) than that of the Pacific Ocean coast.

Localities of younger Late Cretaceous floras are situated in the western states of the

USA, where two roughly coeval floras are reported. The first is associated with the Lance formation (Wyoming) overlying the Fox-Hills suite. The latter has yielded an ammonite *Sphenodiscus lenticularis* implying the Upper Senonian. The age of this flora is regarded as Maastrichtian because its roof includes the Fort Union formation whose lower part is attributed to the Danian while the upper part is Palaeocene (Anon., 1977).

The Medicine Bow formation coeval with the Lance suite ranges through the southern part of Wyoming and the adjoining part of Colorado. This formation also rests on the Fox-Hills. The floras from the Lance and Medicine Bow formations which are very similar in composition were examined by Dorf (1942). They exhibit the ferns *Anemia*, '*Asplenium*', *Woodwardia* and *Salvinia* (?). Gymnosperms are represented by *Ginkgo* ex gr. *adiantoides*, *Metasequoia*, *Sequoia* (?) and *Araucarites*. The majority of imprints belong to angiosperms *Pistia*, '*Salix*', *Myrica*, *Dryophyllum*, *Ficus*, *Platanophyllum*, *Laurophyllum*, *Menispermites*, *Magnoliaephyllum*, *Vitis*, *Grewiopsis*, *Dillenites*, *Myrtophyllum*, *Dombeyopsis*, *Viburnum*, '*Zizyphus*' and a number of other forms. The presumed presence of *Cercidiphyllum arcticus* (= *Trochodendroides arctica*) is doubtful judging from the photograph of this fossil. It is important to note the presence of well-preserved palm leaves (*Sabalites*) in both the suites, corroborating the subtropical aspect of the flora.

The lower half of the Fort Union formation, deposited stratigraphically higher, also features the presence of palms. Brown (1962) in describing these remains related them directly to the contemporary genus *Sabal* but they are indistinguishable from the leaves of palms illustrated by Dorf (1942) from the underlying Medicine Bow and Lance suites and referred to the genus *Sabalites*. Unlike the floras described from these two suites the upper half of the Fort Union formation sees the emergence of *Carya*, *Juglans*, *Betula*, *Corylus*, *Castanea*, *Ulmus*, *Zelkova* and *Acer*. The number of twigs of *Metasequoia* is sharply increased, coupled with the appearance of twigs of *Taxodium* and *Glyptostrobus*. Comparison of the specific composition of Fort Union and Lance suite floras reveals only five common species.

Such a major change of flora composition is doubtless due to the climatic alteration from subtropical to warm-temperate and probably more humid (abundance of ferns). We have already dealt with the differences pointed out, notably between the floras from the Pacific Ocean coast and the central areas of the USA likely to have been caused by varied climatic conditions prevalent within the Potomac province. Its subtropical climate is well recorded by the dominant evolution of plants with smooth-margined, occasionally narrow leaves approximating to the present-day genera common in subtropical areas.

An essential criterion is the finds of palms on the Atlantic coast (Magothy formation, Turonian), on the Pacific Ocean coast (Nanaimo series, Campanian), in New Mexico (Upper Cretaceous of San Juan basin) and in the western states (Lance and Medicine Bow formations, Maastrichtian). Emergence of palms from as late as the Turonian is not necessarily due to climatic change but rather to the evolutionary process because the pre-Turonian deposits prove devoid of undoubted palm remains.

Differences between the floras of the Potomac province and those of the European province considered on the generic level are not major. Europe displays petrified stems of dendroid ferns (*Dicksonia*, *Oncopteris*) missing so far from the USA. *Tempskya* is common for both. *Metasequoia* and *Taxodium* occur in the USA earlier, in the second half of the Late Cretaceous, while in Europe they become well-developed only in the Palaeogene. *Protophyllocladus* and *Libocedrus* (*Androvettia*), as well as such angiosperms as *Dillenites*, *Liriodendron*, *Trochodendroides* and large-leaved *Protophyllum*, are lacking in Europe. Genuine *Credneria* were not found in the Potomac province.

There are slightly more numerous common features between the Potomac province and the Okhotsk–Chukotka and Japanese provinces. This is attributable to the existence of the Bering land accommodating migration of plants and animals between Asia and North America which was especially intense as the northern boundary of the subtropical area advanced northwards.

4.2.3 Greenland province

The localities of the Late Cretaceous plants on the western coast of Greenland are confined to the littoral outcrops of the Nugsuak peninsula and in part Disco island. The Cretaceous continental deposits rest here with unconformity on the gneisses of the Precambrian.

Plant remains were studied by Heer (1882, 1883) and Seward (1926). The latter supplemented investigation of his own collection with revision of the definitions of Heer, significantly reducing the list of species proposed by the earlier researcher. Like the majority of palaeobotanists of the last century, Heer identified new species relying on minute differences which should be considered as intraspecific or related to heterogeneous preservation of fossil material.

Initially Heer recognized three fossil floras, i.e. Kome, Atane and Patut, assigning them to a variety of ages ranging from the upper strata of Lower Cretaceous (Kome) to Senonian (Patut). However, later Seward who visited Greenland came to doubt the stratigraphic sequence of the palaeofloras identified by Heer. Koch (1964) also came to a similar conclusion. Recently, Budantsev (1983) who studied the Greenlandian flora from literature inferred that in systematic composition the Patut flora is younger than the Atane.

To settle this issue it is necessary to conduct a detailed stratigraphic investigation in this area of Greenland with allocation of the localities of Cretaceous floras to sequences followed by correlational analysis. In the first place it is necessary to clarify the correlation of continental formations with the marine horizon containing the Late Santonian *Inoceramus patootensis* and *Scaphites nicoletti*. For our purpose, which is to characterize the peculiarities of the Greenland phytochoria, we shall regard it as a whole since we do not see any possibility of adopting an alternative concept at present. The characterization below relies on the lists of taxa cited by Budanstev (1983) who critically revised the figures of leaves supplied in the works of Heer (1882, 1883), excluding some species based on remains of dubious systematic affinity or forms whose names are synonymous with those of other species. We also make use of the work of Seward (1926).

Presence of the representatives of the following genera has been ascertained among ferns: 'Asplenium', Osmunda, Dicksonia (petrified trunk), frequently encountered Gleichenites (3 species), Phlebopteris, Hausmannia, Onychiopsis, Cladophlebis (5 species) and Sphenopteris (2 species). V. A. Krassilov, who examined the fertile pinnae from the Late Cretaceous of Sakhalin referred to Gleichenia, revealed that the structure of sori in these pinnae implies their affinity to the cyathean ferns. It is likely that in Greenland Gleicheniaceae are present concurrently with Cyatheaceae but the presence of the former (Gleichenites) is borne out by the finds of dichotomizing rachises found in great numbers. Cycads are represented by Nilssonia johnstruppii and Pseudoctenis latipennis, and the bennettitaleans by Pseudocycas streenstrupii, P. insignis, Ptilophyllum sp. and Williamsonia ? sp.; scraps of Taeniopteris were found as well. Gymnosperms feature Podozamites, Protophyllocladus, Cryptomeria, Sequoia, Cupressinocladus (Thuja), Cyparissidium, Widdringtonites, Moriconia, Elatocladus, Pagiophyllum and Pityospermum.

Angiosperms are represented by Magnoliophyllum, Magnoliistrobus, Laurophyllum (4 species), Cinnamomides, Menispermites (3 species), Arctocarpus, Juglandiphyllum, Andromeda, Cassia, Bauhinites, Dalbergites, Diopsyros, Leguminosites, Myrtophyllum, Sapindopsis, Macclintockia (2 species), Debeya (Dewalquea), Cornophyllum, Celastrophyllum, Zizyphoides, Cissus, etc. This list should be supplemented by Ginkgoites pluripartita, established by Seward and omitted for some reason by Budantsev, as well as Phoenicopsis steenstrupii and Baiera ikorfatensis. Ginkgoites feature dissection of the leaf lamella into separate lobes.

Looking at the composition of this flora we perceive at once certain features allowing recognition of it as separate from other phytochoria. This is primarily the abundance of ferns with outstanding Gleichenites having dichotomous branching rachises and Phlebopteris represented mainly by fertile pinnae. Among fern remains a find of a fragment of dendroid fern stem should be noted (Dicksonia punctata) belonging to the species originally described from the Upper Cretaceous of Czechoslovakia and suggesting the existence of a subtropical climate during its growth.

The second peculiarity is the presence of relicts manifested in Ginkgoaceae, i.e. Baiera and Phoenicopsis, cycads (Ctenis) and bennettitaleans (Pseudocycas and Williamsonia). These should be coupled with the fragments of taeniate leaves assigned by Seward (1926)to Taeniopteris. Relicts may also be said to include Ginkgoites with the lamella dissected into narrow lobes as well as Podozamitaceae. Among relicts we also find such subtropical genera as Pseudocycas and Ptilophyllum.

The leaves of angiosperms are often represented by entire forms peculiar to a subtropical climate (Laurophyllum, Cinnamomoides, Andromeda, Bauhinites, etc.). Debeya

(= *Dewalquea*) is also present, being typical for ecotonal zones between the subtropical and warm-temperate climates. Platanoid leaves feature only *Platanus* (*P. latiloba*) while the large leaves of the *Pseudoprotophyllum* type widely disseminated in Siberia, the north-west of the USSR and Alaska were not encountered in the belt of warm-temperate climate of Greenland.

This suggests that Greenland (at least its southern half) of Late Cretaceous times was situated near the northern boundary of the subtropical belt. This is supported by the composition of the Late Triassic and Early Jurassic floras (localities of the Upper–Middle Jurassic floras are absent in Greenland) coupled with palaeomagnetic research data. The climate was moist which is indicated by abundant ferns and a large number of moisture-loving relicts surviving the epoch of the Early Cretaceous.

A conspicuous feature is the similarity between the Late Cretaceous flora of Greenland and Sakhalin permitting Krassilov (1979) to identify the Greenland–Sakhalin phytochoria which he regarded as an 'ecotone of subtropical and warm-temperate zones'. That Greenland and especially Sakhalin (see below) were situated in the boundary zone between the warm-temperate and subtropical belts is a fact, but is it possible to merge these two regions so remote from each other into one phytochoria? In this case it should be stretching throughout the North American continent including at least the southern part of Canada. But data for this are insufficient. Apart from this it is hardly reasonable to regard an ecotonal area between two regions as an independent region. This ecotonal zone is likely to be even narrower than the basic zones thus entailing delineation of two boundaries segregating the ecotonal zone from the basic phytochoria. Recognition of Greenland as an independent province looked upon as an ecotone is in great measure conventional because we cannot draw either its southern or northern boundaries due to lack of data.

4.2.4 Middle Asia province

The Cenomanian floras of this province, as was pointed out above, still had a moderate enough aspect. The warming that set in at the end of the Cenomanian markedly changed the composition of the Turonian flora of the southern Urals and particularly the Aral Sea vicinity making us refer this territory to the Siberian–Canadian region situated in the zone of warm-temperate climate. The province embracing Kazakhstan and Middle Asia in the Late Cenomanian time is termed Middle Asiatic by myself.

The remains of the Turonian plants (Shilin, 1986) are confined to the localities Ayat (eastern slope of the Urals), Jaksybutash, Dyurmen-Tube, Jamanshin, Kankazgan, Tyuratam, Beleuty (north-east Priaralje area), and Kyzyldjar (north-western tip of the Karatau ridge), etc. Although in many localities we find numerous fossil leaf impressions of *Platanus* they are accompanied by considerable numbers of entire, occasionally narrow leaves assigned to *Cocculus, Daphnophyllum, Diopsyros, Laurophyllum, Leguminosites, Lindera, Magnolia, Myrtophyllum, Sophora*, etc. Some are recorded in the Cenomanian floras as well (*Dalbergites, Ficus, Laurophyllum, Lindera, Magnolia*) but in the Turonian they become more numerous. A find of *Liliodendron* from Kankazgan is of special interest. These are concomitant with persisting species of *Cissites* (prevalent in Taldyespe locality), *Menispermites* (*Viburnum*) and *Sapindus*. Rare leaves have been assigned to *Viburnites* (*Viburnum*) and *Trochodendroides*. The affinity of platanoid leaves to the genus *Platanus* is confirmed by a find of collective fruit very closely similar to those from present-day planes but featuring a far smaller diameter (Samsonov, 1966, Table IX, Fig. 2). Conifers continue to be represented by Taxodiaceae (*Sequoia*), Pinaceae (*Pinus*) and emerging *Protophyllocladus polymorphous* (Kangazgan, Kyzyldjar) missing from the older deposits. In comparison with the Cenomanian and Late Albian, fern remains are encountered more and more seldom (*Asplenium dicksonianum*) and in a number of localities they are altogether absent. Difference between the coeval Turonian floras of the Chulym–Yenisei basin and Kazakhstan increase. These changes indicating drying and simultaneous warming during the transition from the Cenomanian to the Turonian are borne out by the changing lithologic composition of host deposits. The deposits of the Altykuduk suite (Upper Albian–Cenomanian) composed mainly of obliquely laminated sands of alluvial or deltaic origin are replaced with red or variegated primarily clayey formations (Jirkendek suite in the northern Priaralje area). It will be also noted that the amount of *Classopollis* pollen in the Turonian is appreciably enhanced (Vakhrameev, 1980), suggestive of climatic change towards drying and warming.

In the Senonian the greater part of the Priaralje area and the Turgay depression were inundated by sea, therefore here we already fail to find localities with plant remains of the land flora.

Localities of Senonian plants associated with continental sediments have been reported from the right bank part of the Syr-Darya river and Chu-Sary region (Shilin and Romanova, 1978; Shilin, 1986). The former includes the Shah-Shah locality related to the red deposits of the Bostobinsk suite attributed to the Santonian. The latter houses the locality Taldysay; its age is assessed as Santonian–Early Campanian because the upper boundary of the continental deposits containing plant remains encloses marine sediments of the Upper Campanian or Maastrichtian. The fossil flora from these two localities as well as from others situated north-west of the Shah-Shah locality (Baybishe, Akkurgan-Boltyk, Irinino) consists of angiosperms and to a lesser degree of conifers. The latter are represented by shoots with mainly short reduced needles, i.e. *Sequoia reichenbachii, Glyptostrobus groenlandicus, Cyparissidium gracile* and *Brachyphyllum* sp. Sporadic cones or their scales were also found, being referred by P. V. Shilin to *Agathis, Pinus* and *Sequoia*. The impressions of elongated lancet-like and apparently narrow leaves were described under the form-generic name *Dammarophyllum*.

Angiosperms are represented by the genera *Andromeda, Araliaephyllum, Aryskumia, Celastrophyllum, Celtidophyllum, Cissites, Dalbergites, Diospyros, Laurophyllum* (*Laurus*), *Magnolia, Myrica, Myrtophyllum, Quercus, Populus, 'Salix'* and *Trochodendroides*. In spite of meticulous collection not a single impression was recorded of plane or platanoid leaves which are so abundant in the Turonian and particularly the Cenomanian floras. Remains of angiosperms are largely manifested in smooth-margined, often very small leaves (*Aryskumia*, *Laurophyllum*, *Myrica*, *Myrtophyllum*, '*Salix*', etc.). The new genus *Aryskumia* Shilin proves to be of particular importance being represented by two species, one of these being similar to *Zelkova* (*A. zelkoviifolia*) and the other to an elm tree (*A. ulmifolia*). Individual representatives of *Celtis*, *Quercophyllum* (*Quercus*), *Ulmus* and *Viburnum* are emerging.

The general composition of the flora and leaf morphology suggest advanced xero-phytization of climate as compared with the Turonian, the same being corroborated by the complete disappearance of ferns. The dry climate is also evidenced by the high proportions of *Classopollis* pollen (usually around 50%) and ephedroid-type pollen assigned to the genus *Gnetaceaepollenites*. In the Cretaceous deposits of more northerly areas (e.g. south of western Siberia) with a moister climate *Gnetaceaepollenites* is encountered in the form of separate grains with the amount of *Classopollis* also sharply reduced. The local dry climate is signalled by the finds of crocodile and dinosaur bones at the Shah–Shah locality. The most favourable crocodile body temperature amounts to about 35 °C while the lower limit of activity is 20 °C. P. V. Shilin (Shilin and Romanova, 1978) justly believes that at the time in question this territory was covered by forest vegetation only in the river valleys.

The presence of *Trochodendroides* in conjunction with the collective fruit of *Trochodendrocarpus* detected in Shah–Shah is apparently due to the relative proximity of the East Siberian province which is part of the Siberian–Canadian region. It is characteristic that remains of *Trochodendroides* were observed in the European province.

A number of common species and the subtropical physiognomy of flora approximate the Senonian vegetation of southern Kazakhstan to the coeval flora of the European province. However, the former proves more xerophilic containing many endemic species (Shilin and Romanova, 1978) as well as the endemic genus *Aryskumia*.

The youngest, possibly, Maastrichtian floras of eastern Kazakhstan represented by the localities of Juvankara (northern shore of the Zaysan lake) and Ulken–Kalkan (basin of the river Ili) are already of a totally different nature. This is moisture-loving vegetation containing *Metasequoia*, *Taxodium*, various *Trochodendroides*, *Platanus* and rare *Corylus* (Ulken–Kalkan). This flora (Makulbekov, 1974; Shilin and Romanova, 1978) signals moistening and cooling of climate in the Maastrichtian causing the advance to the south of the southern boundary of the Siberian–Canadian region encompassing at this time the area of eastern Kazakhstan as well. This flora is very similar to that of Tsagayan of the Far East which we shall deal with in detail when describing the Siberian–Canadian region.

The eastern extension of the Euro-Sinian region is to be sought in China and Mongolia. However, the widespread development of red deposits of this age does not contribute to good preservation of plant remains. There are very scarce data on the composition of the Late Cretaceous floras in China and Mongolia. Krassilov and Martinson (1982) report a find in the Early Cretaceous of Mongolia of fruit imprints of *Nyssoidea mongolica* and *Botrycaryum gobiense*. Cones of Araucariaceae were also found coupled with fragments of petrified woods of gymnosperms.

The existence of vegetation, most probably of the savannah type, is indicated by finds of bones and sometimes whole skeletons of large dinosaurs known mainly from the southern and eastern parts of Mongolia and the adjacent part of north-eastern China. These gigantic reptiles were hardly adapted to living in dense almost impassable forests like those that covered the territory of Siberia and the Far East in the Cretaceous.

The southern part of the Middle Asia province embracing Fergana and Tadjikistan is almost deprived of plant macro-remains. Marine shallow-water formations are developed here, occasionally including gypsum beds (Turonian) and red or variegated continental deposits. In Fergana the latter have yielded plant remains living in a littoral zone of a fresh water lake. *Quereuxia angulata*, *Nelumbites* sp. and *Typha* sp., as well as *Equisetites* sp., were found here. No leaf remains of woody plants were found among these. Probably these were inhabitants of a small lake situated in a forest-free area, in all probability savannah.

The basic difference between the Middle Asia province and the European province caused by different distribution of land and sea was that the former occupied the western projection of the vast continent extending to the east (China) and reaching up to the Pacific Ocean whereas the European province embraced large and small islands existing in the Early Cretaceous time in place of Europe.

The direct contact between the Middle Asia province and the Siberian–Canadian region occurring on this mainland was conducive to favourable conditions for mutual exchange of plant elements from both phytochoria and for formation of ecotonal floras. Thus, the composition of the Late Cretaceous floras of Kazakhstan often reveals the presence of *Trochodendroides*, i.e. the genus which became widespread in the Siberian–Canadian region and was totally missing from the European province. Another genus of this kind was *Cissites*, widely disseminated in Siberia and the Pacific Ocean coastal area of the USSR as well as in Kazakhstan but almost unknown in Europe.

Climatic change led to boundary migrations between these regions. As was already shown, during the moistening and cooling of climate in the Albian (mostly Late) and the Maastrichtian this boundary shifted noticeably southwards while afterwards during warming and drying it moved northwards (Turonian–Campanian).

Within Europe the eastern extension of the Siberian–Canadian region was located chiefly in the recent Arctic basin possibly encompassing only the northern part of Scandinavia separated from the remainder of Europe by a sea basin inhibiting penetration from the north of the elements of the Siberian–Canadian region. The only locality of a flora apparently belonging to the Siberian–Canadian region and found within Europe is situated on the western slope of the northern Urals on the river

Lemve (Baykovskaya, 1956). Leaf impressions were collected from the littoral deposits of Santonian age. Presence of *Macclintockia lyelli*, widespread in the Late Cretaceous floras of Siberia and the North-East implies affinity of the plants of this locality to the Siberian–Canadian region.

The occurrence of flora belonging to this region in the north-west of Asia is borne out by the locality situated on the river Lozva (eastern slope of the northern Urals) hosting a rich complex of plants. An accurate age for this vegetation is not known. Some refer it to the Senonian (Baykovskaya, 1956), others to the Palaeocene (Krishto-fovich, 1933), but its affinity to the Siberian–Canadian or Boreal region (Budanstev, 1983) is unquestionable. Its composition also features two species of *Macclintockia* including *M. lyelli* coupled with *Trochodendroides* ex gr. *arctica* (*T. richardsonii*).

Another difference between the European province and the Middle Asia province is a more moisture-loving aspect of the flora due to the mild marine climate of the former. This stands out noticeably when comparing the Senonian floras of western Europe and Kazakhstan. The composition of the Senonian flora of western Europe features occurrence of the large-leaved *Credneria* and *Debeya* coupled with numerous ferns in addition to the smooth-margined and narrow-leaved forms of *Laurophyllum* and *Myrtophyllum* type while the Senonian of Kazakhstan is devoid of large-leaved angiosperms and ferns. As was shown above, carbonaceous deposits were recorded in the Upper Turonian of Bulgaria.

4.2.5 Japanese province

The flora from the Asuva suite developed in the central part of Japan (Matsuo, 1962) may presumably be referred to the oldest Late Cretaceous floras of this province, namely to the Cenomanian or Turonian. The correlation of rocks enclosing plant remains along with faunistically characterized marine deposits remains obscure. *Osmunda*, several species of *Cladophlebis*, *Nilssonia*, *Zamites*, *Ginkgoites*, *Sequoia*, *Metasequoia*, *Menispermites*, *Nelumbium*, *Viburnites*, etc. were described from here. Matsuo described five species of *Nilssonia*, but a revision of the illustrations suggests reduction to two, at most to three, species. The sparse composition of angiosperms probably points to the affinity of the flora to the lower part of the Upper Cretaceous.

The Omchidani flora which is close to Asuva compositionally belongs to the same stratigraphic level (Matsuo, 1970). Within Omchidani no leaf imprints of angiosperms were found but several fruits were assigned to this group of plants. Younger floras have been carefully studied by Tanai (1979) in the north-east of Honsyu in the Kudzi district.

Three suites are distinguishable here (upwards): Tamagava, Kunitan and Savayama. The lower and the upper ones host plant remains, while the middle Kunitan is com-posed of littoral terrigenous deposits containing ammonites (*Gaudryceras denseplicatus*,

Polyptychoceras subundulatum), inocerams (*Inoceramus japonicus*) and a number of others. Presence of this fauna is suggestive of Late Santonian age for the Kunitan suite.

The underlying Tomagava formation has yielded *Equisetum, Osmunda, Gleichenites, Asplenium dicksonianum,* several species of *Cladophlebis, Sachalinia sachalinensis, Glyptostrobus, Araucarites, Dammarites, Brachyphyllum, Protophyllocladus, Magnolia, Platanus, Dryophyllum, 'Salix', Sapindophyllum, Cissus, Debeya tikhnovichii,* etc. The presence of *Protophyllocladus* and *Debeya* is conspicuous. The former of these does not usually appear earlier than in the Turonian while the latter emerges only after the Santonian. This correlates well with the age assigned by Matsuo to this flora, i.e. Lower Santonian. The Mgachi flora from Sakhalin is similar to the above in composition being assigned by Krassilov (1979) to the Santonian and also containing *Protophyllocladus, Debeya* and *Liriodendron.*

The flora from the Savayama formation covering the Late Santonian deposits of the Kunitan suite comprises *Equisetum, Anemia elongata, Gleichenites, Asplenium dicksonianum, Salvinia, Adiantopteris,* several species of *Cladophlebis, Sachalinia, Zamiopsis, Otozamites schenkii, Nilssonia, Glyptostrobus, Metasequoia, Araucarites, Thuja (Cupressinocladus), Cyparissidium gracile, Protophyllocladus, Liriodendron, Magnolia, Laurophyllum, Cinnamophyllum, Menispermites, Trochodendroides, Platanus, Dryophyllum, Dillenites, 'Salix', Sapindophyllum, 'Zizyphus', Hemitrapa, Debeya,* etc. The deposition of the Savayama formation above the Kunitan formation assigned to the Late Santonian permits reference of the former to the Early Campanian.

The presence in it of the relict *Otozamites* is noteworthy. This relict is typical of the Jurassic and Early Cretaceous of the Euro-Sinian region and is indicative of subtropical climate. However, the above entire forms co-occur with the leaves having a broad lamella with a toothed or cavitous margin, viz. *Platanus, Menispermites, Trochodendroides.* In Sakhalin the Savayama flora may correspond to the Jonkyer suite vegetation which is indicated by the presence of such smooth-margined and often narrow-leaved forms as *Magnolia, Liriodendron, Laurophyllum* and *'Salix'.*

The Oarai flora proves to be the youngest of the Late Cretaceous floras of Japan. This floral locality is situated north-east of Tokyo (Oyama and Matsuo, 1964). Its composition features a palm *Sabalites caraiensis.* The ammonites recovered from the overlying marine deposits imply an Upper Senonian age. Apart from palm remains this suite has yielded *Zamiophyllum, Zamites* and many other plants. However, many definitions cited in the work of Matsuo (1962) are probably outdated.

In Sakhalin this flora seems to correspond in age to the Avgustovsk or Bosnyakovsk vegetation (within the volume suggested by Krassilov, 1979) which feature *Nilssonia* but are devoid of palms. The youngest Danian or even Early Palaeogene flora is one on the river Takhobe (Primorje) with representatives of the genera which become widely developed in the Palaeogene (*Zelkova, Alnus, Corylus, Acer*) over and above the forms disseminated in the Cretaceous. A dubious impression of a *Nilssonia* leaf has also been reported. Fragments of platanoid leaves are numerous.

Giving a general evaluation of the flora of the Japanese province it should be emphasized that it contains such thermophilic plants missing from the adjacent Okhotsk–Chukotka province as palms, *Otozamites*, *Debeya*, *Liriodendron*. Entire and occasionally narrow leaves also prove to be more numerous in the Japanese province. On the other hand, it is deprived of such moderate elements as *Phoenicopsis* as well as representatives of the genera *Macclintockia* and *Pseudoprotophyllum*.

The most thermophilic subtropical floras of the Japanese province are associated with Japan proper whereas the floras of Sakhalin and Primorje are ecotonal in nature. In pre-Santonian time Sakhalin (Figs. 4.13, 4.14) tended more to the Okhotsk–Chukotka province while from the second half of the Cretaceous and especially in the Campanian marked by a warming it, as well as the Primorje area, was situated within the northern fringe of the Japanese province. At the same time Sakhalin received many of the migrating representatives of the subtropical genera (*Cycas*, *Debeya*, *Liriodendron*). Abundance of *Nilssonia* is noted for this time as well.

The elements of largely warm-temperate climate include apparently deciduous tree forms possessing large platanoid leaves which became most widespread in Sakhalin in pre-Santonian time (*Platanus*, *Pseudoprotophyllum*, *Paraprotophyllum*). They are widely distributed in the deposits of the Turonian, Coniacian and Santonian of the Okhotsk–Chukotka province. They are also numerous in Sakhalin embracing, however, a narrower time interval (the Ayn complex of the lower half of the Arkovo suite assigned to the Coniacian).

Platanoid leaves are abundant in the Sabuinsk suite of the Primorje area. Ablayev (1974) thinks it to be younger (Coniacian–Santonian) in comparison with the Partisansk suite deprived, according to him, of these forms. The age of the latter is determined by him to be Cenomanian–Turonian. In the lists of the Partisansk suite vegetation Nevolina (1984) includes such platanoids as *Platanus*, *Protophyllum*, *Pseudoprotophyllum* and *Credneria* referring the Partisan group to the Turonian–Early Coniacian. Going by all data the peak of development of the platanoid forms in the east of Asia coincides with the Turonian–Coniacian which is probably related to the time of southward expansion of the warm-temperate zone whose boundary was situated at this time south of Sakhalin. The numbers and diversity of these forms declined to the south. In Japan (Tanai, 1979) they are represented by sporadic species of planes. The Takhobe flora, just as the Boshnyakovsk suite vegetation (Sakhalin), again highlights the emergence of numerous large-leaved forms implying a certain cooling and simultaneous moistening of climate during this era.

4.2.6 South China province

This province incorporates central and southern China. The northern and northeastern parts of this country belong to the Siberian–Canadian region. In the very south of China (Hsü, 1983) the deposits of the Cenomanian (Boli suite) in Guansi province yielded the shoots of the squamifoliate conifers *Brachyphyllum rhombiomaniform* and

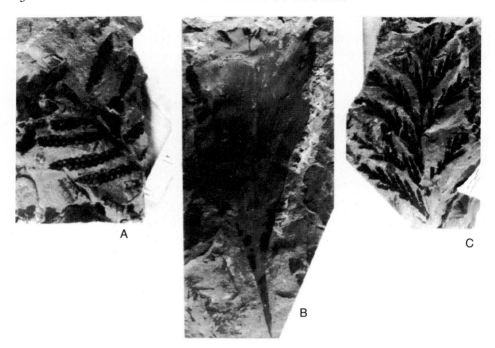

Fig. 4.13 Plant fossils of Late Cretaceous age from Sakhalin; natural size.
A, *Gleichenia crenata* Kryshtofovich; B, *Protophyllocladus polymorphus* (Lesq.)
Berry; C, *Thuja cretacea* (Heer) Newberry.

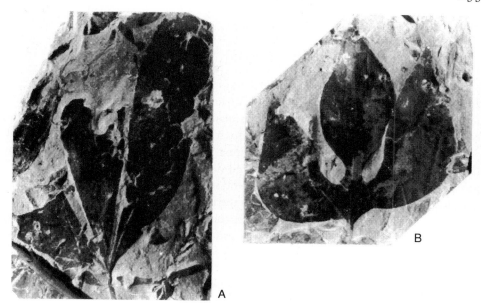

Fig. 4.14 Plant fossils of Late Cretaceous age from Sakhalin; natural size; A, *Debeya tikhonovichii* (Krysht.) Krassilov; B, *Araliaephyllum* (= *Sassafras*) *palevoii* (Krysht.) Krassilov.

entire, possibly, sclerophyllous angiosperms such as *Cinnamomum hesperium*, *C. new-berryi*, *Nectandra prolifera* and *N. guangxiensis*.

Palynologic investigations revealed that the deposits of the upper strata of the Lower Cretaceous and Cenomanian are dominated by *Classopollis* pollen produced by squamifoliate conifers like *Brachyphyllum* and *Pagiophyllum* as well as by the pollen of *Ephedra*, Schizeaceae and Ulmaceae. The bisaccate pollen of the conifers is encountered rarely (Song *et al.*, 1982) while the abundance of *Classopollis* pollen and the shoots of *Brachyphyllum* and *Pagiophyllum* coupled with entire leaves are suggestive of a hot, possibly, tropical dry climate for South China. Climatic dryness is also supported by the broad range of the Late Cretaceous red deposits well developed in central and southern China. In South China gypsum and salt are associated with red formations. This leads to a belief that the whole of this territory may also have been inhabited by xerophytic vegetation. The aspect of the vegetation of the South China province may be presumed similar to that of Middle Asia and south Mongolia.

4.3 Late Cretaceous phytochoria in the Northern Hemisphere according to palynologic data

For the Late Cretaceous epoch correlation of data obtained by studies of macro-remains and by the evidence of palynologic research remains a serious and still unresolved problem. The numerous factual data referred to in the summary works of Samoilovich (1977), Zaklinskaya (1977), Srivastava (1978), Herngreen and Khlonova (1981), Batten (1984) and some other researchers who reviewed these palynofloras shows the Northern Hemisphere to include two basic palynogeographic regions which were respectively dominated by two groups of pollen of the angiosperm group Normapolles and Aquilapollenites.

The first representatives of the group Normapolles (primarily *Complexiopollis* and *Atlantopollis*) occur in the Middle or Late Cenomanian of Bohemia and the Atlantic coast of the USA. Starting in the Turonian this group evolved very rapidly resulting in as many as 50 genera at the end of the Cretaceous; the last representatives of this group survived until the Eocene. This group is apparently of heterogeneous origin. Its distribution encompasses Europe, including the Trans–Ural area, and the east and south of North America. From the south the area of distribution of Normapolles comes into contact with the Equatorial region of Palmae which occupied the tropical belt. In Malaysia mingled forms were detected belonging to these two palynofloras.

The most ancient palynofloras with Normapolles both in Europe and North America had a very similar composition which is consistent with the notion of the North Atlantic as being narrow or even land at this time. Later at the beginning of the Senonian the increasing isolation of North America from Europe concomitant with divergence caused by evolution led to the emergence of genera confined in distribution to one or the other continent. Such genera, for instance, include *Choano-*

pollenites, Minorpollis, Osculapollis, Plicapollis and *Pseudoplicapollis* which are all typical of the Late Cretaceous of North America. There is an emerging difference between generic compositions of Normapolles pollen on the Atlantic coast of the USA and its southern states.

Europe witnessed the spread of *Interporopollenites, Krutzschipollis, Papillopollis* and *Vancampopollis* unknown in the USA (Batten, 1984). This differentiation brought about development of a number of provinces within the vast region of Normapolles distribution. Of these special mention should be made of the Turkmenia–Kazakhstan province featuring, unlike the rest of the area, arid climate and characterized by the presence of *Betpakdalina, Khlonovia, Boreapollis* which do not cross its boundary. A peculiarity of this province was the abundance of *Classopollis* pollen confirming the prevalence of an arid climate within its territory.

Another major region was occupied by the palynoflora 'Aquilapollenites' appearing in the Turonian, i.e. somewhat later as compared with Normapolles pollen and becoming extinct at the end of the Cretaceous. The most widespread genera in this region are *Aquilapollenites, Fubalapollis, Mancicorpus* and *Triprojectus*. These co-occur with *Orbiculapollis, Wodehouseia, Cranwellia, Loranthancites, Beaupreaicidites* and *Expressipollis*. The genus *Proteacidites* is said to belong here, too, but Batten (1984) rightly remarks that the pollen from the Northern Hemisphere incorporated in this genus is hardly identical with that of the same name from the Southern Hemisphere where it is likely to be connected with the family Proteaceae.

Originally the palynoflora 'Aquilapollenites' was assumed to embrace the entire territory north of the Tropic of Cancer excluding the region occupied by the palynoflora Normapolles stretching from the eastern shore of the West Siberian Sea through the Far East and north-east of the USSR, west of Canada and the USA up to the western border of the middle (Cretaceous) sea of North America. The boundary between these two regions was presumed (Zaklinskaya, 1977) to pass meridionally crossing the peripolar space.

Such a position of the boundary was not consistent with the evidence of vegetation remains of the Late Cretaceous nor with the limits of the boundaries between the largest phytochoria of other epochs and periods of geological history. These boundaries considered on the whole had always retained a latitudinal or sublatitudinal course although in places they could well become submeridional. This regularity owed to the existence of latitudinal climatic zonality which, in turn, was associated with the Earth's axis of inclination towards the ecliptic.

Gradually amassed facts, in particular, the discovery of the typical palynoflora 'Aquilapollenites' following the study of cores from boreholes drilled in the North Sea, coupled with finds on the island of Mull, situated between mainland Scotland and Ireland, in Greenland (Batten, 1982) and in the Barents Sea (Bratseva, 1985), revealed that the phytochoria taken up by 'Aquilapollenites' was located north of that with Normapolles, and occupying the circumpolar position. The first to come to this conclusion were Srivastava (1975) and Samoilovich (1977). On and off the boundary between these phytochoria strongly deviated from the latitudinal trend acquiring in certain quite elongated sections a meridional course which was determined by the orientation of the West Siberian Sea and the 'middle seas' of North America which were major barriers in the way of plant migration.

In places the palynoflora 'Aquilapollenites' penetrated far south as was the case in California which was possibly favoured by the mountainous structure of North America along which plants could penetrate from the north. These plants were inherent in warm-temperate climate whose belt largely accommodated the plants producing *Aquilapollenites* pollen. Some elements from this flora have been reported from China, Malaysia, India, Africa, the Atlantic coast of the USA and Brazil where they are represented by sporadic pollen grains.

Within the USSR, S. R. Samoilovich identified two provinces inside the area of 'Aquilapollenites', i.e. the Yenisei–Amur with *Orbiculapollis, Proteacidites, Loranthacites, Wodehouseia* (of the oculata type according to Khlonova) and *Expressipollis*, and the Hattang–Lena province where these forms are absent except for *Wodehouseia* which has been recorded for the southern margin of this province. The Yenisei–Amur province, in turn, is divided into three subprovinces, i.e. Ust–Yenisei, Middle Yenisei and Zeya-Bureya. These subprovinces are most distinctly delineated for the Maastrichtian. The first of them is typified by an abundance of *Expressipollis* pollen coupled with diversity of *Triprojectacites* as well as the pollen of Taxodiaceae and spores of sphagnum mosses. The Middle Yenisei subprovince includes numerous *Orbiculapollis* with occurrences of *Wodehouseia, Proteacidites, Ulmoidepites* and *Cranwellia*. The Zeya-Bureya subprovince is characterized by the same genera but *Expressipollis* is lacking here.

Between the Yenisei–Amur province belonging to the 'Aquilapollenites' region and the region with Normapolles there is an ecotonal, Ural–West Siberian band of mixed palynoflora (Herngreen and Khlonova, 1981). This band mostly occupies the shallow sea which covered the West Siberian lowland in the Late Cretaceous. The pollen and spores buried in the sediments of this sea had been brought both from the Urals, then forming a gentle highland covered with vegetation producing Normapolles pollen, and from the Siberian Platform whence pollen of the *Aquilapollenites* type was transported.

At the present time there are no plants producing Normapolles and *Aquilapollenites* pollen. This does not permit the establishment of the systematic affinity of the plants that produced this pollen or the association of these morphologically expressed but formal pollen genera with those genera established by leaf morphology. The only comparison that is afforded by the contemporary level of development of palaeobotany is between the areas of this or that pollen or spores and the region of distribution of some genera or other according to leaves.

Now that the largely sublatitudinal dissemination of *Aquilapollenites* has been ascertained it is clear that its area of incidence considered as a palynofloristic region roughly coincides with the Siberian–Canadian region established on the basis of leaves. Hence we can presume that the plants, assigned to the genus *Trochodendroides* whose distribution remains within the confines of the Siberian–Canadian region, correspond to some genus of pollen constituting part of 'Aquilapollenites' group.

Superimposing the area of *Trochodendroides,* which is very widespread in the Siberian–Canadian region, on the area of pollens from the various genera of 'Aquila-pollenites' we find that *Wodehouseia* pollen (or the oculata type, according to A. F. Khlonova) proved to be most broadly disseminated and virtually to coincide with the area of *Trochodendroides.* The temporal intervals of their distribution also roughly coincide. It is highly probable that the plants with the leaves *Trochodendroides* and fructifications *Trochodendrocarpus* constantly attending *Trochodendroides* produced *Wodehouseia* pollen. Attention should be paid to the fact that the leaves *Macclintockia* disseminated in the northern part of Eurasia have the area coinciding with that of *Expressipollis* pollen.

On the other hand, plants with leaves belonging to *Debeya, Laurophyllum, Cinnamomoides, Myricaephyllum,* etc., which continued within the Patapsco–Euro-Sinian region might have produced some pollen or other of Normapolles type. Thus, this leads to a way of establishing correlation between the form-genera of the dispersed pollen (and spores) and the taxa established by leaves on the generic level.

This method may be termed comparative arealogical. Of course, its application requires that both the stratigraphic position of leaf remains and spore (pollen) and the pattern of their areas by stages be well studied. Material has started coming in which will allow resort to this technique in the near future. The area of Normapolles pollen by and large coincides with the limits of the Euro-Sinian region.

4.4 Equatorial (palm) region

Virtually no macro-remains of Late Cretaceous plants are known here, therefore the characterization of this region relies upon pollen composition (Herngreen and Khlonova, 1981). It should be added that the palynologic complexes of the Equatorial region have been reported largely from western Africa and South America. The Cenomanian complexes still contain much *Classopollis* pollen but on the boundary with the Turonian both this pollen and *Afropollis* (coupled with the pollen with elaters) disappears. *Ephedripites* pollen continues into the Turonian.

The Turonian complexes are typified by periporous (*Cretacaeiporites*) and multi-colpate pollen. Tricolpate grains prove to be dominant (30 to 60%) being represented by various species of *Tricolpites*. In the Lower Senonian the upper section includes an increased variety and content (10 to 15%) of monocolpate pollen (*Psilamonocolpites*, less *Retimonocolpites*) referred to the palms. The amount of periporous pollen declines (5% and less) coupled with decreasing proportions of multicolpate form-taxa. Ferns become fewer which is supplemented with a reduced variety of ephedroid pollen. The first pollen type that may be referred to the currently living palms was discovered in the Senonian, i.e. *Spinizonocolpites echinatus*. It is identified with *Nipa fruticans* pollen. *Proxaperites* pollen may well belong to the extant group of palms related to *Nipa*.

According to the data of Herngreen (Herngreen and Khlonova, 1981) the first differentiation relying on the palynologic data between western Africa and Brazil was manifested at the very beginning of the Cenomanian or at the outset of the Turonian. This is indicated by the distribution in the north of South America of *Steevesipollenites nativensis* and *S. amphoriformis* pollen lacking in Africa. The upper Senonian deposits feature some endemic thick-walled species of pollen as, for instance, *Crassitricolporites brasiliensis*. Endemic species have been reported from Gabon which are also known from eastern India including *Andreisporis*, *Constantisporis* and *Victorisporis*.

However, certain parallelism is observed between the developments of the palynofloras of both the continents amounting to reduced variety of *Ephedripites*, increased numbers and, in part, diversity of monocolpate types of pollen (viz. *Psila-Retimonocolpites*) and emergence of the tricolpate pollen *Proteacidites*. The Senonian witnesses appearance on both the sides of common species such as *Buttinia andreevii*, *Auriculiidites reticulatus* and *Proxapertites operculatus*, as well as representatives of *Spinizonocolpites*.

Transition from the Equatorial to the Euro-Sinian region has been observed by Herngreen in the Middle East (Israel, Saudi Arabia, Egypt) (Herngreen and Khlonova, 1981). Here the representatives of Normapolles peculiar to the Euro-Sinian region, such as *Basopollis*, *Oculopollis* and *Trudopollis*, are encountered in conjunction with the forms of the Equatorial region referred to Palmaceae, i.e. *Echitriporites* and *Proxapertites*.

4.5 Austral (Notal) region

This region encompasses the southern part of South America, Australia, New Zealand, the Antarctic and India (Indostan). A very restricted palynologic characterization of the Indian province points to its close relation to the Equatorial region. A typical feature of the Austral region is the wide distribution of Podocarpaceae and Araucariaceae coupled with the dissemination of the southern beech *Nothofagus* from the Senonian. It comprises the Indian, Patagonian and the Antarctic provinces, the latter being separated from the Senonian onwards. Localities of the Late Cretaceous floras of the Austral region are not numerous.

4.5.1 Patagonian province

Just as in the case of the Early Cretaceous, the Patagonian province is limited to the southern half of Argentina and the adjoining part of Chile. Rather scanty evidence on the Late Cretaceous flora of this phytochoria is summed up in the work of Menendez

(1969) and supplied in a number of his earlier articles. All known localities of this flora in Argentina (Chubut, Santa Cruz) and Chile belong to the Maastrichtian or are closer to the undivided Senonian. The richest locality is situated in Chubut province. Its composition includes horse-tails *Equisetites*, ferns *Adiantus*, *Blechnum*, *Dryopteris*, *Gleichenites*, *Hymenophyllum* and *Thyrsopteris*. The remains of *Thyrsopteris antiqua* found in the Maastrichtian of southern Chile both morphologically and in fructification are similar to the endemic genus *Thyrsopteris* growing currently only on Juan Fernandez Island. Among relicts *Nilssonia*, whose remains were found both in Argentina (Chubut) and Chile, is noteworthy. The diversity of conifers is great, they are represented by the genera *Agathis* (?), *Protophyllocladus*, *Pseudoaraucaria*, *Podocarpus* and *Acmopyle*.

Angiosperms are most varied, which is not surprising since their remains descend, as has already been pointed out, from the upper strata of the Upper Cretaceous. Leaves have been identified that refer to the genera *Araliaephyllum*, *Rignonites*, *Cissites*, *Euphorbiatheca*, *Ficus*, *Laurophyllum* (3 species), *Ribes* and *Sterculia*. Leaves of aquatic or semi-aquatic plants (*Scirpites*, *Potamogeton*, *Paranymphaea*) have been detected. The Upper Cretaceous of Tierra del Fuego has yielded *Nothofacidites* pollen (5 species) (Menendez and Caccavari de Filice, 1975). The pollen of this genus is close to the present-day *Nothofagus* pollen and is widespread in the Tertiary deposits of the southern part of South America, Antarctica and Australia as well as in the sediments of this age exposed by deep-sea drilling. It has not been recorded for the Tertiary deposits of Africa which might be due to the earlier drift of this continent northwards whereas contact among South America, Antarctica and Australia was not severed even at the beginning of the Palaeogene.

The composition of this flora highlights abundance of ferns and representatives of the genera *Laurophyllum* and *Sterculia* which implies a moist subtropical climate prevalent in the Senonian in Patagonia. Southwards the climate becomes cooler which is evidenced by *Nothofacidites* pollen found on Tierra del Fuego and, possibly, warm-temperate. No remains of Late Cretaceous vegetation have so far been recorded from the Antarctic.

4.5.2 Australian province

Apart from Australia this province includes New Zealand. Evidence on the composition of the Late Cretaceous flora of these countries is very scarce being represented almost exclusively by palynologic data obtained both from New Zealand (Mildenhall, 1980) and Australia (Dettmann and Playford, 1968). Mildenhall emphasized the similarity of the composition of spores and pollen from both places which indicates their proximity in the Mesozoic.

Appearance of angiosperms is noted in the Albian of both the countries. They were represented by the pollen *Clavatipollenites*, *Tricolpites pannosus*, *Phimopollenites angathelaensis, Liliacidites peroroticulatus, Asteropollis asteroides*, etc. Judging from pollen structure, the Late Cretaceous saw the development of many recent families such as Fagaceae, Proteaceae (*Proteacidites*), Loranthaceae (*Cranwellia*), Gunneraceae (*Tricolpites waiparaensis*), Liliaceae, Chloranthaceae, Caryophyllaceae (*Caryophyllidites polyporatus*) and Winteraceae. Also encountered were the spores of mosses, lycopods and a variety of ferns as follows: Gleicheniaceae (*Clavifera, Ornamentifera*), Schizeaceae (*Appendicisporites*, rarer *Cicatricosisporites*), Dicksoniaceae and Pteridaceae. *Podocarpus* pollen appears to be commonly represented by several morphologically different species such as the pollen *Dacrydiumites*. There is a great deal of trisaccate pollen of *Microcachryidites* and *Trisaccites microsaccatus*. At the beginning of the Late Cretaceous, *Classopollis* (from the Turonian) and *Callialasporites* pollen disappears.

Nothofagidites emerges from the Senonian. This pollen apparently belongs to the southern beech (*Nothofagus*). Its appearance is due to climatic cooling pervading the southern fringes of southern mainlands. According to the distribution of this pollen the independent Antarctic province is identified (see below). Certain ferns (*Cladophlebis australis*) and a number of other species including *Sphenopteris* spp. and *Coniopteris* (?) *lobata*, and cycadophytes (*Taeniopteris spatulata, T. stipulata, Pterophyllum clarencianicum* and *Ptilophyllum seymouricum*) were described (McQueen, 1956) from the Upper Cretaceous of New Zealand. Some of the above species have been reported from Australia as well.

From the Late Cretaceous onwards the Australian province started separating from the other continents of the Southern Hemisphere with its flora acquiring its special features (notably a broad range of Proteaceae) becoming still more clear-cut in the Palaeogene and Neogene. At the end of the Cretaceous Australia became separated from New Zealand by the Tasman Sea located on oceanic crust.

4.5.3 Indian province

Detailed palynologic research has been conducted only for the basin of Cauvery in eastern India (Herngreen and Khlonova, 1981). While the palynofloras of the Early Cretaceous of India compositionally belong to the Austral region, the Late Cretaceous epoch witnessed the emergence in them of forms signalling development of relations with the floras of Africa and the north of South America. The representatives of the genera *Rouseisporites, Cupaniedites, Turonipollis, Gothanipollis, Proteacidites* and *Liliacidites* are concomitant with the appearance of the forms belonging to the genera *Andreisporis, Constantinsporis* and *Victorisporis* reported from Gabon. The uppermost strata of the Cretaceous reveal such pantropical forms as *Spinizonocolpites* and *Proxapertites* assigned to the palms. Certain species of *Aquilapollenites* also penetrate here. These are similar to the forms known from the Maastrichtian of north-eastern Brazil. *Turonopollenites* is familiar from the palynofloras with Normapolles, and *Cranwellia* and *Scollardia* from the palynofloras with *Aquilapollenites*.

It may be presumed that the palynofloras of the Late Cretaceous of India experienced a strong influence on the part of the vegetation of the Northern Hemisphere. By this time India must have advanced farther north in comparison with its position in the Early Cretaceous. It is believed that in the Late Cretaceous epoch (probably from the Senonian) India was already part of the Equatorial region. No Late Cretaceous floras represented by macro-remains of plants have so far been found in India. All the localities with palm remains are confined to the intra-trappean deposits referring to the Palaeogene.

4.5.4 Antarctic province

From the start of the Santonian the palynoflora of south-eastern Australia, New Zealand, Antarctica and Patagonia witness the appearance of *Nothofagidites* pollen whose distribution permits the delineation of a belt of moderate-warm climate emerging in the Southern Hemisphere as well as the Antarctic phytogeographic province corresponding to it. With *Nothofagidites* are found *Proteacidites amolosexinus, Proteacidites* sp., *Triorites edwardsii, Tricolpites sabulosus, T. gillii* and *Tricolporites lieeiei*. The pollen of angiosperms co-occurs with that of conifers probably making part of the family Podocarpaceae including *Podocarpites ellipticus, Microcachryidites antarcticus, Podosporites imicrosaccatus* and *Phyllocladidites mawsonii*. The latter form is close to the pollen of the genus *Dacrydium* (Herngreen and Khlonova, 1981).

The present-day distribution of the southern beech *Nothofagus* is restricted mostly by the southern margins of Australia, New Zealand and South America. In more northerly areas with subtropical or even tropical climates (New Guinea, New Caledonia) the southern beech is part of the composition of the mountainous rain forests situated at altitudes of 1500 to 3000 m in the belt of cool climate. In more southerly areas its distribution descends to sea level.

In South America of both the Senonian and of today *Nothofagus* is absent. This is corroborated by the separation of the African continent from South America in the pre-Senonian and its more northerly position as against other mainlands of the Southern Hemisphere. This is again borne out by absence of the remains of marsupials in Africa. Absence of *Nothofagidites* pollen in the Upper Cretaceous of India also proves its more northerly position at the time as compared with Australia.

Phytogeography, palaeoclimates and position of continents in the Mesozoic

There are three main techniques of climatic reconstruction. The first technique amounts to a study of the distribution of terrigenous rocks which are climatic indicators. For a moist (humid) climate these include coals, sedimentary iron ores, kaolin weathering crusts and bauxites. For a dry (arid) climate these are carbonate red beds, gypsums, rock salt and, from the Jurassic period, primary dolomites. The maps of palaeoclimates in the works of Strakhov (1960) and Ronov and Balukhovsky (1981) are based mainly on the distribution of these rocks.

The second method is palaeoecologic and is founded on the distribution of various groups of animals and plants sensitively responding to climatic change. Of these the most important indicators for climatic reconstructions of continents are the remains of land plants (Vakhrameev, 1975). They are found far more often than the remains of land animals. Furthermore, the latter might hibernate in the wake of temperature changes or migrate seasonally with adverse conditions arising. Thus, for instance, finds of the remains of large-sized marine reptiles in high latitudes does not imply that they inhabited this area in winter.

It is common knowledge that among the marine Mesozoic invertebrates attached to the substrate, colonial corals and thick-valved molluscs (rudists, *Diceras*), as well as buchias are reliable indicators of the temperature of sea water. However, the number of these groups is insignificant and they point only to water temperature, naturally being unable to respond to air humidity.

The third method is physical, meaning determination of absolute annual temperatures by studying the ratios of the isotopes ^{16}O and ^{18}O or the Ca–Mg ratio in shell valves (Yasamanov, 1980). Accurate enough data are provided only by analysing the shells of molluscs, foraminifera or nannoplankton, living under conditions of normal salinity and not subjected to compositional change as a result of diagenesis. However, for this technique to yield sufficiently accurate and comparable results it is necessary to analyse shells belonging to the same group of organisms whose remains should be collected stratum by stratum from a single sequence or adjacent sequences. Otherwise, constructing temperature curves on the basis of different authors' materials lacking in systematic selection is conducive to major disagreement (Krasheninnikov and Bassov, 1985, Fig. 5, p. 16).

It should be noted that absolute temperature changes help to obtain the results for a comparatively short period of time while the study of distribution of terrigenous rocks or climatic indicators and their changes in space and time as well as the analysis of the history of floras and faunas permit the tracing of climatic changes over a long period almost throughout the Phanerozoic. However, we should not neglect the significance of measuring absolute temperatures because only these measurements allow direct comparison of the temperature conditions of the past and present.

It is certain that in order to reconstruct a climate of some earlier geological epoch it is necessary to apply all the above techniques and that is the prevalent practice nowadays. In this work emphasis is laid on the study of structural features and history of the areas of land Mesozoic plants.

The Mesozoic (Table 5.1) was an era of warm climate. Absence of the polar ice-caps and even belts of cold climate is indicated by finds of thermophilic plant remains in the Arctic and Antarctic. Thus, remains of a temperate warmth-loving flora of Early Cretaceous age were found in the Arctic on the Spitsbergen islands and Franz Joseph Land. In the Antarctic in the region of the Beardmore glacier (approximately 82° S latitude) remains of Permian and Triassic plants including cycads were detected, suggesting a subtropical climate. In the area of the Queen Alexandra range trunks of trees of Triassic age were found reaching 23 m in length. Lava-covered silicified trunks from 0.5 to 1.0 m diameter that were not subjected to transportation were found in the south of Victoria Land. Abundant remains of plants of Jurassic and Early Cretaceous age have also been reported from the Antarctic peninsula.

In the Mesozoic annual average temperatures in the north of Siberia measured from the oxygen isotope ratios fluctuated at various epochs from 15 to 17 °C, and in the vicinity of the Mediterranean belt from 18 to 24 °C. The temperature gradient was half that of today.

The onset of the Triassic period coincided with one of the turning points in climatic history causing one of the most major alterations of floristic composition in the history of the Earth (Dobruskina, 1982). The variety of plants was dramatically reduced resulting in the representation of the extremely impoverished Early Triassic flora mainly of the derivatives of the Permian vegetation. The principal cause of this change was abrupt climatic drying that set in in the Permian and reached its climax in the Early Triassic. Climatic drying was facilitated by a considerable regression of marine basins. It should be noted that the only spot on the Earth witnessing deposition of coals which is indicative of a moist climate was south-eastern Australia.

Three palaeofloristic regions take shape in the first half of the Early Triassic. Pteridosperms, ferns and conifers were chiefly growing in the north of Asia now the Angara region. This flora is associated with the characteristic lycopsid 'Pseudoaraucarites'. The flora suggests a semi-arid warm climate probably close to subtropical. In the Euroamerican region which, just as in the Permian, embraced Europe, Middle Asia and

Table 5.1. *Geochronological scale of Mesozoic time (Harland et al., 1982). Ma = million years*

Period	Epoch	Age (and duration)
Cretaceous	Late (K2) 97.5–65 Ma	Maastrichtian (8 Ma) Campanian (10) Santonian (4.5) Coniacian (1) Turonian (2.5) Cenomanian (6.5)
	Early (K1) 144–97.5 Ma	Albian (15.5) Aptian (6) Barremian (6) Hauterivian (6) Valanginian (7) Berriasian (6)
Jurassic (J)	Late (J3) 163–144 Ma	Tithonian (6) Kimmeridgian (6) Oxfordian (7)
	Middle (J2) 188–163 Ma	Callovian (6) Bathonian (6) Bajocian (6) Aalenian (7)
	Early (J1) 213–188 Ma	Toarcian (6) Pliensbachian (6) Sinemurian (6) Hettangian (7)
Triassic	Late (Tr3) 231–213 Ma	Rhaetian–Norian (12) Carnian (6)
	Middle (Tr2) 243–231 Ma	Ladinian (7) Anisian (5)
	Early (Tr1) 248–243 Ma	Olenekian (?) Induan (?)

North America, there was the so-called Voltzia flora named for the characteristic representative of the coniferous genus *Voltzia*. No identifiable plant remains in the first half of the Triassic were found in North America. The Voltzia flora grew in the belt of tropical arid and semi-arid climate inhabiting individual oases and apparently formed no continuous cover. The southern region named Gondwana, just as for the Palaeozoic, is typified by the representatives of corystosperms most often featuring the genus *Dicroidium* by whose name the entire flora is referred to as the Dicroidium flora.

The Early and beginning of the Middle Triassic is characterized by the wide distribution of the lycopsid *Pleuromeia* possessing a relatively short (about 1 m) unramified trunk. These plants as a rule formed monodominant thickets on the sea coasts and inland lakes. *Pleuromeia* and its related forms were extremely widely disseminated throughout the Earth occurring from Australia and India to the Far East and Taimyr, i.e. in all phytogeographic regions of the first half of the Triassic. Such a distribution encompassing both high and low latitudes testifies to the absence of any more or less distinct climatic zonality and sufficiently uniform warm climate.

From the second half of the Triassic the flora was gradually replenished by emerging basic groups of the Mesozoic plants, i.e. Dipteridaceae, Matoniaceae, Marattiaceae, ferns and gymnosperms (Peltasperms, cycads, Bennettitales, Ginkgoaceae, Czekanowskiaceae).

Dobruskina (1982) believes that emergence of several centres of species formation caused a restructuring of phytochoria from sublatitudinal or latitudinal into sectorial. In the map presented by her, floral composition changes latitudinally more significantly than meridionally which permits recognition of a number of sectors extending from the Arctic latitudes to the tropics. Large changes of floristic composition took place in the meridional direction for a number of epochs (Neocomian, Late Cretaceous) but they were always considered on the level of provincial differences whereas latitudinal floristic alterations remained crucial.

Consideration of the map of I. A. Dobruskina reveals extreme scarcity of evidence on the composition of the Middle Triassic and Carnian floras of the northern regions of Eurasia. This impedes the solution of the problem of the extent of differences between the floras of warm-temperate and subtropical or even equatorial belts. Without denying the possibility of existing centres of rapid generic or specific development and hence the emerging discrepancies between the floras of the same climatic belt, one may doubt the assumption of their homogeneity within one and the same sector when moving from the Arctic region to the subtropical or tropical areas.

Towards the end of the Triassic the flora became a typically Mesozoic one. Its composition was balanced owing to migration and its distribution was distinctly governed by latitudinal zonality.

Comparing the distribution of the Dicroidium flora of the Southern Hemisphere on a map with the current position of the continents and on a map compiled on the palaeomagnetic basis taking into account continental drift we see that localities of this flora are scattered not only across all continents of the Southern Hemisphere but are present in the Northern Hemisphere, notably in India. On the map of the Triassic time based on palaeomagnetic evidence the distribution of this group of plants is confined to the ancient mainland of Gondwana which embraced Africa, South America, Australia, Antarctica and India and looks far more compact and natural than in the first case. It will also be noted that in the Permian period which preceded the Triassic the so-called Glossopteris floras were located, just as are the Dicroidium floras, exclusively within Gondwana.

The end of the Triassic and the following Early and Middle Jurassic epochs are marked by major climatic moistening signalled by the large-scale coal-bed formation. Due to a certain cooling, the end of the Triassic sees the emergence in the north of Eurasia of a moderate-warm belt with well-expressed zonality. This belt is distinctly traceable in Eurasia while data on North America are much more scanty. The arid climate region is dramatically reduced.

Phytogeographically and climatically the Jurassic period clearly falls into two parts. The first embraces the Early and Middle Jurassic while the second encompasses the Late Jurassic. The Jurassic period is marked by the emerging distinctly latitudinal position of climatic and vegetational zones.

In characterizing the climates of the Mesozoic we use the terminology applied for present-day climate although there are some reservations for such usage because in the Mesozoic there were no polar or even moderate-cold climatic belts. For the Mesozoic we recognize three thermal belts, i.e. temperate (moderate)-warm, subtropical and tropical (equatorial). According to the degree of humidity, division is into humid (moist) and arid (or occasionally semi-arid).

Deciduous trees with pycnoxylic-type wood with well manifested annual rings were widespread in the belt of moderate-warm climate. Evergreen wood forms were encountered more rarely, notably the forms with the manoxylic wood type rich in parenchyma. We assume that in the northern part of the moderate-warm belt occupying the position of the contemporary Arctic area of the Northern Hemisphere, winter temperature might temporarily have declined below zero. This might have caused snowfalls without, however, permanent snow cover.

Snow formation may be assumed only for the extreme north-east of Asia (Chukotka and adjacent areas) which on mobilistic maps, based on palaeomagnetic data, were situated farther north than the current position since Eurasia of that time was turned counterclockwise relative to its present-day position. Some authors (e.g. Epstein, 1982) recognize occurrence of glacial-marine sediments in the Jurassic and Cretaceous that were deposited in shore-ice formation areas. But other researchers think them to be ordinary pebble-beds forming in the littoral zone.

A very gradual cooling when moving from the southern boundary of the warm-temperate belt northwards is corroborated by major similarities of floras whose remains were gathered from the Lower Cretaceous deposits of Spitsbergen, Franz Joseph Land and the Novosibirsk Isles now situated far beyond the Polar Circle, to the age-equivalent floras of central and southern Yakutia. Such a nearness indicates that these floras belong to the same climatic belt although the meridionally more northern and southern localities are separated from each other by more than 30° latitude.

The belt of the most subtropical climate comprises the areas of growth of cycads and Bennettitales with manoxylic type of stem structure implying warm and frost-free weather. Dendritic ferns become widely disseminated here and from the Late Cretaceous this includes palms and evergreen angiosperms, the latter dominated by entire leaves. The systematic variety of plants increases almost twofold in comparison with the belt of moderate-warm climate. Many genera are confined in distribution to the northern boundary of this belt. Arid or semi-arid areas within the subtropical and tropical zones are established on the basis of a sharp reduction of diversity of such hygrophilic plants as ferns, *Nilssonia*, etc., by the broad range of Cheirolepidiaceae (*Frenelopsis, Pseudofrenelopsis* and other plants producing *Classopollis* pollen) contributing to vegetation composition, appearance of evaporites, red-coloured carbonate rocks and the lack of coals. According to the composition of plant remains the southern boundary of subtropical climate in the Southern Hemisphere reached up to the fringes of the Antarctic until the middle of the Late Cretaceous.

In the present work an attempt is made to identify the belt of tropical or equatorial climate proper which in the Mesozoic took up Central America, the greater part of South America, Africa and, perhaps, south-east Asia. The remaining part of this belt passed through the Pacific Ocean whose width at the time might have exceeded the present. A boundary between the tropical and subtropical belts can be drawn only for the Cretaceous period in the first place going by changes in the composition of spores and pollen. Jurassic deposits are little known within this belt while the Cretaceous deposits have been recorded in a number of sites by deep-sea drilling primarily in the Atlantic Ocean. Macrofossils of vegetation origin are very rare in the Equatorial region.

Below we shall move on to a detailed characterization of the floras of the belts and phytochoria outlined above across separate geological epochs.

The belt of warm-temperate seasonal climate of the Northern Hemisphere in the Early and Middle Jurassic is particularly well expressed in Siberia and Kazakhstan which in large measure were land areas. They were dominated by forest vegetation represented by deciduous gymnosperms including Ginkgoaceae, Czekanowskiaceae (see Fig. 3.15), Podozamitaceae and ancient Pinaceae. (Leaf-falls of gymnosperms growing in Siberia were caused not so much by low temperatures as by prolonged nights typical of these latitudes in winter time.) Their annual rings are well manifested suggesting distinct seasonality. The ground flora accommodated various herbaceous ferns as well as creeping herbaceous lycopsids. Thickets of short horse-tails inhabited swampy areas. Wide development of bogs and associated peat formation led to the emergence of many large-sized coal basins, i.e. Msykubensk, Kansk-Achinsk, Irkutsk, etc., related to the big intracontinental depressions. Remains of Jurassic plants in Canada situated in the same belt are almost unknown.

Southwards the Euro-Sinian region was situated corresponding to the belt of subtropical, mainly humid climate. Localities of remains of this region are widespread in Eurasia from England and France to China and Japan. They are also recorded on the western shore of the USA. The gymnosperms forming wood vegetation were dominated by Cheirolepidiaceae (the group of conifers becoming extinct at the beginning of the Palaeogene), ancient Araucariaceae and Ginkgoales. Ancient Pinaceae and particularly Czekanowskiales became rare.

Czekanowskiales attained the widest distribution in the Middle Asia province of the Euro-Sinian region. This province was situated on the southern projection of a vast land encompassing Siberia and Kazakhstan whence the representatives of these groups of plants typical of the Siberian region might also have penetrated into Middle Asia (which was located in the subtropical belt).

Cycads and Bennettitales became very diverse in the Euro-Sinian region featuring both the weakly ramified column-shaped forms covered with a wide crown of palm-like leaves (see Fig. 3.15) and the comparatively thin-stemmed branching trees. In size they were inferior to the majority of conifers and Ginkgoales.

Representatives of the genera *Dictyozamites*, *Otozamites*, *Sphenozamites*, *Zamites* and *Zamiophyllum* are, as was shown above, excellent climatic indicators because their distribution keeps within the equatorial belt and the subtropical areas of both hemi-spheres. *Ptilophyllum* could also be added to this list but it was apparently capable of withstanding a somewhat more temperate climate because during the Toarcian warming *Ptilophyllum* penetrated into Siberia far to the north without being accompanied by the above-mentioned other Bennettitales. The cycads, notably *Nilssonia*, *Ctenis* and *Pseudoctenis*, coupled with such bennettitaleans as *Pterophyllum* and its relative *Anomozamites*, inhabited not only the tropical and subtropical areas but also penetrated into the moderate-warm climate region. But there they were less var-ied and far less numerous.

Marattiaceae, Matoniaceae and Dipteridaceae became most diverse among the ferns here. Some of the ferns were dendrolithic which is supported by the finds of petrified stems. Just as earlier the above bennettitaleans and ferns represented by the families already enumerated are widespread not only in the Euro-Sinian region but also in the Austral region, i.e. in the subtropics of the Mesozoic of the Southern Hemi-sphere. All these groups could probably migrate in the meridional direction crossing the Equator. This is indicated by finds of plant remains belonging to these taxa in the few localities of the Equatorial region.

The genus *Piazopteris*, belonging to the Matoniaceae (see Fig. 3.15, p. 129), is widespread in southern Mexico, Cuba, Brazil and northern Africa. In the Upper Jurassic of Madagascar remains of Dipteridaceae included *Dictyophyllum* sp. and *Thaumatopteris* sp.

Another group characteristic both of the Euro-Sinian and the Austral region was the Caytoniales represented both by reproductive organs (*Caytonia*, *Caytonianthus*) and leaves (*Sagenopteris*) widely disseminated in the subtropical areas of the Jurassic of both the Northern and Southern Hemispheres (see Fig. 2.2, p. 000). It is not until the Cretaceous that they encroach upon the southern tip of the Siberian–Canadian region. Thus, *Sagenopteris* were found in the basin of the river Turukhan (Western Siberia), in the north of the Verkhoyansk area and in the lower reaches of the river Aldan, and in the Upper Cretaceous in the basin of the river Anadyr (Grebenka river) as well as in Alaska.

Representatives of the genus *Pachypteris* also serve as good subtropical climate indicators. The localities containing *Pachypteris* are mostly confined to south-western Eurasia stretching from southern England to Iran and Tadjikistan (Fig. 5.1). They are primarily found in the Jurassic deposits and are far less frequently encountered in the Lower Cretaceous (Wealden of southern England). *Pachypteris* has not been reported from the belt of moderate-warm climate (Siberian region).

In the Southern Hemisphere the representatives of this genus were distributed both in the Jurassic (Antarctic peninsula) and the Early Cretaceous of Australia (Queensland, Victoria) and Patagonia. The leaves of this genus were also found in India (the basin of Satpura) situated in the Mesozoic in the Southern Hemisphere. In the Equatorial belt no remains of *Pachypteris* have so far been found but the possibility of its distribution here is quite plausible since finds of this kind have been reported from the Mesozoic subtropical areas on both sides of the Equator.

The analysis of the distribution of the genus *Pachypteris* plants shows their relation either with littoral sediments (Lower Jurassic of Romania, Middle Jurassic of Yorkshire, England, Upper Jurassic of Georgia and the Hissar ridge) containing remains of marine fauna or with continental deposits that formed near the coastline (delta deposits).

The majority of palaeobotanists think that, ecologically, thickets of *Pachypteris* were analogues of contemporary mangroves. This conjecture accounts for the absence of the remains of these plants in the deposits that formed in the intra-continental basins of Fergana, Kirgizia, Kazakhstan and China. In the Jurassic and Cretaceous of western Europe, the southern part of the USSR, the Caucasus and the south-western part of Middle Asia, major areas of land surface were occupied by the sea with islands being scattered about. The coastline of the land was also rugged and abounded in peninsulas.

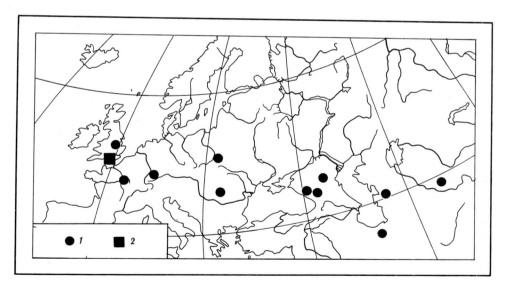

Fig. 5.1 Distribution of the genus *Pachypteris* in the Jurassic and Lower Cretaceous deposits of Eurasia. 1, localities of Jurassic age; 2, localities of Early Cretaceous age.

Absence of remains of *Pachypteris* in the moderate-warm belt (Siberian region) shows that they were sensitive to temperature fluctuations.

Two types of vegetation assemblages were forming in the Euro-Sinian region. The first one took up moist, occasionally marshy littoral or intra-montane lowlands being dominated by more moisture-loving plants, i.e. various ferns, horse-tails and Caytoniales. Conifers proved to be far less numerous here. Generally the remains of such types of vegetation are associated with coal-bearing deposits. The second type of vegetation occurred under conditions of a drier microclimate on the slopes mostly adjoining marine basins. It was dominated by Cheirolepidiaceae and Araucariaceae as well as Bennettitales and cycads. Ferns were subordinate here. This type is often related to the littoral or continental sediments containing no coals.

The Early or Middle Jurassic deposits of the Equatorial region located between the tropical areas are developed on a restricted scale because within the vast Afro–South-American mainland processes of erosion and uplift were prevalent. Only its peripheral part hosted marine and continental deposition with these sediments being for the most part devoid of plant remains. Rare finds point to the prevalence of Araucariaceae, Bennettitales, cycads and especially Cheirolepidiaceae supplemented with certain unique ferns referred to the genera *Piazopteris* and *Weichselia*.

The subtropical region of the Southern Hemisphere referred to as the Austral or Notal embracing the south of South America and the extreme south of Africa as well as India and Australia differed from the Euro-Sinian region by total absence of ancient Pinaceae and Czekanowskiales, and the weak development of Ginkgoales represented chiefly by the genus *Ginkgo*. Cheirolepidiaceae, Araucariaceae and Podocarpaceae were widespread here. Bennettitales are represented mainly by the same genera as in the Euro-Sinian region (*Ptilophyllum* is particularly numerous); the same applies to ferns. Representatives of these two groups must have been able to migrate across the tropical area.

Until recently, the climate was thought to be virtually unchanged throughout the Early and Middle Jurassic. But a more detailed study of the remains of plants (Ilyina, 1985; Kirichkova, 1985) revealed many fluctuations. From the Norian age temperature began slowly declining reaching a relative minimum in the Pliensbachian. The latter age is associated in Siberia with extensive coal formation and maximal development of sphagnum mosses. The latter is ascertained by the maximal number of the spores *Stereisporites* as well as of fern and lycopsid remains. Annual average temperature at this time was somewhat lower than 20 °C.

The Early Toarcian age was concomitant with the onset of a warming (see Fig. 2.1, p. 13) which is evidenced by increased amount of *Classopollis* pollen produced by the thermophilic conifers belonging to the family Cheirolepidiaceae. This pollen co-occurred with the spores, ferns and leaves of bennettitaleans (*Ptilophyllum*) characteristic of the southern Euro-Sinian region. A short-term cessation of coal accumulation is linked with a phase of this warming. This interruption was apparently due to the drying of peat swamps. The advance of the thermophilic vegetation northwards followed chiefly the coastline of the transgressing Toarcian sea occupying the marginal areas of the USSR and opening into the Pacific Ocean. Intercalations of limestone with admixtures of glauconite have been reported from marine deposits of the lower part of the Toarcian.

The Arctic basin of Toarcian age (Saks and Halnyayeva, 1975) was invaded by belemnites coming from Europe and settling there. The average annual temperature of the Arctic basin in the first half of the Toarcian age was about 21 to 23 °C which led to temporary replacement of the moderate-warm climate with temperate subtropical.

Cooling set in from the second half of the Toarcian. The average annual temperature dropped in the Bajocian down to 10 to 12 °C. This was the age of a possibly maximal cooling and humidization of climate. Again, just as in the Pliensbachian, coal formation is resumed, occasionally evolving into superthick beds (Kansk–Achinsk and Irkutsk basins). *Classopollis* pollen practically disappears from the spore–pollen complexes but, as in the Pliensbachian, a drastic increase in the number of sphagnum moss spores is observed (Ilyina, 1985). Herbaceous ferns become widespread along with Czekanowskiales, Ginkgoales and ancient Pinaceae making up a core of the Jurassic flora of Siberia. The invertebrate fauna including ammonites of the Arctic basin comes to be endemic which is probably due to lost contact with the seas of Europe and to water temperature decrease (Figs. 5.2, 5.3).

At the very outset of the Late Jurassic a prolonged warming set in embracing the whole of the Late Jurassic epoch and attaining its peak in the Oxfordian. This was accompanied by the shifting of the northern boundary of the subtropical belt (along with the boundary of the Euro-Sinian region) by about 10 to 15°. Average annual temperature rose to 17 to 22 °C in the northern part of Europe and to 22 to 26 °C in Middle Asia and adjacent areas (Yasamanov, 1980). It is also noteworthy that the *Classopollis* content increased in the southern parts of the European area of the USSR and, probably, throughout southern Europe and Middle Asia from the second half of the Bathonian indicating a warming already at this time whereas more northern areas witnessed a tangible warming from as late as the Callovian.

Simultaneously with the warming, aridization of climate followed, emerging from North and Central America and being particularly manifested in the southern part of Eurasia where it encompassed almost all the Euro-Sinian region from southern England and the Iberian peninsula up to China. Its magnitude on this mainland was enhanced from west to east reaching its maximum in the areas of Middle and Central Asia within which the Late Cretaceous deposits are represented not only by carbonate red beds but also by thick gypsums occasionally accompanied by rock and potassium salts. This is well seen from the altered composition of the flora of the Euro-Sinian region as well. Bennettitaleans with thinnish leaves become widespread. Czekanowskiales virtually vanish, while diversity of ferns, Ginkgoales and *Nilssonia* is drastically reduced. *Classopollis* pollen content, on the other hand, is increased. This easily identifiable pollen is produced by conifers of the family Cheirolepidiaceae capable of withstanding drought conditions.

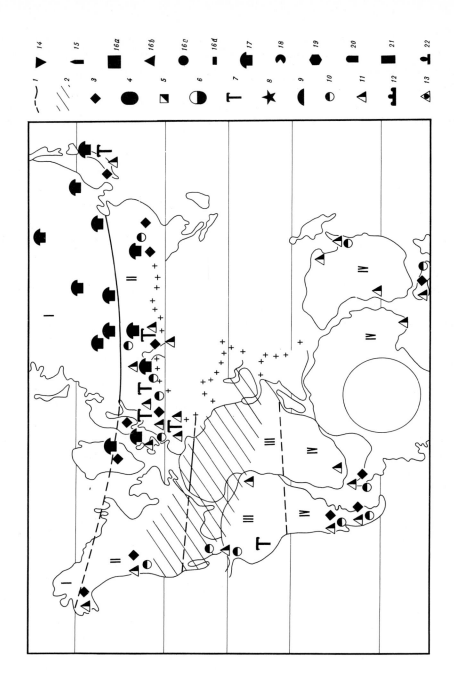

The curves (see Fig. 2.12, p. 53) showing the percentage content of *Classopollis* pollen in the coeval deposits of the Jurassic and Cretaceous at various latitudes within the territory of the USSR show distinct peaks of varying heights. In the north (Yenisei–Hattang depression) warming spells are marked by low peaks while in the south (Middle Asia) these peaks attain high values which is suggestive of a sharp increase in *Classopollis* content (up to 80 to 90%) associated with temperature rise and climatic drying (Vakhrameev, 1980).

In Eurasia the confines of the Siberian region witnessed the emergence of a vast Amur province taking up north-eastern and northern China, the left bank of the Amur river, Transbaikalia and, possibly, northern Mongolia. Farther west the boundary of this province is not traceable. It is probable that climatic warming enhanced the migration of thermophilic elements from south to north from the East Asia province. However, the basic aspect of this flora remained warm-temperate which is characteristic of the Siberian region.

In the Equatorial region embracing Africa and South America except for their southern margins localities of plant remains are known to be few. They are mainly concentrated in Israel, Egypt, Libya, Mexico and Brazil. Hence are known the ferns *Weichselia*, *Piazopteris* possessing thinnish leaves, various bennettitaleans, and conifers belonging to the Araucariaceae and Cheirolepidiaceae. Judging from the nature of plant remains the prevalent climate within the gigantic, still partially undivided, Afro–South-American mainland was semi-arid or arid. As in the Early Cretaceous, moist tropics must have existed only in the north-west and north of South America (Pons, 1982a, b). In the Late Jurassic and Early Cretaceous epochs the arid zones of the Northern and Southern Hemispheres were almost in contact with each other in the region of Brazil and the northern parts of Africa. Only the Equator proper passing through Morocco, Tunisia and Egypt was apparently dry, as is corroborated by a decline in *Classopollis* pollen content and increased amounts of the spores of ferns coupled with thin coal intercalations.

Fig. 5.2 (*opposite*) Climatic belts and basic phytochoria in Early and Middle Jurassic time (legends given for Figs. 5.2 through 5.6). I, belt of moderate-warm climate of the Northern Hemisphere; II, subtropical belt of the Northern Hemisphere (Euro-Sinian region); III, tropical belt (Equatorial region); IV, subtropical belt of the Southern Hemisphere (Austral or Notal region); V, belt of moderate-warm climate of the Southern Hemisphere (Antarctic province) expressed for the second half of Late Cretaceous time. 1, boundaries between climatic belts and corresponding phytochoria; 2, distribution of semi-arid or arid climate; 3, localities of *Dictyophyllum*; 4, *Phlebopteris*, *Matonidium*; 5, *Piazopteris*; 6, *Weichselia*; 7, *Klukia*; 8, *Tempskya*; 9, other tree ferns; 10, *Ptilophyllum*; 11, *Otozamites*; 12, *Ptilophyllum*, *Otozamites*, *Dictyozamites*; 13, *Cycadeoidea*; 14, *Nilssonia*; 15, *Frenelopsis*; 16, *Classopollis* pollen percentages (*a*) over 50%, (*b*) over 25%, (*c*) over 10%, (*d*) rare grains; 17, *Czekanowskiales*; 18, *Dewalquea*; 19, *Palmae*; 20, *Trochodendroides*; 21, *Pseudoprotophyllum*; 22, *Nothofagus* pollen.

The Austral (Notal) region occupying the southern parts of South America and Africa as well as India and Australia had certain features in common with the Euro-Sinian region of the Northern Hemisphere. Key differences included absence of ancient Pinaceae, and wide development of Podocarpaceae and Araucariaceae. At the end of the Jurassic and beginning of the Cretaceous Pentoxylales occur here (these are a special group of plants similar to bennettitaleans). Pentoxylales have been reported from India, New Zealand and the south of Australia. Their distribution suggests relations between India and the mainland of the Southern Hemisphere. The climate of the Austral region in the Late Cretaceous proved subtropical, being semi-arid in the south of South America and moist in Australia and India. No features of the existence of a warm-temperate belt in the Southern Hemisphere up to the second half of the Late Cretaceous on the basis of plant remains have so far been discovered.

The onset of the Early Cretaceous (Neocomian) did not see any substantial changes in the composition of vegetation of the Siberian–Canadian region. Several endemic cycads and bennettitaleans came into existence, possibly indicating a temporary climatic warming (Kirichkova, 1985). The flora of the Euro-Sinian region situated in the subtropical zone of the Northern Hemisphere was twice as rich as the Siberian–Canadian region in the number of species and genera. It completely lost the Czekanowskiales, with Ginkgoales becoming rare. Cheirolepidiaceae played a noticeable role among conifers although their number had been gradually declining towards the end of the Early Cretaceous epoch. As in the Late Cretaceous epoch, climate was changing from west to east up to the onset of the Albian age ranging from moderate moist (western Ukraine) to arid (Central and Middle Asia). Near the shore of the Pacific Ocean it again became humid. From the start of the Berriasian or even, perhaps, Late Volgian time the Northern Hemisphere sees some indications of climatic cooling and certain humidization. Thus, in southern England the gypsum-bearing deposits of the Purbeckian are supplanted by the carbonaceous Wealden (Fig. 5.4).

Average annual temperatures in the Arctic basin in the Neocomian failed to exceed 15 °C whereas in the Tethys Ocean in the Berriasian they reached 22 °C. It is quite striking that Greenland's temperature in the middle of the Volgian age attained 22 °C which indirectly implies a far more southern position of this large island as compared with the present day.

Interesting data on average annual temperature change throughout the Lower Cretaceous and on *Classopollis* pollen content fluctuation for the same time intervals within the Crimean–Caucasian region are cited by Yasamanov and Petrosyants (1983). They show a gradual although non-uniform decline of temperature throughout the Early Cretaceous from the Berriasian (20 to 25 °C, and in the Caucasus up to 27 °C) to the Barremian (northern Caucasus 17.5 to 20 °C, the Transcaucasian area 17 to 21 °C). Temperature fall was concomitant with *Classopollis* pollen content reduction from 75 to 90% in some samples in the Valanginian to the Barremian maximal values at 20 to 40%.

A new minor warming occurred apparently somewhere at the close of the Aptian or onset of the Albian. The cycadophytes index increase in the continental deposits of the Southern Primorje area whose age is presumably Aptian as shown by Krassilov (1973b). The palaeotemperature curves constructed from the oxygen isotope composition in the carbonate shells of various organisms for certain regions of the Pacific Ocean (Krasheninnikov and Bassov, 1985) reveal a temperature rise in the Aptian only for benthonic foraminifera. Climatic moistening concomitant with cooling ensues in the Albian, probably in its second half (the duration of this age is currently estimated at 15 million years). For the first time in the Early Cretaceous epoch Mongolia witnessed appearance of coal-bearing deposits crowning the sequence of the Lower Cretaceous (the Khukhtyk suite). A carbonaceous formation was also observed in the upper strata of the Lower Cretaceous of Transbaikalia. In western Kazakhstan, the Chulym-Yenisei basin and the Vilyui trough sands of alluvial origin testify to climatic humidization which came to be widely developed at the end of the Albian. This rapid alluvial development was facilitated by displacement area uplifts.

Consideration of *Classopollis* content curves for the various regions of the European part of the USSR, the Caucasus and Middle Asia shows its proportions to decrease from the upper strata of the upper Jurassic to the Albian (see Fig. 2.12, p. 53). It reaches minimal values in the deposits of this age (less than 5%). In the Cenomanian and

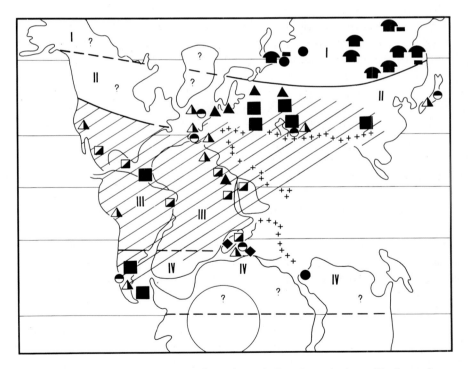

Fig. 5.3 Climatic belts and basic phytochoria in Late Jurassic time. (For legends see Fig. 5.2.)

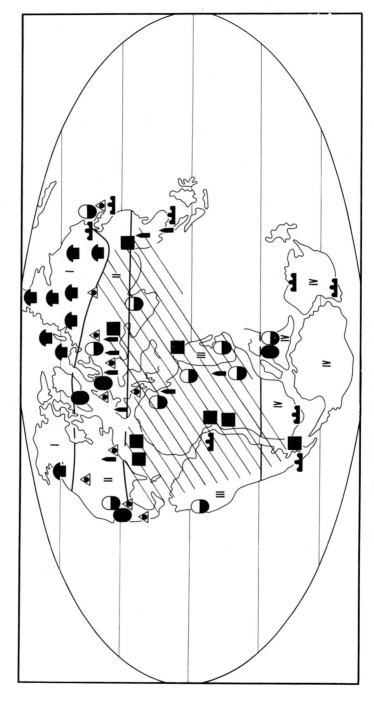

Fig. 5.4 Climatic belts and basic phytochoria in Neocomian time (including Barremian). (For legends see Fig. 5.2.)

Turonian of Middle Asia and western Kazakhstan *Classopollis* content again shows a gradual rise up to 30 to 40% but in the Upper Cretaceous of the remaining part of the USSR this pollen is represented by sporadic grains. The declining *Classopollis* pollen content from the Berriasian to the Albian is concurrent with a rise in the proportion of spores of ferns (primarily Gleicheniaceae and Schizeaceae).

The above correlations certainly suggest a gradual climatic humidization and cooling which was particularly pronounced in the subtropical belt (wherein Middle Asia was situated). This is supported among other things by replacement of red deposits with grey beds and disappearance of gypsum layers upward in the sequence.

In North America the northern boundary of humid climate shifted considerably northwards coupled with the expansion in the same direction of the area of distribution of coal-bearing deposits and the displacement of growth sites of tree-ferns *Tempskya*. At this time the arid climate areas vanish from Eurasia persisting, however, in South America and Africa. Their drastic reduction is related to the emergence in the Albian deposits of both Middle Asia (Kyzylkum, etc.) and the west of the USA of a number of localities of plant megafossils including ferns (Fig. 5.5).

Climatic cooling is also consistent with palaeotemperature changes. Thus, according to the data of Yasamanov (1980), the average annual temperature in the western Transcaucasian area fell from 20–21 °C in the Early Aptian down to 14–18 °C in the Late Aptian to 15–17 °C in the Early and Middle Albian and, finally, to 11–12 °C in the Late Albian. The temperature of the Albian in Middle Asia was maintained at 9 to 10 °C. The temperature curve built on the basis of plankton foraminifera in the region of the Falkland plateau (Krasheninnikov and Bassov, 1985) reveals a cooling in the Middle and Late Albian. On the other hand, the same authors' data on nanno-plankton from the northern part of the Pacific Ocean are indicative of a certain warming in the Albian.

However, the bulk of the data show the Late Albian to be marked by temperature decrease throughout. In particular, a temperature fall has been recorded for the northern part of the English–Parisian basin. The onset of the Albian is concomitant here with disappearance of sponges and is noted for the paucity of stromatopores and Chaetetidae which corroborates sea cooling covering the northern part of this basin. Emergence in the Cenomanian of diverse sponges points to a climatic warming. The above-mentioned cooling must have been instrumental in bringing about rapid extinction of the bennettitaleans and cycads over the greater part of the Earth in the second half of the Albian.

The Equatorial region for the Early Cretaceous epoch has been identified mainly on the basis of palynologic evidence (Doyle *et al.*, 1982). It was devoid of ancient Pinaceae but witnessed, however, the emergence of very characteristic pollen forms including those supplied with processes (elaters) typifying the Albian age. High *Classopollis* pollen content testifies to the hot and dry climate of Brazil and western and central Africa; this is also borne out by deposition of salts in the narrow strait located between the two mainlands. This is quite consistent with the notion of the existence

Fig. 5.5 Climatic belts and basic phytochoria in Albian time. (For legends see Fig. 5.2.)

at the beginning of the Cretaceous of a united continent formed by Africa and South America.

This is corroborated not simply by the similarity of the palynologic assemblages at one of the stratigraphic levels traceable in Brazil and western Africa but also by the same evolution of these complexes and consequently by the production of vegetation assemblages throughout the Early Cretaceous. The palynofloras of the Neocomian, Aptian and Albian prove to be in common.

The deposits of the Cenomanian on the western shore of central Africa (Gabon, Cameroon) see a considerable reduction in *Classopollis* pollen content while in the Turonian it virtually disappears (Boltenhagen and Salard-Cheboldaeff, 1980) which is indicative of climatic humidization. This must have been caused by the expanding South Atlantic completely merging with the ocean's central part, as is corroborated by identical composition of marine fauna (see Figs. 5.4 and 5.5). Due to the opening of the South Atlantic, differentiation of floras of these mainlands there ensues concomitant with the emergence of a belt of moist tropical forest dominated by angiosperms. Floristic differentiation attains its peak in Tertiary time witnessing the rise of the independent Palaeotropical and Neotropical vegetation kingdoms. The earlier separation of Africa from the other continents of the Southern Hemisphere is indicated by the absence in it of the remains of marsupials as well as of *Nothofagidites* pollen known from the sediments of the second half of the Upper Cretaceous of South America, Antarctica and Australia.

The Austral or Notal region including India highlighted paucity of Ginkgoales and *Nilssonia*, absence of Czekanowskiales and ancient Pinaceae. The proportion of Cheirolepidiaceae was moderate. Bennettitaleans, notably *Ptilophyllum*, as well as conifers (Podocarpaceae and Araucariaceae) were widespread. Presence of carbonate rocks containing spores and pollen of plants, at oceanic bottoms notably in the Atlantic, provided another possibility for reconstruction of floras inhabiting the coasts.

The Late Cretaceous epoch, a period of wide distribution and rapid evolution of angiosperms is marked by a certain climatic cooling most strongly pronounced in the Southern Hemisphere which was previously warmer than the northern one. A warm-temperate zone embracing the margin of South America, southern Australia and Antarctica (Herngreen and Khlonova, 1981) emerged in its southern part from the onset of the second half of the Late Cretaceous epoch. This is supported by the finds of the southern beech pollen (*Nothofagidites*) in the areas mentioned above. This beech typifies the moderate zone of the Southern Hemisphere at present. Absence of the southern beech either in the present-day vegetation or of its remains in the Upper Cretaceous deposits of this region confirms an earlier separation of South Africa (probably in pre-Cretaceous time) from the remainder of Gondwana (Fig. 5.6).

In the Late Cenomanian–Turonian there are signs of a warming in the Northern Hemisphere followed by a certain temperature drop in the Santonian and a new rise in the Campanian. The latter rise is concomitant with the advance northwards of evergreen elements with entire leaves including the palm (Krassilov, 1975b) reaching on

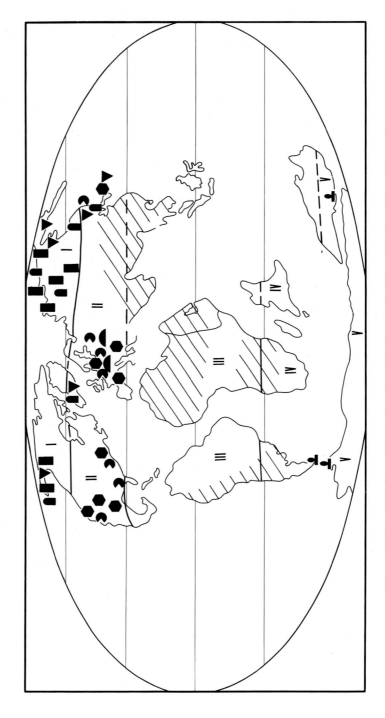

Fig. 5.6 Climatic belts and basic phytochoria in Campanian time. (For legends see Fig. 5.2.)

the western shore of North America up to the USA–Canada boundary (Vancouver Island). The warming is noted in the seas covering Middle Asia by the wide distribution of rudists (average annual temperature of 18 to 20 °C). The increased humidity of the Albian conformed to the reduction of areas occupied by the arid belt in the Northern Hemisphere at the expense of the southward advance of the southern boundary of the humid climate. In the Cenomanian the arid belt began expanding and this increased in the Turonian and Campanian. The maps of Ronov and Balukhovsky (1981) compiled on the basis of the distribution of terrigenous rock-climatic indicators of the Northern Hemisphere show a wide belt for the Cenomanian–Campanian encompassing Mexico, all of Northern Africa, the greater part of the Middle East, Middle Asia and southern China. The arid belt of the Southern Hemisphere takes up less space. In the Western Hemisphere it stretches meridionally along the coast of South America passing on to its eastern coast embracing here Argentina. In Africa it occupies the entire southern part of this continent save for its extreme southern tip. Indications of arid climate in Australia are lacking. ·

The conifer *Protophyllocladus* occurred in the Turonian. The role of leaves in this conifer was performed by cladodii. Its remains have been reported from the Euro-Sinian (Eurasia, Greenland, North America) and Siberian–Canadian regions (Fig. 5.7). *Protophyllocladus* was growing both under conditions of moist warm-temperate climate and in humid and semi-arid subtropical sites (southern Kazakhstan).

The north-east of Asia and western Canada of the Late Cretaceous saw the appearance of a vast refuge (Vakhrameev, 1981b) retaining many typical Mesozoic gymnosperms which vanished at the Early–Late Cretaceous boundary elsewhere on the globe. Varied ferns also continued to grow here in abundance in the Late Cretaceous.

Czekanowskiales represented by the genus *Phoenicopsis* are confined in their distribution to the north-east of the USSR. According to the observations of Lebedev (1983) they were particularly widely disseminated within the Okhotsk–Chukotka belt which was then a mountainous volcanic structure. In the more eastern littoral zone and in western Canada no remains of *Phoenicopsis* have so far been reported (see Fig. 4.11, p. 209).

The bennettitaleans, whose composition is believed to comprise among others the remains with unstudied cuticle determined as *Pseudocycas*, appear in the north-east of the USSR only during the periods of warming (Late Cenomanian–Turonian, Campanian). They are lacking within the Okhotsk–Chukotka belt itself save for rare exceptions being, however, encountered in the coastal plain. More often the remains of these plants represented by *Zamites, Zamiophyllum, Otozamites, Dictyozamites Pseudocycas* (?), etc. have been recorded for the Upper Cretaceous of the Primorje area, Sakhalin, Japan and south-western Canada, i.e. already within the subtropical belt (Euro-Sinian region).

Nilssonia are less sensitive to temperature. Therefore in the Late Cretaceous their areas are elongated meridionally along the two coasts of the Pacific Ocean invading

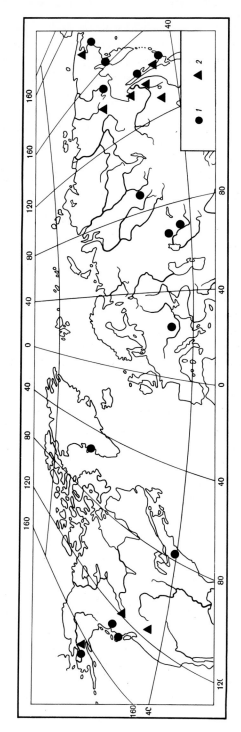

Fig. 5.7 Localities: 1, with *Protophyllocladus*; 2, with *Quereuxia*.

both the warm-temperate and subtropical climatic sites. In the Early and Middle Jurassic the Caytoniales occupied vast spaces largely within the subtropical belt of the Northern Hemisphere (Euro-Sinian region), in places penetrating the Siberian and, less frequently, the Equatorial (Mexico) regions. Climatic drying in the Late Jurassic must have resulted in the rupture of the continuous area and disintegration into smaller ones. In the Early Cretaceous they again became widespread. In the Later Cretaceous the Caytoniales are preserved along with other ancient gymnosperms in Alaska, western Canada, Sakhalin and the north-east of the USSR (see Fig. 2.2, p. 18).

Let us try to elucidate the climatic peculiarities permitting the mesophytic elements to persist on the Pacific coast of the USSR and in part in Canada throughout or almost throughout the Late Cretaceous. Considering the palaeogeographic maps of various ages of the Cretaceous period starting with the Valanginian we notice that land areas increase progressively during the entire Early Cretaceous in the north-east of Eurasia. While in the Valanginian there were marine basins with scattered groups of islands, the onset of the Hauterivian witnessed the emergence of a significant land area in place of Chukotka in the east connected with Alaska and separated from the remaining part of Eurasia by the narrow Anuy strait linking the Arctic basin with the Pacific Ocean roughly along the line passing through the lower reaches of the Kolyma river to the lower reaches of the Penzhina river according to present-day maps.

In the Albian the Anuy strait disappears. In the Aptian the north-east of the USSR sees the appearance of a number of elevations (Chukotkan, Omolonskian, Cherskian) continuing to rise towards the end of the Early Cretaceous and merging with one another in the Albian. The same process occurred in the Primorje area. The uplifts in the north of Eurasia resulted in the boundary line shifting to the northern latitudes and the Northern Land and Novosibirsk Isles being supplanted by the emerging lowland of great extent constituting a unity with the mainland.

But the most important event ultimately determining the climatic difference between the Pacific coast and the inner regions of Siberia was the rise in the Albian–Cenomanian of the volcanic-montane Okhotsk–Chukotka belt. Somewhat later a montane belt was also formed in Sikhote-Alin. Both the belts continued developing towards the end of the Late Cretaceous. Their prevalent effusive rocks largely of acidic or middle composition suggest that the volcanoes of the fissure or central type formed by them were as high as 2000 to 3000 m. The resultant mountainous chains held up the moist winds blowing from the Pacific Ocean conducive to precipitation fall-outs both in the littoral zone and in the areas of volcanic mountainous chains.

The humid climate of the montane belt and the littoral lowland adjacent to the Pacific Ocean is indicated by the occurrence of carbonaceous members intercalated with marine formations in continental deposits. Temperatures in the coastal part of the Pacific Ocean retaining certain bennettitaleans whose number and diversity increased southwards were probably higher than in the belt because we fail to find any remains of bennettitaleans in the belt's Late Cretaceous deposits. On the other hand, *Phoeni-copsis* were found only within the belt. *Nilssonia* were largely growing on the coastal plain east of the Okhotsk–Chukotka montane country as well as in Sakhalin, Alaska, Japan and western Canada where they were rather numerous.

The Late Cretaceous floras situated west of the present-day Verkhoyansk ridge have a quite different composition. While in the Early Cretaceous the aspect of the floras of the Lena basin and the more easterly Zyryanka, Omsukchan and Uda basins was similar especially on the level of relations between major plant taxa, the onset of the Late Cretaceous epoch coincided with another state of things. With the onset of the Late Cretaceous the composition of floras of the Vilyui and Hattang troughs as well as western Siberia (rivers Kas and Sym) became deprived of bennettitaleans, cycads, Czekanowskiales, Ginkgoales except for the genus *Ginkgo* proper, and Caytoniales. The variety and number of ferns is sharply reduced; conifers and angiosperms take the foreground. Carbonaceous deposits are virtually absent. It should be pointed out that small-leaved forms prevailed among angiosperms at certain stages of development of the Lena basin Late Cretaceous flora. This dramatic difference between the compositions of the floras cannot be attributed to the various latitudes of the sites of growth since they were actually located in the same latitudinal belt.

Such a composition points to the continentalization of the climate west of the Verkhoyansk ridge. Hot summers with drought led to the extinction of such moisture-loving forms as ferns, Czekanowskiales, the majority of Ginkgoales, *Nilssonia* and Caytoniales. The extinction of the bennettitaleans was facilitated by cool winters. Thus, latitudinal zonality gave way to meridional zonality in the Late Cretaceous epoch in the east of northern Eurasia.

The survival of the moisture-loving groups of plants in Canada and Alaska, which was mentioned above, was due to the montane ridges coming up in the west of North America holding up the moisture emanating from the Pacific Ocean. Coal formation was also continuing here sporadically. The Late Cretaceous floras of the central part of the USA (Dakota, Kansas, etc.) remote from the Pacific Ocean were totally devoid of the above-mentioned groups of moisture-loving gymnosperms.

The climatic humidity of north-eastern Asia, Alaska, western Canada and the north-east of the USA that made up the refuge outlined above is corroborated by the distribution here of the aquatic angiosperm *Quereuxia* in the Late Cretaceous. It exhibited a rosette of small leaves in part overlapping and situated on elongated radially diverging petioles. The leaf rosette floated on the water surface, and the plant must have inhabited the banks of lakes and river reaches. Finds of this plant within the region delineated are rather numerous. At the same time outside this area *Quereuxia* is almost unknown (see Fig. 5.7). It should be emphasized that the growth area of this plant coincides with the region of coal-formation in the Late Cretaceous in the Northern Hemisphere.

A new major cooling of climate started in the second half of the Maastrichtian–Danian. We shall not dwell in more detail on the peculiarities of this spell because it was analysed in many works including the collective contribution 'Development of Floras at the Mesozoic–Cenozoic boundary' (Anon., 1977). We shall merely point out such facts as disappearance of palms from the Fort-Union suite flora (Danian–Lower Palaeocene) in the western part of the USA and their absence in the floras situated to the north as well as the progression towards the south of the deciduous trocho-dendroidal forests. The latter became widely disseminated in the basin of the river Amur penetrating to Mongolia and reaching the Zaysan trough in eastern Kazakhstan.

The Equatorial zone of Africa drifting to the region of the present-day Equator underwent a humidization accompanied by coal deposition (Nigeria). Throughout the Late Cretaceous the distance between Africa and South America was increasing and this brought about the differentiation of the floras of the diverging continents according to palynologic data.

Touching on continental drift we should dwell on the boundary separating the Siberian–Canadian and the Euro-Sinian region. Plotting on a map of the present-day position of the mainlands the sites of the localities of fossil floras we see that the boundary between these regions approaching Scandinavia rises far northwards passing through Greenland towards North America then descending southwards.

It will be noted that the Greenlandian floras are very close compositionally to those of western Europe although now they are situated far more to the north. This is true of the floras of the end of the Triassic and beginning of the Jurassic as well as those of the Early and Late Cretaceous. It is curious that temperatures measured by isotopes in Greenland are far more similar to those of western Europe of the corresponding periods of geological time than to the temperatures of Siberia although these parts of Siberia are now lying at the same latitude as Greenland. This is one of the convincing arguments of the change of the position of Greenland, i.e. its northward drift, that had occurred already in the Tertiary time.

In the work of Krassilov (1985, pp. 184–185) devoted to the history of the Cretaceous period and particularly climatic evolution of this period, the boundary between the moderate-warm and the subtropical climates in the Northern Hemisphere reflected on the charts of phytoclimatic zones of the Early and Late Cretaceous coincides with ours (Vakhrameev, 1985). However, V. A. Krassilov draws this boundary only within the limits of Eurasia, Greenland and North America without interconnecting these sections which enables him to avoid the issue of the existence of this bend inevitably showing when linking these sections.

Judging by the text Krassilov does not think it necessary to resort to the hypothesis of continental drift or, as he puts it 'the plate-tectonic scheme', for elucidating the position of phytoclimatic belts and peculiarities of development of vegetation in Africa and South America. It appears to me that the data on palaeomagnetism and distribution of plants and land animals both in the Cretaceous and preceding periods confirm continental drift as they are inexplicable without this hypothesis.

Absence in the Southern Hemisphere in the Jurassic and, perhaps, Early Cretaceous of an analogue of the Siberian–Canadian region with its deciduous forests devoid of bennettitaleans is conspicuous. A few localities of the Jurassic and Early Cretaceous floras known from the littoral areas of the Antarctic do contain remains of

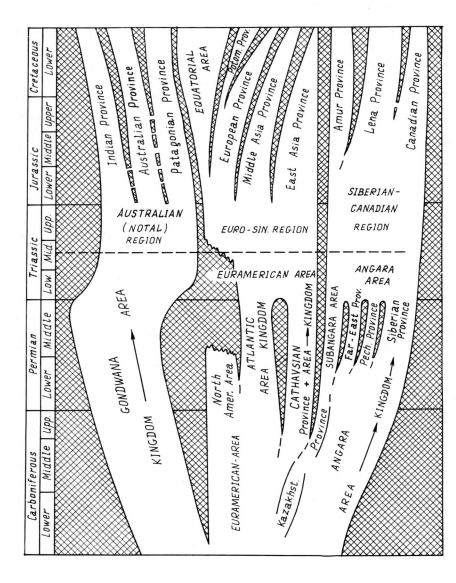

Fig. 5.8 Evolution of phytochoria in Late Palaeozoic and Mesozoic time.
Compiled by S. V. Meyen for Late Palaeozoic; V. A. Vakhrameev for Mesozoic.

bennettitaleans such as *Ptilophyllum* and *Otozamites* characteristic of the subtropical belt of the Northern Hemisphere (Fig. 5.2).

This asymmetry is also indicated by the distribution of dinosaurs (Charig, 1973). While in the Southern Hemisphere localities of their remains are encountered in Patagonia, South Africa and the south of Australia, in the Northern Hemisphere they were not reported from the confines of the northern part of Asia and North America. The exception is Spitsbergen.[1] It is not excluded that they might have swum there in summer time. The asymmetry is also confirmed by the distribution of belemnites typical of temperate-warm and tropical basins (Stevens, 1973). In the Jurassic the belemnites used to attain 75° S, while in the north they never crossed the 45° latitude. This suggests that the climate of the Northern Hemisphere until the middle of the Cretaceous period was warmer than that of the Southern Hemisphere.

The warmer climate of the Mesozoic area was due to the absence of the peripolar glacial caps and another circulation of oceanic water featuring prevalent latitudinal currents which, in turn, depended on the position of the continents and their configuration. Frakes (1979) thinks that in the Mesozoic there was a strong equatorial current (the Pacific was then wider) which was split as it approached the eastern coasts of Asia. One of its ramifications proceeded northwards by-passing the polar regions. The Antarctic of the time was shifted to the north of the Pole. Then both the branches of this gigantic current merged west of South America in the vicinity of the Equator. This system provided rather high temperatures even in the now near-Polar parts. The emergence of the South Atlantic connection in post-Albian time leading to contact between the equatorial waters and the Antarctic ones contributed to the disturbance of the circulation of water masses thus causing change of climatic conditions.

Summing up it can be said that the greater part of the Mesozoic is characterized by the asymmetry in the position of climatic belts (Fig. 5.8) caused by the drastic reduction in the Southern Hemisphere of the belt of the warm-moderate climate. This reduction was due to the peculiarities of currents and non-uniform distribution of the continental masses in the two hemispheres.

A few words on the peculiarities of migration of certain plant groups should be added. The arid belt of the Northern Hemisphere coupled with the tropical areas served as a barrier for southward migration of ancient Pinaceae and Czekanowskiales, the inhabitants of the warm-moderate climate. In contrast, the thermophilic bennettitaleans as well as Cheirolepidiaceae could freely migrate along the subtropics crossing the Equatorial belt from the Northern to the Southern Hemisphere and back. The arid belts were no obstacles for them since they withstood drought conditions which are corroborated by the distribution of the remains of these plants. Cheirolepidiaceae, more than others adapted to the drought conditions, gained advantageous distribution capabilities in arid climate. These characters enabled the representatives

---

[1] According to the data of W. Clemens and J. Archibald (1980) hadrosaurs and plesiosaurs lived at the end of the Cretaceous period in Alaska (70° N latitude) in moderate-warm climate.

of the same genera of Bennettitales and Cheirolepidiaceae to live in western Europe, China, Africa, Mexico and Argentina.

The northward distribution of many of them was impeded by the cooler climate of Siberia. However, during warming spells certain representatives of bennettitaleans such as *Ptilophyllum* and Cheirolepidiaceae got as far north as 60° to 70° latitude. The subtropical climate of southern latitudes enabled them to move southwards as far as the fringes of the Antarctic. *Nilssonia* and, possibly, other cycads were more responsive to drought environment preferring a mild marine climate without abrupt fluctuations of temperatures.

The greatest variety of *Nilssonia* is observed in the moist subtropical areas of Asia whence they must have originated for that matter. Hence they easily migrated northwards and to the north-east into the zone of warm-temperate climate. In the Southern Hemisphere *Nilssonia* are seldom encountered, the arid belt being a formidable obstacle on their way southwards. Many remains initially assigned to this genus as, for instance, in India were subsequently re-estimated and referred to the Bennettitales (*Nilssoniopteris*). In the Southern Hemisphere no reproductive organs of *Beania* and *Androstrobus* associated with the genus *Nilssonia* established by leaves have so far been reported.

It is probable that the arid belts also formed a serious hindrance for the ferns. But among these, too, certain forms arose adapted to arid climate. These forms include *Weichselia reticulata* and *Piazopteris branneri* that were growing not only in moist but also in dry localities. This enabled them to gain ground almost globally in the tropical and subtropical areas. The exception was the warm-moderate zone of the Northern Hemisphere.

The similarities among many Northern and Southern Hemisphere genera which included an abundance of bennettitaleans in the first place makes one seek for ways of migration. The path along the eastern and northern coasts of Africa washed by the Tethys Ocean seems to be one of the most critical. Furthermore, it crossed the Iberian peninsula at times touching northern Africa and then passing into Europe. The subsequent migratory path eastwards followed the northern coast of the Tethys Ocean up to China and south-east Asia. The connection between Asia and North America was provided by the Bering bridge.

During the Mesozoic period two major phases may be identified, characterized by expansion and even merging to some extent of the arid zones of both the hemispheres (the first half of the Triassic, Late Jurassic–beginning of Early Cretaceous) followed by two phases alternating with the preceding two and being typified by moistened and reduced areas of arid climate (second half of the Triassic, Early and Middle Jurassic, end of the Early and Late Cretaceous); minor climatic fluctuations are traceable against the background of these major phases. The distribution on a contemporary map of fossil plants whose composition determined some phytogeographic region and the position of the corresponding climatic belt prove that in the Mesozoic the continents took up positions different to those of today. In particular, almost until the end of the

Early Cretaceous there was no South Atlantic. Throughout the Mesozoic India was situated in the Southern Hemisphere while Greenland was located south of its present position.

The study of phytogeography provides major support for the continental drift theory being worked out by tectonists. This theory makes it possible to trace changes in position and configuration of humid belts and related coal-formation areas. Revealing distribution of phytochoria on the surface of the globe and the composition of respective floras permits correlation of sequences of continental sediments deposited in mutually remote areas on different mainlands, thus providing a biostratigraphic basis needed for completion of geological maps.

Postscript

V. A. Vakhrameev's book devoted to Jurassic and Cretaceous floras and climates was finished by the year 1986 and published in the USSR in 1988, after the death of the author.

Since that time, many new works on different aspects of the problem, new facts on available Mesozoic floras of different parts of the Earth with descriptions of new taxa of Jurassic and Cretaceous flora, palaeoecological and palaeoclimatological conditions of the environment have appeared. Below, there is a short review of several of them. Information on new taxa is given only for the territory of the USSR because the others are better known to the Western reader.

Some methodological aspects of phytogeographical zoning of the Earth during the Mesozoic era and proposals on how this or that phytochoria should be defined in accordance with palynological facts are described in the works of Batten and Wenben (1987), Kedves (1985) and Frederiksen (1987). The first shows the influence of some taxa on the determination of the more fractional phytochoria which depend on climatological environment of existing provinces. For example, for the Senonian era, the European part of the Normapolles province has been divided into several subprovinces. M. Kedves describes in his book the variant of phytogeographical zoning of the Earth for the Upper Cretaceous, pointing out the importance of detailed assessment of stratigeographical distribution of some pollen and spore taxa.

N. Frederiksen, for the Santonian and Campanian of northern America, besides previously known *Aquilapollenites* and Normapolles provinces, has defined a new one for the western margin of the continent with various endemic pollen, domination of *Proteacidites* group and a smaller content of *Aquilapollenites*, etc. In Vakhrameev's opinion, the province initially included British Columbia and California, but later, at the end of Cretaceous time, it was separated from other provinces by water and mountain barriers and, apparently, was formed under different climatic conditions which were close to tropical. It is supposed that, since the Campanian, this region moved 2000 km to the north. However, it is desirable to draw attention to the fact that, so far, in accordance with known palaeobotanic assessments of the neighbouring regions, the

270

presence of tropical flora has not been confirmed. Thus, the Lower Campanian flora of Vancouver Island (Nanaimo flora) in the south of western Canada (Bell, 1956) is of subtropical type and that is why V. A. Vakhrameev included it in the Potomac province. But, the fact of presence in the Vancouver flora, besides subtropical forms (palms, *Debeya*, etc.), of plentiful *Trochodendroides* and *Metasequoia* typical for warm-temperate flora characterizes the status of the considered flora in the northern outskirts of subtropics, i.e. to the south from the Canadian–Alaska province. The Cenomanian Dunvegan flora (Peace River) and the Maastrichtian flora of the Edmonton group (Alberta province) are typical for the warm-temperate Canadian–Alaska province.

Besides palaeontological works specifying the palaeogeographical zoning of the Earth during the Mesozoic, several works based on the results of macroflora studies and devoted to this problem have been published.

Makulbekov and Sodov (1989) gave their own understanding of a zoning method for Mongolia. The main part of its territory is attributed to the Siberian region (Lower and Upper Jurassic) and to the Euro-Sinian region (Upper Jurassic and Lower Cretaceous); this fact coincides with the conclusions of V. A. Vakhrameev. However, N. M. Makulbekov changed the names of the phytochoria (for the Upper Cretaceous) and related the northern part of Mongolia to the warm-temperate Boreal kingdom, and the southern part to the Central Asian palaeofloristic region (term of N. M. Makulbekov). The author renamed the Siberian region as Boreal because of the fact that during the Upper Cretaceous the composition of the flora abruptly changed in favour of angiosperms, whereas during the Lower Cretaceous mesophytic plants still prevailed. Meanwhile, V. A. Vakhrameev in his introduction to his book considered the reasons against the term 'Boreal kingdom' and stood for the term 'Siberian region' or, to be more exact, 'Siberian–Canadian region'. The problem of zoning a Central Asian region, and its correspondence and distinctions from those phytochoria which were located further to western and eastern parts of the subtropic floral area (Euro-Sinian region) demand further studies. V. A. Vakhrameev pointed out that due to the wide spread of Upper Cretaceous continental red beds unfavourable for the development of fossil vegetation it became difficult to define the composition and the type of flora of that age.

There are several works which V. A. Vakhrameev could not take into account devoted to palaeoclimatical aspects. Thus, Koeniguer (1986) characterized Cretaceous palaeoclimates and vegetation of that period. His conclusion that by the middle of the Senonian the Southern Hemisphere's warm-temperate zone was distinctively formed corresponds to the conclusions made by V. A. Vakhrameev on the necessity of zoning out (since the Santonian) the Antarctic province of the Southern Hemisphere with warm-temperate climate because of the presence of the *Nothofagidites* pollen (south of Southern America, south-eastern part of Australia, New Zealand, Antarctica). The same conclusion about the warm temperate climate of the Antarctic regions, during Upper Cretaceous time, was confirmed by Francis (1986) by the results of the investigation of the *Nothofagus* woods with the expressly

seen age rings. At the same time Creber and Chaloner (1985) thought the materials of
the studies of the fossil woods of tropics and subtropics (between 32° N and S latitudes)
showed no sign of climatic seasonal changes. There are several ideas which are of great
interest based on palaeodendrological and palaeomagnetic studies according to which
the northern part of Alaska was located further towards the North Pole (80°–85° N
latitude) than today. Probably, the sizes of age rings were caused by the light regime
changes from winter to summer with the average temperature of 10° ± 3 °C (Spicer
and Parrish, 1987).

V. A. Vakhrameev also pointed out that the location of Chukotka and its neighbouring territories has changed further to the north and supposed the temperature
could provisionally fall to a negative mark.

An important regional generalization on the Cretaceous flora of the Okhotsk–
Chukotka volcanogenic belt has been made by E. L. Lebedev (1987). The specific feature of the majority of the known belt's flora is its mountain nature. Thus its composition and development are slightly different from the flat country flora of the same
age. For example, some Lower Cretaceous relicts (*Sphenobaiera*, *Phoenicopsis*) of the
belt existed until Senonian time. The analysis of the belt's flora pointed out the
extreme fall of temperature on the boundary of the Lower and Upper Cretaceous that
influenced deeply the development of the Albian–Cenomanian flora of the
Siberian–Canadian region where the synchronous climate changes of different groups
could be observed. It is supposed that this temperature fall was caused by the disastrous
outburst of volcanic activity in the Okhotsk—Chukotka volcanogenic belt which
affected the climate.

Lately, publications dealing with those aspects of studies that were of great interest
to V. A. Vakhrameev have appeared; most important, on *Classopollis* pollen and its role
in reconstruction of palaeoclimatological conditions. It is proposed that the plants that
produced the *Classopollis* pollen should be united into the Frenelopsideaceae family
(Hluštik, 1987) which became extinct by the end of the Cretaceous. The domination
of *Classopollis* pollen and participation of the *Frenelopsis* conifer genus has been pointed
out for the optimum climate phases for many regions of the USSR (Sheiko, 1984;
Schilkina and Doludenko, 1985; Ediger and Papulov, 1987).

Among others published lately, the distribution maps of the most interesting taxa of
Eurasian Lower and Middle Jurassic plants (Vozenin-Serra and Taugourdeau-Lantz,
1985) and the report of Kimura *et al.* (1986) on the Middle Jurassic Utano flora from
south-western Japan should be mentioned. It is interesting to note that the Middle
Jurassic *Acrostichopteris* ferns found here could also be found among the Albian flora of
north-eastern Asia and Northern America.

The work of A. F. Khlonova is devoted to the appearance of some gymnospermae
and their circumboreal distribution (Khlonova, 1985; Khlonova and Lebedeva,
1988).

In conclusion, we mention the works dealing with the description of new genera
of fossil plants of the USSR and the systematization of the known plants.

V. A. Samylina (1988) described the Lower Cenomanian Arkagala flora of the north-eastern part of Asia totalling of 108 (26 new) fossil plants. Others deal with the description of new fern taxa (Vassilevskaya, Pavlov and Lodkina, 1985; Samylina and Schepetov, 1988); ginkgophyta (Samylina and Srebrodolskaya, 1986; Bystritskaya, 1987; Gomolitzky, 1987; Orlovskaya, 1987); conifers (Abramova, 1985; Doludenko and Kostina, 1985, 1986, 1987, 1988; Golovneva, 1988); angiosperm plants (Lebedev, 1986; Golovneva, 1987; Herman, 1987, 1988; Herman and Golovneva, 1988; Filippova, 1988; Lebedev and Herman, 1989); and other plants (Bystritskaya and Tatjanin, 1983; Kirichkova and Khramova, 1984; Krassilov, 1986; Doludenko and Teslenko, 1987; Krassilov and Bugdaeva, 1988).

M. A. Akhmetiev

References (Postscript)

Abramova, L. N. 1985. Cretaceous conifers from Bungin formation, Fadeevsky island (New Siberian islands). In: *Stratigraphy and Palaeontology of Mesozoic Sedimentary Basins in North of USSR*. Leningrad, pp. 104–107. (In Russian.)

Akhmetiev, M. A. 1990. Mesozoic paleofloristics, phytogeography and climates by V. A. Vakhrameev. Selected transactions, Geological Institute. Moscow: Nauka, 292 pp. (In Russian.)

Batten, D. J. and Wenben, L. 1987. Aspects of palynomorph distribution, floral provinces and climate during the Cretaceous. *Geol. Jahrb.*, Series A **96**, 219–237.

Bell, W. A. 1956. Lower Cretaceous floras of Western Canada. *Geol. Surv. Canada Mem.* **285**, 1–331.

Bystritskaya, L. I. 1987. Some *Phoenicopsis* from Western Siberia. *Materials of Stratigraphy and Palaeontology of Siberia*. Tomsk, pp. 87–96. (In Russian.)

Bystritskaya, L. I. and Tatjanin, G. M. 1983. New data on stratigraphy of Jurassic deposits in southeastern part of Western Siberia. *Materials of Geology of Siberia*. Tomsk, pp. 85–97. (In Russian.)

Creber, G. T. and Chaloner, W. G. 1985. Tree growth in the Mesozoic and Early Tertiary and the reconstruction of palaeoclimates. *Palaeogeogr., Palaeoclimatol., Palaeoecol.* **52** (1–2), 35–53.

Doludenko, M. P. and Kostina, E. I. 1985. *Schizolepis fanica* – a Mesozoic species of Pinaceae family. *Botan. Journ.* **70** (4), 464–471. (In Russian.)

    1986. New genus of conifers *Kantia* from the Jurassic of Tadjikistan. *Palaeontol. Journ.* **1**, 105–112. (In Russian.)

    1987. On conifers of the genus *Elatides*. *Palaeontol. Journ.* **1**, 110–114. (In Russian.)

    1988. New conifers genus *Kanevia* (Taxodiaceae) from the Late Albian of the Ukraine. *Botan. Journ.* **73** (4), 465–476. (In Russian.)

Doludenko, M. P. and Teslenko, Yu. V. 1987. New data on the Late Jurassic flora of the Ukraine (the environs of Kanev). *Palaeontol. Journ.* **3**, 114–118. (In Russian.)

Ediger, I. S. and Papulov, G. N. 1987. Palynocomplexes composition and stratigraphy of Jurassic continental deposits in South Trans-Urals region. New stratigraphic data on Phanerozoic Urals and correlated region. In: *Informational Material*. Sveldlovsk, pp. 92–99. (In Russian.)

Filippova, G. G. 1988. On some plants from Cretaceous deposits in the Arman River basin (North Okhotsk region). *Palaeontol. Journ.* **4**, 88–95. (In Russian.)

Francis, J. 1986. Growth rings in Cretaceous and Tertiary wood from Antarctica and their palaeoclimatic implications. *Palaeontology*, **29** (4), 665–684.

Frederiksen, N. O. 1987. Tectonic and paleographic setting of a new latest Cretaceous floristic province in North America. *Palaios*. **2** (6), 533–542.

Golovneva, L. B. 1987. New species of *Haemanthophyllum* genus from Rarytkin formation of the Koryak upland. *Botan Journ*. **72** (8), 1127–1131 (In Russian.)

    1988. New genus *Microconium* (Cupressaceae) from Late Cretaceous deposits of northeastern USSR. *Botan Journ*. **73** (8), 1179–1183. (In Russian.)

Gomolitsky, N. P. 1987. A new species of *Eretmophyllum* from the Jurassic of Middle Asia. *Rev. Palaeobot. and Palynol*. **51** (1–3), 127–131.

Herman, A. B. 1987. New Turonian angiosperms from north-western Kamchatka. *Palaeontol. Journ*. **4**, 96–105. (In Russian.)

    1988. Cenomanian flora from Ugol'naya bay (northeastern USSR). *Proceedings of the USSR Acad. Sci. Geol. Ser*. **11**, 110–114. (In Russian.)

Herman, A. B. and Goloneva, L. B. 1988. New genus of the Late Cretaceous platanoids from northeastern USSR. *Botan Journ*. **73** (10), 1456–1467. (In Russian.)

Hluštik, A. 1987. Frenelopsidaceae fam. nov., a group of highly specialized *Classopollis*-producing conifers. *Acta Palaeobot*. **27** (2), 3–20.

Kedves, M. 1985. The present-day state of Upper Cretaceous palaeophytogeography on palynological evidence. *Acta Biol. Szeged*. **31** (1–4), 115–127.

Khlonova, A. F. 1985. First evidence and stratigraphical levels of main types of angiosperm pollen. *Trans. Geol. and Geophys. Inst., Siberian branch the USSR Acad. Sci*. **620**, 21–34. (In Russian.)

Khlonova, A. F. and Lebedeva, N. K. 1988. Peculiarities of circumboreal correlation of Upper Cretaceous deposits as evidenced by palynological data. *Geol. and Geophys*. **2**, 13–19. (In Russian.)

Kimura, T., Ohaka, T. and Kurihara, Y. 1986. Middle Jurassic Utano flora and its significance for biostratigraphy and palaeophytogeography in East Asia. *Proc. Jap. Acad*. **62** (9), 341–344.

Kirichkova, A. I. and Khramova, A. N. 1984. Genus *Uralophyllum* and its taxonomic status. *Palaeontol. Journ*. **3**, 138–143. (In Russian.)

Koeniguer, J.-C. 1986. Les paleoflores comparées du Crétacé d'Europe et d'Afrique dans leurs rapports avec la paleoclimatologia. *C. r. Acad. Sci. Ser*. 2, **303** (9), 869–872.

Krassilov, V. A. 1986. New floral structure from the Lower Cretaceous of lake Baikal area. *Rev. Palaeobot. and Palynol*. **47** (1–2), 9–16.

Krassilov, V. A. and Bugdaeva, E. V. 1988. Gnetalean plants from the Jurassic of Ust-Baly, East Siberia. *Rev. Palaeobot. and Palynol*. **53** (3–4), 359–374.

Lebedev, E. L. 1986. New genus of Cretaceous angiosperms of North-East Asia – *Grebenkia*. *Palaeontol. Journ*. **1**, 134–139. (In Russian.)

    1987. Stratigraphy and age of the Okhotsk–Chukotian volcanic belt. *Trans. GIN, the USSR Acad. Sci*. **421**, 175 pp. (In Russian.)

Lebedev, E. L. and Herman, A. B. 1989. A new genus of Cretaceous angiosperms – *Dalembia*. *Rev. Palaeobot. and Palynol*. **59**, 77–91.

Makulbekov, N. M. and Sodov, G. 1989. Mesozoic and Cenozoic flora of Mongolia and phytogeography of Central Asia. *Main Results of Joint Investigations of Soviet–Mongolian Palaeontological Expedition during 1969–1988. Abstracts of Reports*. Moscow, pp. 25–27. (In Russian.)

Orlovskaya, E. R. 1987. On the Jurassic flora of Nizhne–Iliyan coalfield. In: *Materials on the History of the Fauna and Flora of Kazakhstan*. Alma-Ata, pp. 121–131. (In Russian.)

Petrescu, I. and Dusa, A. 1985. Paleoflora din senomanul bazinului Rusca Montana. *Dari Seame Sedinct Lust. Geol. si Geofiz. Paleontol. 1982*, **69**, 107–124.

Samylina, V. A. 1988. *Arkagalian Stratoflora of North-Eastern Asia*. Leningrad: Nauka, 131 pp. (In Russian.)

Samylina, V. A. and Shchepetov, S. V. 1988. New fern species of *Hausmannia* genus and its distribution in the Cretaceous of northeastern USSR. *Palaeontol. Journ.* **2**, 128–133. (In Russian.)

Samylina, V. A. and Srebrodolskaya, I. N. 1986. New species of *Phoenicopsis* genus from the Mesozoic of the Asian part of the USSR. *Botan. Journ.* **71** (9), 1262–1266. (In Russian.)

Sheiko, L. N. 1984. The role of *Classopollis* pollen in palynological complexes from Jurassic deposits of West-Siberian plain. *Trans. West-Siberian Sci. Research. Geol. Institut*, **187**, 10–14. (In Russian.)

Shilkina, I. A. and Doludenko, M. P. 1985. *Frenelopsis* and *Cryptomeria* – dominants of Upper Albian flora of the Ukraine. *Botan. Journ.* **70** (8), 1019–1030. (In Russian.)

Spicer, R. A. and Parrish, J. T. 1987. Plant megafossils, vertebrates and paleoclimate of the Kogosukruk Tongue (late Cretaceous) north slope Alaska. *US Geol. Surv. Circ.* **998**, 47–48.

Vakhrameev, V. A. 1987. Climate and the distribution of some gymnosperms during the Jurassic and Cretaceous. *Rev. Palaeobot. and Palynol.* **51** (1–3), 205–212.

        1988. Jurassic and Cretaceous floras and climates of the world. *Trans. Geol. Inst.*, Moscow, 430, 214 pp. (In Russian.)

Vassilevskaya, N. D., Pavlov, V. V. and Lodkina, L. B. 1985. New Early Cretaceous fern *Hausmannia* from Pronchishchev ridge. *Stratigraphy and Palaeontology of Mesozoic Sedimentary Basins of Northern USSR*. Leningrad, pp. 108–112. (In Russian.)

Vozenin-Serra, C. and Taugourdeau-Lantz, J. 1985. La flore de la formation Schemshak (Rhetien a Bajocien, Iran): rapport avec les flores contemporaines. Implications paleo-geographiques. *Bull. Soc. Fr.* **1** (5), 663–678.

Wolfe, J. A. and Upchurch, G. R. Jr. 1987. North American nonmarine climates and vegetation during the Late Cretageous. *Palaeogeogr., Palaeoclimatol. Palaeoecol.* **61** (1–2), 33–77.

References (Main text)

Ablayev, A. G. 1974. *Late Cretaceous Flora of Eastern Sikhote–Alin and its Significance for Stratigraphy.* Novosibirsk: Nauka, 180 pp. (In Russian.)

Aboul, Ela. 1979. Lower Cretaceous microflora from the north-western desert of Egypt. *Neues Jb. Geol. und Paläontol. Monatsh,* **10**, 586–595.

Abramov, L. N. 1983. Late Cretaceous flora of Hattanga river basin. *Paleontologicheskoye Obosnovaniye Raschleneniya Paleozoya i Mezozoya Arkticheskikh Rayonov SSSR.* Leningrad: PGO Sevmor-geologiya, pp. 118–127. (In Russian.)

Akhmetyev, M. A., Bratseva, G. M. and Vakhrameev, V. A. 1976. On Cretaceous-Palaeogene boundary in lower reacher of Amur River. *Ocherki Geologii i Paleontologii Dalnyego Vostoka SSSR.* Vladivostok: DVNTS AN SSSR, pp. 46–50. (In Russian.)

Aliyev, O. B. 1977. New data on age of copal-bearing suite in Minor Caucasus. *Izv. An AzSSSR. Nauki o Zemlye,* **1**, 3–10. (In Russian.)

Alvin, K. L. 1971. *Weichselia reticulata* (Stokes et Webb) Font. from the Wealden of Belgium. *Mem. Inst. Roy. sci. nat. Belg.* **166**, 1–33.

　　1982. Cheirolepidiaceae: biology, structure and paleoecology. *Rev. Palaeobot. and Palynol.* **37**, 71–98.

Anon. 1960. *Regional Stratigraphy of China.* Moscow: Izd-vo Inostr. Lit. **1**, 657 pp. (In Russian.)

　　1963. *Regional Stratigraphy of China.* Moscow: Izd-vo Inostr. Lit. **2**, 272 pp. (In Russian.)

　　1977. *Development of Floras at the Mesozoic–Cenozoic Boundary.* Moscow: Nauka, 130 pp. (In Russian.)

Appert, O. 1973. Die Pteridophyten aus dem Oberen Jura des Manamana in südwest Madagaskar. *Schweiz. Paläntol. Abh.* **94**, 62.

Archangelsky, S. 1963. A new Mesozoic flora from Ticó, Santa Cruz province, Argentina. *Bull. Brit. Mus. (Natur. Hist.) Geol.* **8** (2), 45–92.

　　1965. Fossil Ginkgoales from the Ticó flora, Santa Cruz province, Argentina. *Bull. Brit. Mus. (Natur. Hist.) Geol.* **10** (5), 119–137.

　　1966. New gymnosperms from the Ticó flora, Santz Cruz province, Argentina. *Bull. Brit. Mus. (Natur. Hist.) Geol.* **13** (5), 259–295.

　　1967. Estudio de la formacion Baquero. Cretacico inferior de Santa Cruz, Argentina. *Rev. Mus. La Plata. N.S. Paleontol.* **5**, 63–171.

　　1968. On the genus *Tomaxellia* (Coniferae) from the Lower Cretaceous of Patagonia (Argentina) and its male and female cones. *Bot. J. Linn. Soc.* **61** (384), 153–165.

1976. Vegetales fosiles de la formacion Springhill, Cretacico, en el subsuelo de la Cuenca Magallanica, Chile. *Ameghiniana*, **13** (2), 141–158.

1980. Palynology of the Lower Cretaceous in Argentina. *Proc. IV Intern. Palynol. Conf., Lucknow, 1976/1977,* **2**, 425–428.

Archangelsky, S. and Baldoni, A. 1972. Revision de la Bennettitales de la formacion Baguero (Cretacico inferior). Provincia de Santa Cruz, I. Hojas, *Rev. Mus. La Plata. N.S. Palaeontol.* **7**, 195–265.

1981. Palinologia estratigrafica del Cretacico de Argentina Austral. *VIII Congr. Geol., Argentino. San Luis,* **4**, 719–742.

Archangelsky, S. and Gamerro, J. 1966. Spore and pollen types of the Lower Cretaceous in Patagonia (Argentina). *Rev. Palaeobot. and Palynol.* **1**, 211–217.

Archangelsky, S. and Seiler, J. 1978. Algunos resultados palinologicos de la Perforacion UN OIL OS–1, del so de la Provincia de Chubut, Argentina. *Act. II Congr. Argent. Paleontol. y Biostratigr. y I Congr. Latinoamer. Paleontol. Buenos Aires,* **5**, 215–225.

Arkell, W. J. 1956. *Jurassic Geology of the World.* Edinburgh: Oliver and Boyd, 806 pp.

Arrondo, O. G. and Petriella, B. 1980. Alicura, nueva localidad plantifera liasica de la Provincia de Neuquén, Argentina. *Ameghiniana*, **17** (3), 200–215.

Asama, K., Nakornsri, N., Hinthong, C. and Sinsakul, S. 1981. Some younger Mesozoic plants from Trang, Southern Thailand. *Geol. and Palaeontol. Southeast Asia,* **13**, 39–46.

Ash, S. and Read, Ch. B. 1976. North American species of *Tempskya* and their stratigraphic significance. *Geol. Surv. Profess. Pap.* **874**, 1–42.

Baldoni, A. M. 1980. Plantas fósiles jurasicas de una nueva localidad en la Provincia del Neuquén. *Ameghiniana*, **17** (3), 255–272.

Baldoni, A. M. and Archangelsky, S. 1983. Palinologia de la formacion Springhill (Cretacico inferior) subsuelo de Argentina y Chile Austral. *Rev. Esp. Micropaleontol.* **15** (1), 47–101.

Barale, G. 1970. *Contribution a l'étude de la flore Jurassique de France: la paleoflore du Gisement Kimméridgien de Creys (Isère).* Lyon: Univ. Lyon Fac. Sci., 134 pp.

1981. *La paléoflore Jurassique du Jura Français: Etude systématique: aspects stratigraphiques et paléoécologiques.* Lyon, 467 pp. (Doc. Lab. Géol. Lyon; No. 81.)

Barale, G. and Contini, D. 1976. La paléoflore continentale de l'Oxfordien superieur du Jura septentrional: Le gisement de l'Hôpital Saint-Lieffroy (Doubs). *C.r. Soc. géol. France,* **1**, 7–9.

Barale, G. and Doludenko, M. P. 1985. Une nouvelle espèce de Cheirolepidiaceae de l'Albien supérieur d'URSS: *Frenelopsis kaneviensis* nov. sp. *Act. 110ᵉ Congr. Nat. Soc. sav. (Montpellier). Sec. sci.* Fasc. 5, pp. 99–114.

Baranova, Z. I. and Kirichkova, A. I. 1972. New data on stratigraphy and flora of Middle Jurassic deposits of Embensk area. *Dokl. AN SSSR,* **203** (5), 1139–1142. (In Russian.)

Baranova, Z. Y., Kirichkova, A. I. and Zauer, V. V. 1975. *Stratigraphy and Flora of Jurassic Deposits of East of Near Caspian Basin.* Leningrad: Nedra. 191 pp. (Tr. VNIGRI, **332**). (In Russian.)

Barnard, P. D. W. 1965. The geology of the Upper Djadjerud and Lar Valleys (North Iran). II. Palaeontology. Flora of the Shemshak Formation. Part 1. Liassic plants from Dorud. Milano. *Riv. Ital. Paleontol.* **71** (4), 1123–1168.

1967. The geology of the Upper Djadjerud and Lar Valleys (North Iran). II. Palaeontology. Flora of the Shemshak Formation. Part 2. Liassic plants from Shemshak and Ashtar. Milano. *Riv. Ital. Paleontol.* **73** (2), 539–588.

Barron, E. J., Harrison, S. G. A., Sloan, J. L. and Hay, W. N. 1981. Paleogeography 180 million years ago to present. *Eclog. Geol. Helv.* **74** (2), 443–470.

Batten, D. J. 1975. Wealden palaeoecology from the distribution of plant fossils. *Proc. Geol. Assoc.* **85** (4), 433–458.

1982. Palynology of shales associated with the Kap Washington Group volcanics, central north Greenland. *Rapp. Grølands Geol. Unders.* **108**, 15–23.

1984. Palynology, climate and the development of Late Cretaceous floral provinces in the Northern Hemisphere: A review. In: *Fossils and Climate*, ed. P. Brenchley. John Wiley and Sons Ltd, pp. 127–164.

Baykovskaya, T. N. 1956. Early Cretaceous floras of northern Asia. *Tr. Botan. In-ta AN SSSR. Ser. 8, Paleobotanika*, **2**, 49–181. (In Russian.)

Bell, W. A. 1949. Uppermost Cretaceous and Paleocene floras of Western Alberta. *Bull. Geol. Surv. Canada*, **13**, 1–231.

1956. Lower Cretaceous floras of Western Canada. *Geol. Surv. Canada Mem.* **285**, 1–331.

1957. Flora of the Upper Cretaceous Nanaimo group of Vancouver island, British Columbia. *Geol. Surv. Canada Mem.* **293**, 1–84.

1963. Upper Cretaceous floras of the Dunvegan, Bad Heart, and Milk River Formations of Western Canada. *Bull. Geol. Surv. Canada*, **94**, 1–76.

Berry, E. W. 1911. *Lower Cretaceous*. Maryland: Geol. Surv. 622 pp.

1916. *Upper Cretaceous*. Maryland: Geol. Surv. 578 pp.

Bogolepov, K. V. 1961. *Mesozoic and Tertiary Deposits of Eastern Fringe of West-Siberian Lowland and Yenisei Ridge*. Moscow: Gosgeoltekhizdat, 150 pp. (In Russian.)

Bolkhovitina, N. A. 1953. Spore-pollen characterization of Cretaceous deposits in central regions of USSR. *Tr. Inst. Geol. Nauk Acad. Nauk SSSR. Ser. Geol.* **145** (61), 1–185. (In Russian.)

1968. *Spores of Gleicheniaceae Ferns and their Stratigraphic Significance*. Moscow: Nauka, 116 pp. (Tr. GIN AN SSSR, **207**). (In Russian.)

Boltenhagen, E. B. and Salard-Cheboldaeff, M. 1980. Essai de reconstitution climatique crétacé et tertiaire du Gabon et du Cameroun d'après la palynologie. *Mem. Mus. Nat. Hist. Natur. N.S. B.*, **27**, 203–210.

Bose, M. N. 1960. Fossil flora of Jabalpur series. 2. Filicales. *Palaeobotanist*, **7** (2), 90–92.

Bose, M. N. and Sukn-Dev. 1959. Occurrence of two characteristic Wealden ferns in the Jabalpur series. *Nature*, **183** (4654), 130–131.

1961. Studies in the fossil flora of the Jabalpur series from the South Rewa, Gondwana Basin. 2. *Onychiopsis paradoxus* n. sp. *Palaeobotanist*, **8** (1/2), 57–64.

Bratseva, G. M. 1985. Palynological studies of Upper Cretaceous Barents Sea deposits. *Rev. Palaeobot. and Palynol.*, **44**, 293–302.

Bratseva, G. M. and Novodvorskaya, I. M. 1969. Palynologic characterization of Lower Cretaceous deposits of Buylasutin–Khuduk and Anda–Khuduk, Bayan–Khongor Aimak, Mongolian Peoples' Republic. *Fauna Mezozoya i Kainozoya. Tr. Sov.-Mong. Paleont. Exspeditsii*, **8**, 98–99. (In Russian.)

Brenner, G. J. 1976. Middle Cretaceous floral provinces and early migrations of angiosperms. In: *Origin and Early Evolution of Angiosperms*, ed. C. B. Beck. NY: Columbia Univ. Press, pp. 23–47.

Brown, R. W. 1950. Cretaceous plants from Southwestern Colorado. *Geol. Surv. Profess. Pap. D.*, **221**, 1–65.

1962. Paleocene flora of the Rocky Mountains and Great Plains. *US Geol. Surv. Profess. Pap.*, **375**, 1–119.

Budantsev, L. Y. 1968. Late Cretaceous flora of Vilyui Trough. *Botan. Journ.*, **53** (1), 3–16. (In Russian.)

1979. Phytostratigraphic complexes of Late Cretaceous of Lena–Vilyui and Chulym–Yenisei basins as a groundwork for correlating continental deposits. In: *Stratigrafiya*

Nizhnemelovikh Otlozheniy Neftegazonosnikh Oblastey SSSR. Leningrad: Nedra, pp. 149–162. (In Russian.)

　　　1983. *History of Arctic Flora of Early Cenophytic Epoch.* Leningrad: Nauka, 156 pp. (In Russian.)

Bugdayeva, Y. V. 1983. Interpretation of floristic paleosuccession in Late Mesozoic of Eastern Transbaikalia. In: *Paleontologiya i Rekonstruktsiya Geologicheskoy Istorii Paleobasseynov: Tez. Dolk. XXIX ses. VPO.* Leningrad: VPO, pp. 11–12. (In Russian.)

　　　1984. Flora and correlation of Tirgin strata in Transbaikalia. *Geol. i. Geofizika,* **11**, 22–27. (In Russian.)

Burakova, A. T. 1963. Flora of Jurassic deposits of Tuarkyr. *Problema Neftegazonosnosti Sredney Azii.* Leningrad: Gostoptekhizdat, **13**, pp. 117–232. (In Russian.)

Cai Zheng-yao. 1982. On the occurrence of *Scoresbya* from Jiangsu and *Weichselia* from Zhejiang. *Acta Palaeontol. Sin.,* **13**, 343–348.

Chandler, M. E. J. and Axelrod, D. I. 1961. An Early Cretaceous (Hauterivian) angiosperm fruit from California. *Amer. J. Sci.,* **259** (6), 441–446.

Charig, A. J. 1973. Jurassic and Cretaceous dinosaurs. In: *Atlas of Palaéobiogeography*, ed. A. Hallam. Amsterdam: Elsevier, pp. 339–352.

Chen Fen, Yang Guan-xiu. 1982. Lower Cretaceous plants from Pingquan, Hebei Province and Beijing , China. *Acta. Bot. Sin.,* **24** (6), 575–580.

Chen Fen, Yang Guan-xiu and Chow Huiqin. 1981. Lower Cretaceous flora in Fuxin basin, Liaoning province, China. *J. Wuhan Coll. Geol.,* **15** (2), 50–51.

Chen-Fen, Yang Guan-xiu and Dou Ya. 1980. The Jurassic Mentougou–Yudaishan flora from Western Yanshan, North China. *Acta Palaeontol. Sin.,* **9** (6), 431.

Chow Tseyen, Tsao Chengyao. 1977. On eight new species of conifers from the Cretaceous of East China with reference to their taxonomic position and phylogenetic relationship. *Acta Palaeontol. Sin.,* **16** (2), 165–181.

Clemens, W. A. and Archibald, J. D. 1980. Evolution of terrestrial faunas during the Cretaceous–Tertiary transition. *Soc. Geol. France Mem.,* **139**, 67–74.

Crane, P. and Dilcher, D. L. 1984. *Lesqueria*: an early angiosperm fruiting axis from the Middle Cretaceous. *Ann. Mo Bot. Garden,* **71** (2), 384–402.

Davydova, T. N. and Goldshtein, T. L. 1949. *Lithologic Research in Bureya Basin.* Moscow: Gosgeolizdat, 306 pp. (In Russian.)

Delle, G. V., Doludenko, M. P. and Krassilov, V. A. 1986. First find in USSR of Jurassic *Angiopteris* Hoffman (Marattiaceae). In: *Problemi Paleobotaniki.* Leningrad: Nauka, pp. 38–44. (In Russian.)

Dettmann, M. E. 1963. Upper Mesozoic microfloras from South-Eastern Australia. *Proc. Roy. Soc. Victoria,* **77** (1), 1–148.

Dettmann, M. E. and Playford, G. 1968. Taxonomy of some Cretaceous spores and pollen grains from Eastern Australia. *Proc. roy. Soc. Victoria,* **81** (2), 69–94.

Devyatilova, A. D., Nevretdinov, E. B. and Filippova, G. G. 1980. Stratigraphy of Upper Cretaceous deposits of basin of middle reaches of Anadyr River. *Geologiya i Geofizika,* **12**, 62–70. (In Russian).

Dilcher, D. L. and Crane, P. 1984. *Archaeanthus*: an early angiosperm from Cenomanian of the Western Interior of North America. *Ann. Mo Bot. Garden,* **71** (2), 351–383.

Dobruskina, I. A. 1961. On Mesozoic flora of Upper Amur. *Vestn. MGU. Ser. 4. Geologiya,* **6**, 29–35. (In Russian.)

　　　1965a. New data towards characterization of Tolbuzin paleofloristic complex (Upper Amur). *Vestn. MGU. Ser. 4. Geologiya,* **2**, 62–74. (In Russian.)

　　　1965b. Revision of Jurassic flora described by O. Heer. *Paleontol. Journ.,* **3**, 110–118. (In Russian.)

1982. *Triassic Floras of Eurasia*. Moscow: Nauka, 196 pp. (In Russian.)

Doludenko, M. P. 1969. On correlation of genera *Pachypteris* and *Thinnfeldia*. In: *Pterispermy Verkhnyego Paleozoya i Mezozoya*. Moscow: Nauka, pp. 14–34. (Tr. GIN AN SSSR, **190**.) (In Russian.)

1978. Genus *Frenelopsis* (Coniferales) and its finds in Cretaceous of USSR. *Paleontol. Journ.*, **3**, 107–121. (In Russian.)

1984. *Late Triassic Floras of South-Western Eurasia*. Moscow: Nauka, 110 pp. (In Russian.)

Doludenko, M. P. and Kostina, Y. N. 1987. On conifers of genus *Eltaides*, *Paleontol. Journ.*, **1**, 110-114. (In Russian.)

Doludenko, M. P. and Orlovskaya, E. R. 1976. Jurassic floras of the Karatau Range, southern Kazakhstan. *Palaeontology*, **19** (4), 627–640.

Doludenko, M. P. and Pons. D. 1986. Silicification remarquable chez une Cheirolepidiaceae d'Ukraine (URSS) a l'Albien supérieur. *Act. 111ᵉ Congr. Nat. Soc. Sav. Poitiers*, Fasc. 2, 23–35.

Doludenko, M. P. and Rasskazova, Y. S. 1972. Ginkgoaceae and Czekanowskiaceae of Irkutsk Basin. In: *Mezozoyskiye Rasteniya (Ginkgoviye i Chekanowskiyeviye) Vostochnoy Sibiri*. Moscow: Nauka, pp. 7–43. (Tr. GIN AN SSSR, **230**.) (In Russian.)

Doludenko, M. P. and Reymanówna, M. 1978. *Frenelopsis harrisii* sp. nov., from the Cretaceous of Tajikistan, USSR. *Acta Palaeobot.*, **1**, 3–11.

Doludenko, M. P. and Svanidze, T. I. 1969. *Late Jurassic Flora of Georgia*. Moscow: Nauka, 116 pp. Tr. GIN AN SSSR, **178**.) (In Russian.)

Dorf, E. 1942. Upper Cretaceous floras of the Rocky Mountain region. *Carnegie Inst. Wash. Publ.*, **508**, 1–168.

Douglas, J. G. 1969. The Mesozoic floras of Victoria. Pt 1, 2. *Geol. Surv. Victoria Mem.*, **28**, 1–310.

1970. *Ginkgoites multiloba*, a new Ginkgo-like leaf. *Mining and Geol. J.*, **6** (6), 27–32.

1973. The Mesozoic floras of Victoria. Part 3. *Geol. Surv. Victoria Mem.*, **29**, 1–185.

Douglas, J. G. and Williams, G. E. 1982. Southern polar forests: the Early Cretaceous floras of Victoria and their palaeoclimatic significance. *Palaeogeogr., Palaeoclimatol. Palaeoecol.*, **39**, 171–185.

Doyle, J. A. 1983. Palynological evidence for Berriasian age of basal Potomac Group sediments, Crisfield Well, Eastern Maryland. *Pollen et Spores*, **25** (3/4), 499–530.

Doyle, J. A. and Hickey, L. J. 1976. Pollen and leaves from the Mid-Cretaceous Potomac Group and their bearing on early angiosperm evolution. In: *Origin and Early Evolution of Angiosperms*, ed. Ch. B. Beck. NY: Columbia Univ. Press, pp. 139–206.

Doyle, J. A., Biens, P., Doerenkamp, A. and Jardine, S. 1977. Angiosperm pollen from the pre-Albian Lower Cretaceous of Equatorial Africa. *Bull. Cent. Rech. Explor.-Product. Elf-Aquitaine*, **1** (2), 451–473.

Doyle, J. A., Biens, P., Jardine S. and Doerenkamp. A. 1982. *Afropollis*, a new genus of early angiosperm pollen, with notes on the Cretaceous palynostratigraphy and paleoenvironments of Northern Gondwana. *Bull. Cent. Rech. Explor.-Product. Elf-Aquitaine*, **6** (1), 39–117.

Dragastan, O. and Barbulescu, A. 1977–1978. La flora Medior Jurassique de la Dobracea centrale. *Inst. Geol. Georg. Paleontol.*, **65**, 77–78.

Drinnan, A. N. and Chambers, T. C. 1985. A reassessment of *Taeniopteris daintreei* from the Victorian Early Cretaceous: a member of the Pentoxylales and a significant Gondwanaland plant. *Austral. J. Bot.*, **33**, 89–100.

Ducreux, J., Gaillard, M. G. and Samuel, E. 1982. Un gisement à plantes du Turonien supérieur la carrière de Sabran, à l'ouest de Bagnols-sur-Cèze (Gard, France). *Mem. Mus. Nat. Hist. Natur. N.S. Ser. C*, **49**, 71–80.

Du Toit, A. 1953. *Geology of South Africa*, 3rd edn, Edinburgh and London.

Epshtein, O. G. 1982. Climates of Mesozoic and Cenozoic of northern Asia and methods of paleoclimatic reconstructions. *AN SSSR. Ser. Geol.*, **5**, 59–68. (In Russian.)

Filippova, G. G. 1975. Fossil angiospermous plants from Arman River Basin. In: *Iskopayemiye Flori Dalnyego Vostoka*. Vladivostok: DVNTS AN SSSR, pp. 60–75. (In Russian.)

    1978. New Cretaceous angiosperms from Anadyr River Basin. *Paleontol. Journ.*, **1**, 138–144. (In Russian.)

    1979. Cenomanian flora of Grebenka River and its meaning for stratigraphy. In: *Dalnevostochnaya Paleofloristika*. Vladivostok: DVNTS AN SSSR, pp. 91-115. (Tr. Biol.-Pochv. In-ta DVNTS AN SSSR. N.S., **53** (156).) (In Russian.)

    1982. New Cretaceous angiosperms from basin of middle reaches of Anadyr River. In: *Materiali po Geologii i Poleznim Iskopayemim Severo-Vostoka SSSR*. Magadan: Kn. Izd-vo, **26**, pp. 69–75. (In Russian.)

Florin, R. 1936. Die fossilen Ginkgophyten von Franz-Joseph Land. 1. Spezieller Teil. *Palaeontographica B*, **81** (1/4), 71–173.

Fontaine, W. M. 1889. The Potomac of Younger Mesozoic Flora. *US Geol. Surv. Monograph*, **15**, 1–377.

Frakes, L. A. 1979. *Climates through Geologic Time*. Amsterdam: Elsevier, 310 pp.

Frenguelli, E. 1937. La florula Jurasica de Paso Flores en el Neuquen con referencias a la de Piedra Pintada y otras floras jurasicas argentinas. *Rev. Mus. La Plata. N.S.*, **1**, 67–108.

Fry, W. L. 1964. Jurassic floras of Western North America and their relationships to other Circum Pacific Jurassic floras. *Palaeobot. Soc. Spec. Ses., India*. Lucknow: Birbal Sahni Inst., 125.

Fuenzalida, H., Araya, R. and Herve, F. 1972. Middle Jurassic flora from North-Eastern Snow Island, South Shetland Islands. *Antarctic Geology and Geophysik*. Oslo, 93–97.

Gee, C. T. 1984. Preliminary studies of a fossil flora from the Orville Coast–eastern Ellsworth Land, Antarctic peninsula. *Antarct. Journal US*, **19** (5), 33–37.

Genkina, R. Z. 1960. Fossil flora and stratigraphy of carbon-bearing deposits of Severo–Sosvin Basin. *Izv. AZN SSSR, Ser. Geol*, **10**, 70–76. (In Russian.)

    1966. *Fossil Flora and Stratigraphy of Lower Mesozoic Deposits of Issykkul Trough*. Moscow: Nauka, 148 pp. (In Russian.)

    1977. Stratigraphy of Jurassic continental deposits of Fergana Ridge and paleontological rationale for their age. *Sov. Geologiya*, **9**, 63–79.

    1979. Division of continental deposits of Upper Triassic and Jurassic in east of Middle Asia. *Sov. Geologiya*, **4**, 27–39. (In Russian.)

Gomolitsky, N. P. 1968. On stratigraphy of Jurassic continental deposits of Yakkabak Mountains. *Izv. AN SSSR. Ser. Geol.*, **2**, 110–116. (In Russian.)

Gorbachev, I. F. and Timofeyev, A. A. 1965. Stratigraphy of Cretaceous deposits of Zeya–Bureya Trough. In: *Geologiya i Palegeograficheskiye Usloviya Formirovaniya Mezo-Kainozoyskikh Kontinentalnikh Vpadin Yuzhnoy Chasti Dalnyego Vostoka*. Moscow: Nauka, pp. 94–106. (In Russian.)

Gothan, W. 1914. Die unterliassische Flora der Umgegend von Nürnberg. *Abh. naturhist. Ges. Nürnberg*, **19**, 91–186.

Gothan, W. 1935. Unterscheidung der Lias uns Rhät Flora. *Ztschr. Dt. geol. Ges.*, **87** (10), 692–695.

Gryazeva, A. S. 1980. Palynologic substantiation of stratigraphy of Lower Cretaceous deposits of Pechora Basin. In: *Mikrofitofossilii v Neftyanoy Geologii*. Leningrad: VNIGRI, pp. 96–112. (In Russian.)

Halle, T. G. 1913. Mesozoic flora of Graham Land. *Wissensch. Ergebn. Schwed. Sudpolar Exped. (1901–1903). Geol. und Paläontol.*, **3** (14), 1–123.

Harland, W. B., Cox, A. B., Llewellyn, P. G., Pickton, C. A. G., Smith A. G. and Walters, P. A. 1982. *Geologic Time Scale*. Cambridge University Press, 131 pp.

Harris, T. M. 1931. The fossil flora of Scoresby Sound East Greenland. 1. Cryptograms (exclusive of Lycopodiales). *Medd. Grønland*, **85** (2), 1–114.

1932a. The fossil flora of Scoresby Sound East Greenland. 2. Description of seed plants *incertae sedis* together with a discussion of certain Cycadophyte cuticles. *Medd. Grønland*, **85** (3), 1–112.

1932b. The fossil flora of Scoresby Sound East Greenland. 3. Caytoniales and Bennettitales. *Medd. Grønland*, **85** (5), 1–133.

1935. The fossil flora of Scoresby Sound East Greenland. 4. Ginkgoales, Coniferales, Lycopodiales and isolated fructifications. *Medd. Grønland*, **112** (1), 1–176.

1937. The fossil flora of Scoresby Sound East Greenland. 5. Stratigraphic relations of the plant beds. *Medd. Grønland*, **112** (2), 1–114.

1961. *The Yorkshire Jurassic flora. I. Thallophyta–Pteridophyta*. London: Brit. Mus. Natur. Hist., 212 pp.

1962. The occurrence of the fructification *Carnoconites* in New Zealand. *Trans. Roy. Soc. N.Z. Geol.*, **1** (4), 17–27.

1964. *The Yorkshire Jurassic flora. II. Caytoniales, Cycadales and Pteridosperms*. London: British Mus. Natur. Hist., 191 pp.

1969. *The Yorkshire Jurassic flora. III. Bennettitales*. London: Brit. Mus. Natur. Hist., 186 pp.

1977. Notes on two of Raciborski's Jurassic ferns. *Acta Palaeobot*, **18** (1), 3–12.

1979. *The Yorkshire Jurassic flora. V. Coniferales*. London: Brit. Mus. Natur. Hist., 166 pp.

1983. The stem of *Pachypteris papillosa* (Thomas and Bose) Harris. *Bot. J. Linn. Soc.*, **86**, 149–159.

Harris, T. M., Millington, W. and Miller, J. 1974. *The Yorkshire Jurassic flora. IV*. London: Brit. Mus. Natur. Hist., 150 pp.

Heer, O. 1882. Die fossile Flora Grönlands. *Flora Fossilis Arct. Zürich*, **6** (2), 1–112.

1883. Die fossile Flora Grönlands. *Flora Fossilis Arct. Zürich*, **7**, 1–275.

Herman, A. B. 1984a. A new genus of platanoid angiosperms from Upper Cretaceous deposits of Kamchatka. *Paleontol. Journ.*, **1**, 71–79. (In Russian.)

1984b. On the age of the Valizhgensk Suite in Kamchatka and Yelistratov Peninsula by paleobotanical data. *Sov. Geologiya*, **11**, 60–69. (In Russian.)

1985. First find of angiosperms of genus *Ternstroemites* in USSR. (Late Cretaceous of Kamchatka). *Paleontol. Journ.*, **1**, 138–141. (In Russian.)

Hernandez, P., Pedro, J. and Azcarate, M. V. 1971. Estudio paleobotanico preliminar sobre restos de una tafoflora de la peninsula Byers (Cerro Negro), isla Livingston; islas Shetland del Sur, Antartica. *INACH. Ser. cient.*, **2** (1), 15–50.

Herngreen, G. F. W. 1974. Middle Cretaceous palynomorphs from northeastern Brazil. *Sci. Géol. Bull.*, **27** (1/2), 101–116.

1975. Palynology of Middle and Upper Cretaceous strata in Brazil. *Med. R.G.D. N.S.*, **26** (3), 39–91.

Herngreen, G. F. W. and Khlonova, A. F. 1981. Cretaceous microfossil provinces. *Pollen et Spores*, **23**, 441–555.

Hickey, L. J. 1973. Classification of the architecture of dicotyledonous leaves. *Amer. J. Bot.*, **60**, 17–33.

Hickey, L. J. and Doyle, J. A. 1977. Early Cretaceous fossil evidence for angiosperm evolution. *Bot. Rev.*, **43** (1), 1–104.

Hill, D., Playford G. and Woods, J. 1966. Jurassic fossils of Queensland. *Queensl. Palaeontogr. Soc. Brisbane*, pp. 203–228.

    1968. Cretaceous fossils of Queensland. *Queensl. Palaeontogr. Soc. Brisbane*, pp. 1–35.

Hluštik, A. 1974a. Contribution to the systematic and leaf anatomy of the genus Dammarites Presl in Sternberg. *Acta Mus. Nat. Pragae. B.*, **30** (1/2), 49–64.

    1974b. The nature of *Podozamites obtusus* Velenovsky. *Sb. Nar. Muz. Praze B.*, **30** (4/5), 173–186.

    1977. Remark on *Dammarites albens* Presl. *Vestn. Ustred. Ustavú Geol.*, **52**, 359–366.

Hollick, A. 1930. The Upper Cretaceous floras of Alaska. *Geol. Surv. Profess. Pap. Wash.*, **139**, 1–124.

Hsü, J. 1983. Late Cretaceous and Cenozoic vegetation in China, emphasizing their connections with North America. *Ann. Mo Bot. Garden*, **70** (3), 490–508.

Hughes, N. F. 1975. Plant succession in the English Wealden strata. *Proc. Geol. Assoc.*, **86** (4), 439–455.

Hughes, N. F., Drewry, G. E. and Laing, J. F. 1979. Barremian earliest angiosperm pollen. *Palaeontology*, **22** (3), 513–535.

Ilyina, V. I. 1985. *Jurassic Palynology of Siberia*. Moscow: Nauka, 237 pp. (In Russian.)

Imkhanitskaya, N. N. 1968. On issue of authenticity of finds of Sassafras in Cretaceous deposits of USSR. *Botan. Journ.*, **53** (5), 639–652. (In Russian.)

Jefferson, T. H. 1982. Fossil forests from the Lower Cretaceous of Alexander Island, Antarctica. *Palaeontology*, **25** (4), 681–708.

Kalugin, A. K. and Kirichkova, A. I. 1968. On stratigraphy of Jurassic continental formation of Mangyshlak. In: *Bul. NTI. Ser. Geologiya Mestorozhdeniy Polez. Iskopayemikh, Region. Geologiya*, Moscow: ONTI VIEMS, **19**, pp. 15–23. (In Russian.)

Kapitsa, A. A. and Ablayev, A. G. 1984. Additional materials on Albian flower plants of Amur vicinity. In: *Materiali po Stratigrafii i Paleografii Vostochnoy Azii*. Vladivostok: DVNTS AN SSSR, pp. 113–115. (In Russian.)

Karczmarz, K. and Popiel, S. 1971. W. sprawie gornokredowych flor Wyzyny Lubelskiej i Roztocza. *Kwart Geol.*, **15** (3), 643–650.

Khain, V. E., Ronov, A. B. and Balukhovsky, A. N. 1975. World's Cretaceous lithologic formations. *Sov. Geologiya*, **11**, 10–39.

Kilpper, K. 1964. Uber eine Rät–Lias Flora aus dem nordlichen Abfall des Alburs-Gebirges in Nordiran. I. Bryophyta und Pteridophyta. *Palaeontographica B.*, **114**, 1–78.

Kimura, T. 1975a. Middle–Late Early Cretaceous plants newly found from the upper course of the Kuzuryu river area, Fukui prefecture, Japan. *Trans. Proc. Palaeontol. Soc. Jap. N.S.*, **98**. 55–93.

    1975b. Notes on the Early Cretaceous floras of Japan. *Bull. Tokyo Gakugei Univ. Ser. IV*, **27**, 218–257.

    1976. Mesozoic plants from the Yatsushiro Formation (Albian), Kumamoto Prefecture, Kyushu, Southwest Japan. *Bull. Nat. Sci. Mus. Ser. C. Geol. Paleontol*, **2** (4), 179–208.

    1979. Late Mesozoic palaeofloristic provinces in East Asia. *Proc. Jap. Acad. Ser. B.*, **55** (9), 425–430.

    1980. The present status of the Mesozoic land floras of Japan. *Prof. S. Kanno Mem. Vol. Tsukuba Univ.*, pp. 379–413.

Kimura, T. and Hirata, M. 1975. Early Cretaceous plants from Kochi Prefecture, Southwest Japan. *Mem. Nat. Sci. Mus. Tokyo*, **8**, 67–90.

Kimura, T. and Sekido, S. 1976. Mesozoic plants from the Akaiwa Formation (Upper Neocomian), the Itoshiro group, Central Honshu, Japan. *Trans. Proc. Palaeontol. Soc. Jap. N.S.*, **103**, 343–378.

1978. Addition to the Mesozoic plants from the Akaiwa Formation (Upper Neocomian), the Itoshiro group, Central Honshu, Inner Zone of Japan. *Trans. Proc. Palaeontol. Soc. Jap. N.S.*, **109**, 259–279.

Kimura, T. and Tsujii, M. 1980a. Early Jurassic plants in Japan. Pt 1. *Trans. Proc. Palaeontol. Soc. Jap. N.S.*, **119**, 339–358.

1980b. Early Jurassic plants in Japan. Pt 2. *Trans. Proc. Palaeontol. Soc. Jap. N.S.*, **120**, 449–465.

1981. Early Jurassic plants in Japan. Pt 3. *Trans. Proc. Palaeontol. Soc. Jap. N.S.*, **124**, 187–207.

1982. Early Jurassic plants in Japan. Pt 4. *Trans. Proc. Palaeontol. Soc. Jap. N.S.*, **125**, 259–276.

1983. Early Jurassic plants in Japan. Pt 5. *Trans. Proc. Palaeontol. Soc. Jap. N.S.*, **129**, 35–57.

1984. Early Jurassic plants in Japan. Pt 6. *Trans. Proc. Palaeontol. Soc. Jap. N.S.*, **133**, 265–287.

Kirichkova, A. I. 1976a. Paleobotanical characterization and correlation of continental deposits of Upper Jurassic of Western Yakutia. *Geologiya i Geofizika*, **11**, 44–54. (In Russian.)

1976b. Flora of Aalenian of Mangyshlak. *Tr VNIGRI*, **388**, 92–113. (In Russian.)

1979. Paleobotanical rationale for stratigraphy and correlation of Jurassic and Lower Cretaceous continental deposits of Western Yakutia. In: *Stratigrafiya Nizhnemelovikh Otlozheniy Neftegazonosnikh Oblastey SSSR*. Leningrad: Nedra, pp. 123–148. (In Russian.)

1984. Cycadaceae and Bennettitales in Jurassic and Early Cretaceous floras of Lena Basin. *Yezhegodnik Vsesoyuznogo Paleontol. O-va*, **27**, 172–189. (In Russian.)

1985. (ed.) *Phytostratigraphy and Flora of Jurassic and Lower Cretaceous Deposits of Lena Basin*. Leningrad: Nedra, 223 pp. (In Russian.)

Kirichkova, A. I. and Kalugin, A. K. 1973. On Middle–Lower Jurassic boundary in Mangyshlak. *Dokl. AN SSSR*, **113** (2), 410–412. (In Russian.)

Kirichkova, A. I. and Samylina, V. A. 1978. Corrleation of Lower Cretaceous deposits of Lena carbon-bearing basin and north-east of USSR. *Sov. Geologiya*, **12**, 3–18. (In Russian.)

1979. On peculiarities of leaves of some Mesozoic Ginkgoaeceae and Czekanowskiaceae. *Botan. Journ.*, **64** (11), 1529–1538.

1983. On certain controversial issues of Mesozoic Ginkgophytes systematics. *Botan. Journ.*, **68** (3), 302–310. (In Russian.)

1984. Peculiarities of paleofloristic characterization of continental deposits of Upper Jurassic and Neocomian in Siberia. In: *Pogranichniye Yarusi Yurskoy i Melovoy Sistem*. Moscow: Nauka, pp. 161–167. (In Russian.)

Knobloch, E. 1964. Neue Pflanzenfunde aus dem südböhmischen Senon. *Jb. Staatl. Mus. Mineral. und Geol. Dresden*, 113–201.

1971. Neue Pflanzenfunde aus dem böhmischen und mährischen Cenoman. *Neues Jb. Geol. und Paläontol. Abh.*, **139** (1), 43–56.

1973. *Debeya insignis* (Hosius et D. Marck) Knobloch aus dem Senon von Friedersreuth (Oberpfalz). *Geol. Bavarica*, **67**, 172–176.

Knowlton, F. N. 1917. A Lower Jurassic flora from the Upper Matanuska Valley, Alaska. *US Nat. Mus.*, **51** (2158), 451–460.

Koch, E. 1964. Review of fossil floras and nonmarine deposits of West Greenland. *Geol. Soc. Amer. Bull.*, **75**, 535–548.

Koeniguer, J. C. 1980. Essai de reconstitution de quelques environments forestiers du Dogger au Crétacé, en Europe occidentale et au Sahara. *Mém. Soc. Geol. France*, **139**, 117–122.

Kon'no, E. 1967. Some younger Mesozoic plants from Malaya. *Geol. Palaeontol. Southeast Asia*, **3**, 135–164.

1968. Addition to some younger Mesozoic plants from Malaya. *Geol. Palaeontol. Southeast Asia*, **4**, 139–155.

Kon'no, E. and Asama. K. 1975. Younger Mesozoic plants from Ulu Endau, Pahang West Malaysia. *Geol. Palaeontol. Southeast Asia*, **16**, 91–102.

Koshman, M. M. 1969. Principal index paleofloristic complexes of Cretaceous system in Amur vicinity. *Geologiya, Geomorfologiya, Polezniye Iskopayemiye Priamuriya. Khabarovsk*, **3** (74), 221–232. (In Russian.)

1970. New Early Cretaceous ferns and cycadophytes of Udsk Depression (Western Okhotsk Sea vicinity). *Paleontol. Journ.*, **3**, 124–130. (In Russian.)

1973. Angiosperms from Lower Cretaceous deposits of Bureya Basin. *Botan. Journ.*, **58** (8), 1142–1146. (In Russian.)

Kossenkova, A. G. 1975. *Jurassic myospores of southern slope of Hissar Ridge and their significance for stratigraphy.* Abstract of Thesis for Award of Degree of Cand. Geol.-Miner. Science. Moscow, 27 pp. (In Russian.)

Kotova, I. Z. 1978. Spores and pollen from Cretaceous deposits of the Eastern North Atlantic ocean, Deep Sea Drilling Project, Leg 41, Sites 367 and 370. *Init. Rep. DSDP*, **41**, 841–881.

1983. Palynological study of Upper Jurassic and Lower Cretaceous sediments, Site 511, Deep Sea Drilling Project, Leg 71 (Falkland plateau). *Init. Rep. DSDP*, **71**, 879–906.

Krasheninnikov, V. A. and Bassov, I. A. 1985. *Stratigraphy of Cretaceous of Southern Ocean.* Moscow: Nauka, 176 pp. (Tr. GIN AN SSSR, 349.) (In Russian.)

Krassilov, V. A. 1967. *Early Cretaceous Flora of Southern Far East Near Sea Area and its Significance for Stratigraphy.* Moscow: Nauka, 264 pp. (In Russian.)

1972a. *Mesozoic Flora of Bureya River (Ginkgoales and Czekanowskiales).* Moscow: Nauka, 150 pp. (In Russian.)

1972b. On coincidence of lower boundaries of Cenozoic and Cenophytic. *Izv. AN SSSR. Ser. Geol.*, **3**, 9–16. (In Russian.)

1973a. Materials on stratigraphy and paleofloristics of carbon-bearing formation of Bureya Basin. In: *Iskopayemiye Flori i Fitostratigrafiya Dalnyego Vostoka.* Vladivostok: DVNTS AN SSSR, pp. 28–51. (In Russian.)

1973b. Climatic changes in Eastern Asia as indicated by fossil floras. I. Early Cretaceous. *Palaeogeogr., Palaeoclimatol., Palaeoecol.*, **13**, 261–273.

1975a. Development of Late Cretaceous vegetation of western Pacific Ocean coast related to climatic change and tectogenesis. In: *Iskopayemiye Flori Dalnyego Vostoka.* Vladivostok: DVNTS AN SSSR, pp. 30–42. (Tr. Biol.-Pochv. In-ta DVNTS AN SSSR. N.S., **27** (130).) (In Russian.)

1975b. Climatic changes in Eastern Asia as indicated by fossil floras. II. Late Cretaceous and Danian. *Palaeogeogr., Palaeoclimatol., Palaeoecol.*, **17**, 157–172.

1976. *Tsagayan Flora of Amur Region.* Moscow: Nauka, 92 pp. (In Russian.)

1978. Mesozoic lycopods and ferns from the Bureya basin. *Palaeontographica B.*, **166** (1/3), 16–29.

1979. *Cretaceous Flora of Sakhalin.* Moscow: Nauka, 183 pp. (In Russian.)

1982. Early Cretaceous flora of Mongolia. *Palaeontographica B.*, **181** (1/3), 1–43.

1984. Albian–Cenomanian flora of Kachi-Bodrak Interfluve (Crimea). *Bul. MOIP. Otd. Geol.*, **59** (4), 104–112. (In Russian.)

1985. *Cretaceous Period. Evolution of Earth's crust and biosphere.* Moscow: Nauka, 239 pp. (In Russian.)

Krassilov, V. A. and Bugdaeva, E. V. 1982. Achene-like fossils from the Lower Cretaceous of the Lake Baikal area. *Rev. Palaeobot. and Palynol.*, **36**, 279–295.

Krassilov, V. A. and Martinson, G. G. 1982. Fruits from Upper Cretaceous deposits of Mongolia. *Paleontol. Journ.*, **1**, 113–122. (In Russian.)

Krassilov, V. A. and Shorokhov, S. A. 1973. Early Jurassic flora in Pterovka River (Far East Primorje Area). In: *Iskopayemiye Flori i Fitostratigrafiya Dalnyego Vostoka.* Vladivostok; DVNTS AN SSSR, pp. 13–27. (In Russian.)

Krassilov, V. A., Nevolina, S. I. and Filippova, G. G. 1981. Evolution of flora of Far East and geological events of middle of Cretaceous Period. In: *Evolutsiya Organizmov i Biostratigrafiya Seredini Molovogo Perioda.* Vladivostok: DVNTS AN SSSR, pp. 103–115. (In Russian.)

Krishtofovich, A. N. 1933. Fossil flora from River Lozva in Northern Urals with remains of Mcclintockia related to Greenlandian. In: *Tr. Vsesoyuzn. Geol.-Razved. Ob-Niya.*, **291**, 1–33. (In Russian.)

1937. Cretaceous flora of Sakhalin. Mgach and Polovinka. In: *Tr. Dalnyevost. Fil. AN SSSR. Ser. Geol.*, **2**, 1–103. (In Russian.)

1958a. Fossil floras of Penzhina Bay, Lake Tastakh and Rarytkin Ridge. In: *Tr. Botan. In-ta AN SSSR. Ser. 8, Paleobotanika*, **3**, 73–121. (In Russian.)

1958b. Cretaceous flora of River Anadyr Basin. In: *Tr. Botan. In-ta AN SSSR. Ser. 8, Paleobotanika*, **3**, 7–70. (In Russian.)

Krishtofovich, A. N. and Baykovskaya, T. N. 1960. *Cretaceous Flora of Sakhalin.* Moscow, Leningrad: Izd-vo AN SSSR, 122 pp. (In Russian.)

1966. Upper Cretaceous flora of Tsagayan in Amur Region. In: *Izbr. Tr.* Moscow, Leningrad: Izd-vo AN SSSR, **3**, 184–320. (In Russian.)

Langenheim, J. 1961. Late Paleozoic and Early Mesozoic plant fossil from the Cordillera oriental and correlation of the Giron formation. *Bol. Geol. Serv. Geol. Nac. Colombia*, **8** (1/3), 95–132.

Lebedev, E. L. 1965. *Late Jurassic of River Zeya and Jurassic–Cretaceous Boundary.* Moscow: Nauka, 142 pp. (Tr. Geol. In-ta AN SSSR, **125**.) (In Russian.)

1974. *Albian Flora and Stratigraphy of Lower Cretaceous of Western Okhotsk Vicinity.* Moscow, 147 pp. (In Russian.)

1979. Paleobotanical rationale for stratigraphy of Cretaceous volcanogenic formations of Ulyinsk Foredeep (Okhotsk–Chukotka Volcanogenic Belt). In: *Izv. AN SSSR. Ser. Geol.*, **10**, 25–39. (In Russian.)

1982. Recurrent development of floras of Okhotsk–Chukotka Volcanogenic Belt at Early–Middle Cretaceous Boundary. *Paleontol. Journ.*, **2**, 3–14. (In Russian.)

1983. *Development of Cretaceous Floras in North-Eastern Asia and Phytostratigraphy of Okhotsk–Chukotka Volcanogenic Belt.* Abstract of Thesis for Award of Degree of Doctor, Geol.-Miner. Science. Moscow, 46 pp. (In Russian.)

Lebedev, I. V. 1955. Cretaceous System. In: *Atlas Rukovodyashchikh Form Iskopayemikh Fauni i Flori Zapadnoy Sibiri.* Moscow: Gosgeoltekhizdat, pp. 183–210. (In Russian.)

1962. Upper Cretaceous plants. In: *Biostratifragiya Mezozoyskikh i Tretichnikh Otlozheniy Zapadnoy Sibiri: Tr. SNIIGGiMS*, **22**, 237–281. (In Russian.)

Lemoigne, Y. 1984. Données nouvelles sur la paleoflora de Colombie. *Geobios*, **17** (6), 607–690.

Lesquereux, L. 1892. The flora of the Dakota group. *Mon. USA Geol. Surv.*, **17**, 1–256.

Li Xingxue and Ye Meina. 1980. Middle–Late Early Cretaceous flora from Jilim, NE China. *I Conf. Intern. Org. Palaeobot., London and Reading, 1980*. Nanjing (China): Inst. Geol. and Palaeontol. Acad. Sin., pp. 1–13.

Loladze, Y. M. 1978. *Gonatosoru dzirulensis* sp. nov. – A new species of fern from Upper Aptian deposits of Georgia. In: *Tr. Gruz. Politekhn. In-ta. im. V. I. Lenina*, **4** (205), 43–48. (In Russian.)

Lopatin, V. M. 1980. Stratigraphy of Lower Cretaceous of Shavokhtin Basin in South-Eastern Mongolia. In: *Rannemelovoye Ozero Manlay*. Moscow: Nauka, pp. 6–19. (In Russian.)

Lorch, J. 1967. Jurassic flora of Makhtesh Ramon, Israel. *J. Bot.*, **16**, 131–180.

Luchnikov, V. S. 1967. Jurassic flora of Darwaz and its stratigraphic significance. *Dokl. AN SSSR*, **176** (2), 406–408. (In Russian.)

1973. Bathonian flora of Darwaz and its stratigraphic significance. *Dokl. AN SSSR*, **209** (3), 662–664. (In Russian.)

1982. Stratigraphy of coal-bearing deposits of Jurassic in Central Tadjikistan. *Sov. Geologiya*, **9**, 75–85. (In Russian.)

1987. Stage-wise division of Lower Jurassic of Middle Asia by floristic data. *Sov. Geologiya*, **3**, 66–75. (In Russian.)

McLachlan, I. and Pieterse, E. 1978. Preliminary palynological results: Site 361, Leg 40, Deep Sea Drilling Project. *Init. Rep. DSDP*, **40**, 857–881.

McQueen, D. R. 1956. Leaves of Middle and Upper Cretaceous Pteridophytes and cycads from New Zealand. *Trans. Roy. Soc. NZ*, **83** (4), 673–686.

Mägdefrau, K. 1956. *Palaeobiologie der Pflanzen*. Jena: Fisher, 443 pp.

Makulbekov, N. M. 1974. Late Cretaceous flora of Ulken-Kalkan (Iliysk Depression). In: *Fauna i Flora iz Mezokaynozoya Yuzhnogo Kazakhstana*. Alma-Ata: Nauka, pp. 108–120. (In Russian.)

Manum, S. 1987. Mesozoic Sciadopitys-like leaves with observations on four species from the Jurassic of Andøya, Northern Norway, and emendation of *Sciadopityoides sveshnikova*. *Rev. Palaeobot. and Palynol*, **51**, 145–168.

Markov, V. A., Trofimuk, A. A. and Shcherbakov, V. S. 1970. On the interrelation between marine and coal-bearing deposits in Bureya Basin. *Dokl. AN SSSR*, **191** (3), 647–649. (In Russian.)

Matsuo, H. 1962. A study of the Asuwa flora (Late Cretaceous age) in the Hokuriku district, Central Japan. *Sci. Rep. Kanazawa Univ.*, **8** (1), 177–250.

1970. On the Omichidani flora (Upper Cretaceous) inner side of Central Japan. *Trans. Proc. Palaeontol. Soc. Jap. N.S.*, **80**, 371–389.

Menendez, C. A. 1966. Bennettitales from the Ticó flora, Santa Cruz province, Argentina. *Bull. Brit. Mus. (Natur. Hist.) Geol.*, **12** (1), 1–42.

1969. Die fossilen Floren Südamericas. *Biogeogr. Ecol. S. Amer.*, **2**, 519–561.

Menendez, C. A. and Caccavari de Filice, M. A. 1975. Las especies de Nothofagidites (polen fosil de Nothofagus) de sedimentos Terciarios y Cretacicos de Estancia La Sara, Norte de Tierra del Fuego, Argentina. *Ameghiniana*, **12** (2), 165–183.

Meyen, S. V. 1984. Philogenia of higher plants and philogenesis. In: *XXVII Mezhdunar. Geol. Kongr. Doklady. Paleontologiya. Sektsiya S.02*. Moscow: Nauka, **2**, p. 146. (In Russian.)

Mildenhall, D. C. 1970. Checklist of valid and invalid plant macrofossils from New Zealand. *Trans. Roy. Soc. NZ Earth Sci.*, **8** (6), 77–89.

1976. Early Cretaceous Podocarp megastrobilus (note). *NZ J. Geol. and Geophys.*, **19** (3), 389–391.

1980. New Zealand Late Cretaceous and Cenozoic plant biogeography. A contribution. *Palaeogeogr., Palaeoclimatol. Palaeoecol.*, **31**, 197–233.

Mildenhall, D. C. and Johnston, M. R. 1971. A megastrobilus belonging to the genus *Araucarites* from the Upper Motuan (Upper Albian), Mairarapa, North Island, New Zealand. *NZ J. Bot.*, **9** (1), 67–79.

Miller, C. N. 1976. Two new pinaceous cones from the Early Cretaceous of California. *J. Paleontol.*, **50** (5), 821–832.

Morgan, R. 1978. Albian to Senonian palynology of site 364, Angola Basin. *Init. Rep. DSDP*, **40**, 915–951.

Muller, J. and Jeletzky, J. A. 1970. Geology of the Upper Cretaceous Nanaimo Group, Vancouver island and Gulf islands, British Columbia. *Geol. Surv. Canada Pap.*, **69–25**, pp. 1–77.

Naryshkina, A. M. 1973. On the boundary between Cretaceous and Palaeogenic deposits in Amur–Zeya Trough. *Sov. Geologiya*, **6**, 148–151. (In Russian.)

Neiburg, M. F. 1932. On a find of Cycadeoidea trunk from South-Eastern Mongolia. *Dokl. AN SSSR. Ser. A.*, **8**, 200–201. (In Russian.)

Němejc, F. 1961. Fossil plants from Klikov in South Bohemia (Senonian). *Rozpr. CZAV. MPV.*, **71** (1), 1–56.

Němejc, F. and Kvaček, Z. 1975. *Senonian plant macrofossils from the region of Zliv and Hluboka (near Ceske Budejovice) in South Bohemia.* Prague: Univ. Karlova, 82 pp.

Nevolina, S. I. 1984. Late Cretaceous flora of Primorje area (Partinzansk Flora of A. N. Krishtofovich). In: *Yezhegodn. Vsesoyuzn. Paleontol., Ob-va.*, **27**, 219–235. (In Russian.)

Nicol-Lejal. A. 1971. Sur trois frondes fossiles de cycadophytes du Lusitanien de Bou Derga (Algérie). *C.r. 94ᵉ Congr. Nat. Soc. Sav. Pau. 1969. Sec sci.*, Fasc. **3**, 219–229.

1981. A propos de nouvelles floras paleozoiques et mesozoiques de l'Egypte du Sud-Ouest. *C.r. Acad. Sci.*, **292**, 1337–1399.

Nikitin, V. G. and Vassilyev, I. V. 1977. Plant remains complexes from Upper Cretaceous deposits of Turan Plate. *Dokl. AN SSSR. Ser. Geol.*, **8**, 53–60. (In Russian.)

Nishida, H. and Nishida, M. 1983. On some petrified plants from the Cretaceous of Choshi, China Prefecture. VII. *Bot. Mag. Tokyo*, **96** (1042), 93–101.

Oyama, T. and Matsuo, H. 1964. Notes on palmean leaf from the Oarai flora (Upper Cretaceous), Oarai Machi, Ibaraki Prefecture, Japan. *Trans. Proc. Palaeontol. Soc. Jap. N.S.*, **55**, 241–246.

Page, V. M. 1984. A possible magnolioid flora axis, *Loishoglia bettencourtii* from the Upper Cretaceous in Central California. *J. Arnold Arboretum*, **65** (1), 95–104.

Pantič, N. 1981. Macroflora and palynomorphs from Lower Jurassic of Bados mountain (Montenegro–Yugoslavia). *Ann. Geol. Pen. Balk*, **45**, 137–171.

Paraketsov, K. V. 1982. Upper Jurassic of Laglykhtansk Trough. In: *Materiali po Geologii i Poleznim Iskopayemym Severo-Vostoka SSSR.* Magadan: In. Izd-vo, **26**, pp. 58–60. (In Russian.)

Patton, W. 1973. Reconnaissance geology of the Northern Yukon-Koynkuk province, Alaska. *Geol. Surv. Profess. Pap. A*, **774**, 1–17.

Peresvetov, A. S. 1947. Floristic characterization of Lower Cretaceous sandstones in Karovo village (Moscow region). In: *Tr. Mosk. Geol-Razved. In-ta*, **22**, 192–208. (In Russian.)

Pergament, M. A. 1961. *Stratigraphy of Upper Cretaceous Deposits of North-Western Kamchatka (Penshina Area).* Moscow: Izd-vo AN SSSR, 147 pp. (Tr. GIN AN SSSR, **39**.) (In Russian.)

1978. *Stratigraphy and Inocerams of Upper Cretaceous in Northern Hemisphere.* Moscow: Nauka, 214 pp. (Tr. GIN AN SSSR, **322**.) (In Russian.)

Petrescu, I. and Dusa, A. 1980. Flora din cretacicul superior de la Rusca Montana–o raritate in patrimoniul paleobotanic national. *Octotirea natur. med. inconj.*, **24** (2), 147–155.

Pimenova, N. V. 1939. Cenomanian flora of environs of Kanev Town. *Geologichniy Zhurn.*, **6** (1/2), 229–243.

Polyansky, B. V., Safronov, D. S. and Sykstel, T. A. 1975. Upper Triassic and Jurassic deposits of South-Eastern Iran (Kerman Area). *Bul. MOIP. Otd. Geol.*, **50** (6), 5–15. (In Russian.)

Pons, D. 1979. Les organes reproducteurs de *Frenelopsis alata* (K. Feistm.) Knobloch. Cheirolepidiaceae de Cénomanien de l'Anjou, France. *Act. 110ᵉ Congr. Nat. Soc. Sav., Bordeaux. Sec. sci.*, Fasc. **1**, 209–231.

    1982a. Decouverte de Crétacé moyen sur le flanc est du Massif de Quetame, Columbie. *C.r. Acad. Sci. Ser. 2*, **294**, 533–536.

    1982b. Etudes paléobotanique et palynologique de la formation Giron (Jurassique moyen–Crétacé inferieur) dans la region de Lebrija, department de Santander, Columbie. *Act. 107ᵉ Congr. Nat. Soc. Sav. Sci. Brest*, Fasc. **1**, 53–78.

Poyarkova, A. I. 1939. On study of fossil floras of Bureya and Amur Tsagayan. In: *Sbornik k 70-letiyu Acad. V. L. Komarova*. Moscow–Leningrad: Izd-vo AN SSSR, pp. 631–682. (In Russian.)

Prynada, V. D. 1937. On study of Lower Cretaceous flora in Voronezh Region. *Yezhegodn. Vseros. Paleontol. Ob-va*, **11**, 71–89. (In Russian.)

    1944. *On Mesozoic Flora of Siberia*. Irkutsk: OGIZ, 44 pp. (In Russian.)

    1962. *Mesozoic Flora of Eastern Siberia and Transbaikalia*. Moscow: Gosgeoltekhizdat, 168 pp. (In Russian.)

Radkevich, G. A. 1895. On fauna of Cretaceous deposits in Kanev and Cherkassy Districts of Kiev Region. *Zap. Kiev. Ob-va Yestestvoispitateley. Kiev*, **14** (1), 95–105. (In Russian.)

Rao, P. V. R., Ramanujam, C. G. K. and Varma, Y. N. R. 1983. Palynology of the Gangapur beds, Pranhita-Godavari basin, Andhra Pradesh. *Geophytology*, **13** (1), 22–45.

Ronov, A. B. and Balukhovsky, A. N. 1981. Climatic zonality of continents and general trends of climatic change in Late Mesozoic and Cenozoic. *Litologiya i Polezn. Izkopayemiye*, **5**, 118–136. (In Russian.).

Rovnina, L. V. 1972. *Stratigraphic Division of Continental Deposits of Triassic and Jurassic in North-West of West-Siberian Lowland*. Moscow: Nauka, 109 pp. (In Russian.)

Sadovnikov, G. N. 1977. Floristic complexes of Mesozoic in Northern Iran. *Bul. MOIP. Otd. Geol.*, **52** (2), 146. (In Russian.)

Saks, V. N. and Halnyayeva, T. I. 1975. *Early and Middle Jurassic Belemnites of North USSR*. Moscow: Nauka, 191 pp. (Tr. IGiG, **239**.) (In Russian.)

Salem, M. I. and Basewia, M. T. (eds.) 1980. *Geology of Libya*. London: Academic Press.

Samoilovich, S. R. 1977. A new pattern of floristic zonation of Northern Hemisphere in Late Senonian. *Paleontol. Journ.*, **3**, 118–127. (In Russian.)

    1980. Microfossils of Upper Jurassic – of Hattang Trough and correlation of precipitation of Late Cretaceous time in North USSR. In: *Mikrofossilii v Neftyanoy Geologii*. Leningrad: VNIGRI, pp. 113–129. (In Russian.)

Samsonov, S. K. 1966. *New Data on Upper Cretaceous Flora of Middle Asia North-East*. Moscow: Nauka, 100 pp. (In Russian.)

Samylina, V. A. 1960. Angiosperms from Lower Cretaceous deposits of Kolyma River. *Botan. Journ.*, **45** (3), 335–352. (In Russian.)

    1963. Mesozoic flora from lower reaches of Aldan River. In: *Paleobotanika*. Moscow–Leningrad: Izd vo AN SSSR, **4**, 57–139. (In Russian.)

1970. Ginkgoaceae and Czekanowskiaceae (certain outcomes and objectives of researchers). *Paleontol. Journ.*, **3**, 114–123. (In Russian.)

1974. Early Cretaceous floras of north-east of USSR (toward problem of emergence of Cenophytic floras). In: *XXVII Komarovskiye Chteniya. (16 Oktyabrya 1972 g.).* Leningrad: Nauka, pp. 1–56.

1976. *Cretaceous Flora of Omsukchan.* Leningrad: Nauka, 206 pp. (In Russian.)

1984. Late Cretaceous Flora of Tap River (Northern Okhotsk vicinity). In: *Yezhegodn. Vsesoyuzn. Paleontol. Ob-va*, **27**, 236–246. (In Russian.)

Samylina, V. A. and Kirichkova, A. I. 1973. Structure of epidermis in leaves of Czekanowskiaceae and Ginkgoaceae and terminological issues. *Paleontol. Journ.*, **4**, 95–100. (In Russian.)

Samylina, V. A. and Yefimova, A. F. 1968. First finds of Early Jurassic flora from Kolyma River Basin. *Dokl. AN SSSR*, **179** (1), 166–168. (In Russian.)

Semyonova, Y. V. 1966. Spore–pollen complexes of Upper Triassic and Lower and Middle Jurassic in north-western margin of Donbass. In: *Znacheniye Palinologicheskogo Analiza dlya Stratigrafii i Paleofloristiki*, Moscow, pp. 104–108. (In Russian.)

Seward, A. C. 1911. The Jurassic flora of Sutherland. *Trans. Roy. Soc. Edinburgh*, **47** (4), 643–709.

1926. The Cretaceous plant-bearing rocks of Western Greenland. *Phil. Trans. Roy. Soc. London. B*, **215**, 57–175.

Shah, S. C. 1977. Jurassic–Lower Cretaceous megaflora in India. A review. *Rec. Geol. Surv. Ind.*, **109** (2), 55–81.

Sharma, B. D. 1974. The Jurassic flora of the Rajmahal Hills, India – Advances and challenges. *Acta palaeobot.*, **15**, 3–15.

Shilin, P. V. 1983. Late Cretaceous floras of Nizhnesyrdarya Uplift. *Paleontol. Journ.*, **2**, 105–112. (In Russian.)

1986. *Late Cretaceous Floras of Kazakhstan.* Alma-Ata: Nauka, 136 pp. (In Russian.)

Shilin, P. V. and Romanova, E. V. 1978. *Senonian Floras of Kazakhstan.* Alma-Ata: Nauka, 176 pp. (In Russian.)

Shilkina, I. A. and Doludenko, M. P. 1985. *Frenelopsis* and *Cryptomeria* – Dominants of Late Albian Flora in Ukraine. *Botan. Journ.*, **70** (8), 1019–1030. (In Russian.)

Shramkova, G. V. 1963. Spore–pollen complexes of Mesozoic deposits in North-Western Donbass and Dniepr-Donets Trough. *Tr. Voronezh. Un-ta*, **62**, 10–25. (In Russian.)

Shrank, E. 1982. Kretazische Pollen und Sporen aus dem 'Nubischen Sandstein' des Dakhla–Beckens (Agypten). *Berliner Geowiss. Abh. A.*, **40**, 87–109.

Shuvalov, V. F. 1982. Paleogeography and history of development of lacustrine systems in Mongolia in Jurassic and Cretaceous. In: *Mezozoysliye Ozerniye Basseyni Mongolii.* Leningrad: Nauka, pp. 3–68. (In Russian.)

Silva-Pineda, A. 1970. Plantas fosiles del Jurassico Medio de la region de Tezoatlan, Oaxaca. *Soc. Geol. Mexicana. Excursion Mexico – Oaxaca*, pp. 129–153.

1978. Plantos del Jurassico Medio del sur de Pueblo y Nordeste de Oaxaca. *Univ. Nac. Auton. Medico, Inst. Geol. Paleontol. Mexicana*, **44** (3), 25–56.

1984. Revision taxonomica y tipificacion de las plantas Jurasicas colectades y estudiadas por Wieland (1914) en la region de el Consuelo, Oaxaca. *Paleontol. Mexicana*, **49**, 1–102.

Singh, C. 1975. Stratigraphic significance of early angiosperm pollen in the mid-Cretaceous strata of Alberta. In: *The Cretaceous System in the Western Interior of North America*, ed. W. G. E. Caldwell. Geol. Assoc. Canada Spec. Pap., **13**, pp. 365–389.

Singh, G. 1974. Demarcation of Jurassic–Cretaceous boundary. *Aspects and Appraisal of Indian Palaeobotany*, Lucknow, pp. 452–466.

Sitholey, R. 1954. Mesozoic and Tertiary floras of India: a review. *Paleobotanist*, **3**, 55–69.

Smiley, C. J. 1969. Cretaceous floras of Chandler Colville region Alaska: stratigraphy and preliminary floristics. *Amer. Assoc. Petrol. Geol. Bull.*, **53** (3), 482–502.

1970. Later Mesozoic flora from Maran, Pahang, West Malaysia. Part 2. Taxonomic consideration. *Bull. Geol. Soc. Malaysia*, **3**, 87–112.

Smith, A. G. and Briden, J. C. 1977. *Mesozoic and Cenozoic Paleocontinental Maps*. Cambridge: Cambridge University Press, 63 pp.

Sodov, J. 1980. New species of *Heilungia* and *Pityospermum* from the Mesozoic of South-East Mongolia. *Paleontol. Journ.*, **4**, 131–133. (In Russian.)

Song, Z., Zheng, Y., Lin, J., Ye, P., Wang, C. and Zhou, S. 1982. *Cretaceous-Tertiary Palynological Assemblages from Jiangsu*. Peking: Geol. Publ. House, 270 pp.

Srivastava, S. K. 1975. Maastrichtian microspore assemblages from the interbasaltic lignites of Mull, Scotland. *Palaeontographica B.*, **150** (5/6), 125–156.

1978. Cretaceous spore–pollen floras: a global evaluation. *Biol. Mem. Palaeopalynol. Ser. 5*, **3** (1), 1–130.

1983. Cretaceous phytogeoprovinces and paleogeography of the Indian plate based on palynological data. *Cretaceous of India*. Lucknow: Ind. Assoc. Palynostratigr., pp. 141–157.

1984. A new elater-bearing Late Albian pollen species from offshore eastern Saudi Arabia. *Bot. J. Linn. Soc.*, **89** (3), 231–238.

Stanislavsky, F. A. and Kiselevich, L. S. 1986. First find of Middle Albian plants in Crimea. *Geol. Journ.*, **46** (5), 121–124. (In Russian.)

Stevens, G. R. 1973. Jurassic Belemnites. In: *Atlas of Paleobiogeography*, ed. A. Hallam. Amsterdam: Elsevier, pp. 259–274.

Stott, D. F. 1961. Summary account of the Cretaceous and equivalent Rocks, Rocky Mountain foothills, Alberta. *Geol. Surv. Canada. Pap.*, **61–2**, 1–34.

1963. Stratigraphy of the Lower Cretaceous Fort St. John group and Gething and Cadomin formations, foothills of Northern Alberta and British Columbia. *Geol. Surv. Canada Pap.*, **62–39**, 1–48.

1968. Lower Cretaceous Bullhead and Fort St. John group between Smoky and Peace rivers, Rocky Mountain foothills, Alberta and British Columbia. *Geol. Surv. Canada. Bull.*, **152**, 1–279.

Strakhov, N. M. 1960. *Main Theories of Lithogenesis*. Vol. 1. Types of Lithogenesis and their distribution on the globe's surface. Moscow. Izd-vo AN SSSR, 212 pp. (In Russian.)

Sultan, I. Z. 1978. Mid-Cretaceous plant microfossils from the northern part of the Western Desert of Egypt. *Rev. Palaeobot. and Palynol.*, **25** (3/4), 259–367.

Surange, K. R. 1966. *Indian Fossil Pteridophytes*. New Delhi: Council of Sci. and Ind. Res., 209 pp.

Svanidze, T. I. 1971. On age of flora-bearing deposits of Lower Jurassic of Dzirul and Lok Crystal Massifs. *Tr. Tbil. Un-ta. Ser. A.*, **2** (141), 163–169.

Sveshnikova, I. N. 1967. Fossil conifers of Vilyui Syneclise. *Tr. Botan. In-ta AN SSSR. Ser. 8, Paleobotanika*, **6**, 177–203. (In Russian.)

Sveshnikova, I. N. and Budantsev, L. Y. 1969. *Fossil Flora of Arctic*. Part 1. Paleozoic and Mesozoic floras of Western Spitzbergen, Franz Joseph Land and Novaya Sibir Island. Leningrad: Nauka, 128 pp. (In Russian.)

Sykstel, T. A. 1954. Certain data on climatic zones of Jurassic period. *Tr. Sredneaz. Un-ta*, **52**, 71–73. (In Russian.)

Sze, H. C. 1945. The Cretaceous flora from the Pantou series in Yungan, Fukien. *J. Paleontol*, **19** (1), 45–59.

1949. Die mesozoische Flora aus der Hsiangchi Kohlen Serie in Westhupeh. *Paleontol. Sin A*, **2**, 1–71.

Takhtajan, A. L. 1966. Main phytochoria of Late Cretaceous and Paleocene on USSR territory and in adjacent countries. *Botan. Journ.*, **51** (9), 1217–1230. (In Russian.)

1978. *Floristic Regions of the Earth*. Leningrad: Nauka, 248 pp. (In Russian.)

Tanai, T. 1979. Late Cretaceous floras from Kuji district, Northeastern Honshu, Japan. *J. Fac. Sci. Hokkaido Univ. Ser. 4*, **19** (1/2), 75–136.

Teixeira, C. and Pais, J. 1976. *Introducao a paleobotanica as grandes fases da evolucao dos vegetais.* Lisboa, 210 pp.

Tenčov, J. and Cernjavska, S. 1965. Paläozoische und mesozoische Floren in Bulgarien. *Ber. Geol. Ges. DDR*, **10** (4), 465–478.

Teslenko, Y. V. 1968. On floras of Late Jurassic and Early Cretaceous in Eastern Transbaikalia. *Dokl. AN SSSR*, **183** (4), 910–913. (In Russian.)

1970. *Stratigraphy and Floras of Jurassic Deposits of Western and Southern Siberia and Tuva.* Moscow: Nedra, 269 pp. (In Russian.)

1975. On stratigraphy of Jurassic deposits in Eastern Transbaikalia. *Geologiya i Geofizika*, **10**, 41–46. (In Russian.)

Thayne, G. F. and Tidwell, W. D. 1983. Flora of the Lower Cretaceous Cedar Mountain formation of Utah and Colorado. Pt I. *Paraphyllanthoxylon utahense. Great Basin Natur.*, **43** (3), 394–402.

Tidwell, W. D. 1966. Cretaceous paleobotany of eastern Utah and western Colorado. *Bull. Utah Geol. and Mineral. Surv.*, **80**, 87–95.

Tidwell, W. D., Ash, S. and Parker, L. R. 1981. Cretaceous and Tertiary floras of the San Juan basin. In: *Advances in San Juan Basin Paleontology*, ed. S. Lucas. Albuquerque: Univ. New Mexico Press, pp. 307–332.

Tidwell, W.D., Rushforth, S. R. and Reveal, J. L. 1967. *Astralopteris*, a new Cretaceous fern genus from Utah and Colorado. *Brigham Yong Univ. Geol. Stud.*, **14**, 237–240.

Trofimov, D. M., Petrosyants, M. A., Gerus, Y. A. and Kamladze, G. A. 1969. On age of Tegama Series in north-western part of Mali–Nigerian Syneclise (Intermediate Continental Formation of Southern Sahara). *Izv. Vuzov. Geologiya i Razvedka*, **5**, 52–61. (In Russian.)

Turutanova-Ketova, A. I. 1962. A new genus of fern from Mesozoic deposits of Kazakhstan. *Paleontol. Journ.*, **2**, 145–158. (In Russian.)

Upchurch, G. R. 1984. Cuticle evolution in Early Cretaceous angiosperms from the Potomac Group of Virginia and Maryland. *Ann. Mo Bot. Garden*, **71**, 522–550.

Upchurch, G. R. and Doyle, J. A. 1981. Paleoecology of the conifers *Frenelopsis* and *Pseudofrenelopsis* (Cheirolepidiaceae) from the Cretaceous Potomac Group of Maryland and Virginia. In: *Geobotany II*, ed. R. C. Romans. NY: Plenum Publ. Corp., pp. 167–202.

Vakhrameev, V. A. 1952. Stratigraphy and fossil flora of Jurassic and Cretaceous deposits of Vilyui Trough and adjacent part of Near Verkhoyansk Foredeep. *Regionalnaya Stratigrafiya SSSR*, Vol. 1, Moscow: AN SSSR, 340 pp. (In Russian.)

1958. Stratigraphy and fossil flora of Cretaceous deposits in Western Kazakhstan. *Regionalnaya Stratigrafiya SSSR*, Vol. 3, Moscow: AN SSSR, 136 pp. (In Russian.)

1964. *Jurassic and Early Cretaceous Floras of Eurasia and Contemporary Paleofloristic Provinces.* Moscow: Nauka, 261 pp. (In Russian.)

1965. First find of Jurassic flora in Cuba. *Paleontol. Journ.*, **3**, 123–126. (In Russian.)

1966. Late Cretaceous floras of the USSR Pacific coast: special features of their composition and stratigraphic sites. *Izv. AN SSSR. Ser. Geol.*, **3**, 76–87. (In Russian.)

1969. Stage-wise division of Middle Jurassic of southern USSR areas by paleobotanical data. *Sov. Geologiya*, **6**, 8–18. (In Russian.)

1970a. Pattern of distribution and palaeoecology of Mesozoic conifers Cheirolepidiaceae. *Paleontol. Journ.*, **1**, 19–34. (In Russian.)

1970b. First find of Bennettitalean *Dictyozamites* in Mesozoic of Siberia. *Paleontol. Journ.*, **4**, 120–123. (In Russian.)

1974. Cretaceous deposits of foothills of Rocky Mountains of Canada (Alberta) and their comparative paleofloristic characterization. In: *Problemi Geologii i Poleznikh Iskopayemikh na XXIV sessii Mezhdunarodnogo Geologicheskogo Kongressa.* Moscow: Nauka, pp. 152–163. (In Russian.)

1975. Main features of phytogeography of the globe in Jurassic and Early Cretaceous time. *Paleontol. Journ.*, **2**, 123–131. (In Russian.)

1976a. Platanoids of Late Cretaceous. In: *Ocherki Geologii i Paleontologii Dalnyego Vostoka SSSR.* Vladivostok: DVNTS AN SSSR, pp. 66–78. (In Russian.)

1976b. Development of Cretaceous floras of northern part of Pacific Belt. In: *Paleontologiya. Morskaya Geologiya.* Moscow: Nauka, pp. 128–137. (In Russian.)

1978. Climates of Northern Hemisphere in Cretaceous Period and paleobotanical data. *Paleontol. Journ.*, **2**, 3–17. (In Russian.)

1980. *Classopollis* pollen as indicator of Jurassic and Cretaceous climates. *Sov. Geologiya*, **8**, 48–56. (In Russian.)

1981a. Time of formation of Atlantic by paleontologic data. In: *Problemi Stroyeniya Zemnoy Kori.* Moscow: Nauka, pp. 29–37. (In Russian.)

1981b. Development of floras in middle part of Cretaceous Period and ancient angiosperms. In: *Paleontol. Journ.*, **2**, 3–14. (In Russian.)

1983. Jurassic and Cretaceous floras of Mongolia and climates of that time. *Izv. AN SSSR, Ser. Geol.*, **11**, 54–58. (In Russian.)

1984. Floras and climate of Earth in Early Cretaceous Epoch. *Sov. Geologiya*, **1**, 41–49. (In Russian.)

1985. Phytogeography, paleoclimates and position of continents in Mesozoic. *Vestn. AN SSSR*, **8**, 30–42. (In Russian.)

Vakhrameev, V. A. and Blinova, Y. V. 1971. New Early Cretaceous flora of Stanovy Ridge. *Paleontol. Journ.*, **1**, 88–94. (In Russian.)

Vakhrameev, V. A. and Doludenko, M. P. 1961. *Upper Jurassic and Lower Cretaceous Flora of Bureya Basin and its Stratigraphic Significance.* Moscow: Izd. AN SSSR, 135 pp. (Tr. GIN AN SSSR, **54**.) (In Russian.)

1976. Middle–Late Jurassic boundary – an important landmark in history of climatic development and vegetation evolution in Northern Hemisphere. *Sov. Geologiya*, **4**, 12–25. (In Russian.)

Vakhrameev, V. A. and Kotova, I. Z. 1977. Ancient angiosperms and their accompanying plants from Lower Cretaceous deposits of Transbaikalia. *Paleontol. Journ.*, **4**, 101–109. (In Russian.)

Vakhrameev, V. A. and Lebedev, E. L. 1967. Paleobotanical characterization and age of carbon-bearing Upper Mesozoic deposits of Far East (between Amur and Uda rivers). *Izv. AN SSSR. Ser. Geol.*, **2**, 120–133. (In Russian.)

Vakhrameev, V. A. and Vassina, R. A. 1959. Lower Jurassic and Aalenian floras of Northern Caucasus. In: *Paleontol. Journ.*, **3**, 125–133. (In Russian.)

Vakhrameev, V. A., Dobruskina, I. A., Zaklinskaya, Y. D. and Meyen, S. V. 1970. *Paleozoic and Mesozoic Floras of Eurasia and the Phytogeography of that Period.* Moscow: Nauka, 424 pp. (In Russian.)

Van der Burgh, J. and Van Konijnenburg-Van Cittert, J. H. 1984. A drifted flora from the

Kimmeridgian (Upper Jurassic) of Lothbeg Point, Sutherland, Scotland. *Rev. Palaeobot. and Palynol.*, **43**, 359–396.

Vassilevskaya, N. D. 1959. Stratigraphy and flora of Mesozoic coal deposits in Sangar Region of Lena carbon-bearing basin. *Tr. NIIGA*, **105** (11), 17–43. (In Russian.)

1966. Certain Early Cretaceous plants of Zhigansk Area (Lena coal-bearing basin). *Uchen. Zap. NIIGA. Paleontologiya i Biostratigrafiya*, **15**, 49–76. (In Russian.)

1977. Early Cretaceous flora of Kotelny Island. In: *Mesozoyskiye Otlozheniya Severo-Vostoka SSSR*. Leningrad, NIIGA, pp. 57–75. (In Russian.)

1980. Early Cretaceous flora of Spitzbergen Island. In: *Geologiya Osadochnogo Chekhla Arkhipelaga Svarlbard*. Leningrad: PGO Sevmorgeologiya, pp. 61–69. (In Russian.)

Vassilevskaya, N. D. and Abramova, L. N. 1966. Materials for cognition of Early Cretaceous flora of Lena Basin. *Uchen. Zap. NIIGA. Paleonologiya i Biostraigrafiya*, **16**, 73–96. (In Russian.)

1974. Floristic complexes of Early Cretaceous deposits of Koryak-Anadyr Region. In: *Stratigrafiya i Litologiya Melovykh, Paleogenovykh i Neogenovykh Otlozheniy Koryaksko-Anadirskoy Oblasti*. Leningrad: NIIGA, pp. 31–37. (In Russian.)

Vassilevskaya, N. D. and Pavlov, V. V. 1963. *Stratigraphy and Flora of Cretaceous Deposits of Lena–Olenyek Area and Lena Carbon-Bearing Basin*. Leningrad: Gostoptekhizdat, 96 pp. (Tr. NIIGA, **128**.) (In Russian.)

1967. On issue of systematic position of Jurassic ferns *Raphaelia*, *Uchen. Zap. NIIGA. Paleontologiya i Biostratigraphiya*, **19**, 41–50. (In Russian.)

Vassina, R. A. and Doludenko, M. P. 1968. Late Aalenian flora of Dagestan. *Paleontol. Journ.*, **3**, 90–98. (In Russian.)

Venkatachala, B. S. 1969. Palynology of the Mesozoic sediments of Kutch. 4. Spores and pollen from the Bhuj exposures near Bhuj, Gujarat district. *Palaeobotanist*, **17** (2), 208–219.

1977. Fossil flora assemblages in the East Coast Gondwanas – a critical review. *J. Geol. Soc. Ind.*, **18**, 378–397.

Vlassov, V. M. and Markovich, Y. M. 1979a. Correlation of Jurassic and Lower Cretaceous deposits of central and eastern parts of South-Yakutsk Coal Basin. *Sov. Geologiya*, **1**, 72–80. (In Russian.)

1979b. On age of Tokinsk and Undytkansk Suites of South-Yakutsk carbon-bearing formation in Tokinsk Trough. In: *Strukturniye Elementy Regiona Baikaloamurskoy Magistrali i Ikh Mineragenicheskiye Osobennosti*. Leningrad: VSEGEI, pp. 33–36. (Tr. VSEGEI: N.S., **303**.) (In Russian.)

Volkheimer, W. and Salas, A. 1975. Die älteste Angiospermen-Palynoflora Argentiniens vor der Typuslokalität der unterkretazischen Huitrin-Folge des Neuquen-Beckens. Mikrofloristische Assoziation und biostratigraphische Bedeutung. *Neues Jb. Geol. und Paläntol. Monatsh*, **7**, 424–436.

Watson, J. 1969. A revision of the English Wealden flora. 1. Charales – Ginkgoales. *Bull. Brit. Mus. (Natur. Hist.) Geol.*, **17** (5), 209–254.

1977. Some Lower Cretaceous conifers of the Cheirolepidiaceae from the USA and England. *Palaeontology*, **20**, 715–749.

Watson, J. and Fisher, H. 1984. A new conifer genus from the Lower Cretaceous Glen Rosa formation, Texas. *Palaeontology*, **27** (4), 719–727.

Weber, R. 1968. Die fossile Flora der Rhät–Lias-Ubergangsschichten von Bayreuth (Oberfranken) unter besonderer Berücksichtigung der Coenologie. *Erlanger geol. Abh.*, **72**, 1–73.

White, D. 1913. A new fossil plant from the state of Bahia, Brazil. *Amer. J. Sci.*, **35** (4), 633.

Yaroshenko, O. P. 1965. *Spore–Pollen Complexes of Jurassic and Lower Cretaceous Deposits of*

Northern Caucasus and their Stratigraphic Significance. Moscow: Nauka, 105 pp. (Tr. GIN AN SSSR, **117**.) (In Russian.)

Yasamanov, N. A. 1980. Paleothermy of Jurassic, Cretaceous and Paleogenic Periods in certain regions of USSR. *Bull. MOIP. Otd. Geol.*, **53** (3), 117–125. (In Russian.)

Yasamanov, N. A. and Petrosyants, M. A. 1983. Climatic conditions of Early Cretaceous in Crimean–Caucasian Region. *Sov. Geologiya*, **4**, 83–85. (In Russian.)

Ye Meina, Li Bauxian. 1980. *Succession of Jurassic Plant Assemblages and Stratigraphic Correlation of China.* Nanjing: Inst. Geol. and Palaeontol. Acad. Sin., 10 pp.

Yefimova, A. F. and Terekhova, G. P. 1966. On age of Ginterovo Suite in Ugolnaya Harbour. In: *Materiali po Geologii i Poleznim Iskopayemym Severo-Vostoka SSSR.* Magadan: Kn. Izd-vo., **19**, pp. 63–76. (In Russian.)

Zaklinskaya, Y. D. 1977. Angiosperms by palynologic data. In: *Razvitiye Flor na Granitse Mezozoya i Kainozoya.* Moscow: Nauka, pp. 66–119. (In Russian.)

Zhou, Z. 1983. Liassic plants from southwest Hunan, China. *Palaeontol. Sin. A.*, **165** (7), 1–115.

Systematic index

Carnoconites compactum Srivastava, 87, 92
  C. cranwellii Harris, 87, 91, 92, 176
  C. laxum Srivastava, 92
Carpolithes, 57, 63, 74
Carya, 221, 222
Caryophyllites polyporatus Couper, 240
Caspiocarpus paniculiger (Krassilov) Vakhrameev, 145
Cassia, 224
Castallites, 210
Castanea, 222
Caytonia, 16, 22, 148, 170, 248
Caytonianthus, 16, 22, 248
Caytonipollenites, 173
Cedrus, 190
Celastrophyllum, 125, 137, 139, 186, 196, 201, 216, 219, 221, 224, 226
Celastrophyllum acutidens Fontaine, 125
  C. kazachstanense Vakhrameev, 145
  C. kolymensis Samylina, 100, 110, 111
  C. oppositifolius Samylina, 100
  C. ovale Vakhrameev, 186
  C. rectinerve Herman, 202
  C. serrulatus Samylina, 100
Celastrus, 207
Celtidphyllum, 226
Celtis, 192, 227
Celtoidophyllum, 215
Cephalotaxopsis, 139
'*Cephalotaxopsis*', 95, 96, 97, 99, 111, 118, 119, 137, 139, 148, 182, 184, 186, 187, 188, 190, 191, 192, 196, 200, 201, 207, 210, 212
Cephalotaxopsis aff. *acuminata* Kryshtofovich and Prynada, 117
  '*C.*' *heterophylla* Hollick, 188, 195
  '*C.*' *intermedia* Hollick, 102, 191, 197, 198
  C. magnifolia Fontaine, 96
Cephalotaxus, 148
Ceratosporites, 177
Ceratostrobus echinatus Velenovsky, 214
Cercidiphyllum, 125
Cercidiphyllum arcticum (Heer) Brown, 222
  '*C.*' *potomacense* (Ward) Vakhrameev, 100
Cheirolepidium, 171
Cheirolepis, 17, 19
Chiaohoella, 121, 122
Chillinia, 121
Choanopollenites, 234
Chomotriletes, 163
Cicatricosisporites, 87, 89, 94, 103, 126, 136, 138, 140, 142, 144, 155, 158, 161, 163, 166, 167, 169, 173, 174, 177, 179, 216, 240
Cicatricosisporites australiensis (Cookson) Potonie, 172, 173
Cinnamomoides, 125, 196, 224, 237
Cinnamomoides elongata Koschman, 110
  C. ievlevii Samylina, 100
Cinnamomophyllum, 216, 217, 230
Cinnamomum, 219, 220, 221
'*Cinnamomum*', 210

Cinnamomum hesperium Knowlton, 234
  C. newberry Berry, 234
Cissites, 110, 121, 133, 145, 148, 184, 186, 191, 195, 200, 201, 207, 210, 211, 219, 226, 228, 239
Cissites comparabilis Hollick, 191
  C. microphylla Budantsev, 186
  C. cf. *parvifolius*, 111
  C. uralensis Kryshtofovich, 145
Cissus, 220, 221, 224, 230
Cissus marginata (Lesquereux), Brown, 193
Cladophlebidium interstifolium (Prynada) Krassilov, 110
Cladophlebis, 11, 22, 32, 33, 35, 36, 37, 38, 39, 40, 41, 42, 43, 46, 65, 68, 69, 75, 77, 79, 80, 82, 84, 87, 88, 95, 97, 101, 119, 120, 121, 124, 131, 133, 136, 137, 138, 144, 147, 149, 152, 153, 154, 155, 158, 170, 172, 175, 178, 184, 196, 200, 201, 210, 216, 221, 224, 229, 230
Cladophlebis aktaschensis Turutanova-Ketova, 33, 38, 42
  C. alberta (Dawson) Bell, 125
  C. albertsii (Dunker) Brongniart, 131
  C. aldanensis Vakhrameev, 49, 51, 56, 59, 63
  C. ankazoaboensis, 81
  C. cf. *arctica* (Heer) Kryshtofovich, 193
  C. argutula (Heer) Fontaine, 49, 96, 101, 124
  C. australis (Morris) Arber, 86, 175, 177, 240
  C. bidentata, 42
  C. browniana Dunker, 79, 131, 133, 144
  C. denticulata (Brongniart) Fontaine, 49,51, 160
  C. exiliformis (Geyler) Oishi, 81, 154, 160
  C. haiburnensis (L and H) Brongniart, 60
  C. ex gr. *haiburnensis* (Lindley and Hutton), 49, 60
  C. heterophylla Fontaine, 123, 124
  C. impressa Bell, 133
  C. ketovae Vassilevskaya, 96
  C. (*Klukia?*) *koraiensis* Yale, 160
  C. laxipinnata Prynada, 56, 58, 63
  C. lenaensis Vakhrameev, 96, 120
  C. lobifolia (Phillips) Brongniart, 63
  C. magnifica Brick, 38
  C. nebbensis (Brongniart) Nathorst, 51
  C. nifica, 42
  C. novopokrovskii Prynada, 105, 110
  C. oblonga, 84
  C. oerstedtii (Heer) Seward, 187
  C. orientalis Prynada, 49, 56, 58, 110
  C. cf. *parva* Fontaine, 125
  C. cf. *patagonica* Frenguelli, 171
  C. pseudolobifolia Vakhrameev, 63, 124, 155
  C. rigida, 125
  C. sangarensis Vakhrameev, 96
  C. saportana (Heer) Vakhrameev, 58
  C. septentrionalis Hollick, 196
  C. serrulata Samylina, 49, 58, 105
  C. sokolovii Teslenko, 63
  C. stenolopha, 42
  C. stricta, 81
  C. sulkata, 42
  C. suluktensis Brick, 33, 38, 42
  C. takezaki Oishi, 152
  C. toungusorum Prynada, 63

Geographic index